Lecture Notes in Mathematics 1946

Editors:
J.-M. Morel, Cachan
F. Takens, Groningen
B. Teissier, Paris

FONDAZIONE
CIME
ROBERTO CONTI
CENTRO INTERNAZIONALE MATEMATICO ESTIVO
INTERNATIONAL MATHEMATICAL SUMMER CENTER

C.I.M.E. means Centro Internazionale Matematico Estivo, that is, International Mathematical Summer Center. Conceived in the early fifties, it was born in 1954 and made welcome by the world mathematical community where it remains in good health and spirit. Many mathematicians from all over the world have been involved in a way or another in C.I.M.E.'s activities during the past years.

So they already know what the C.I.M.E. is all about. For the benefit of future potential users and cooperators the main purposes and the functioning of the Centre may be summarized as follows: every year, during the summer, Sessions (three or four as a rule) on different themes from pure and applied mathematics arc offered by application to mathematicians from all countries. Each session is generally based on three or four main courses $(24-30$ hours over a period of $6\text{-}8$ working days) held from specialists of international renown, plus a certain number of seminars.

A C.I.M.E. Session, therefore, is neither a Symposium, nor just a School, but maybe a blend of both. The aim is that of bringing to the attention of younger researchers the origins, later developments, and perspectives of some branch of live mathematics.

The topics of the courses are generally of international resonance and the participation of the courses cover the expertise of different countries and continents. Such combination, gave an excellent opportunity to young participants to be acquainted with the most advance research in the topics of the courses and the possibility of an interchange with the world famous specialists. The full immersion atmosphere of the courses and the daily exchange among participants are a first building brick in the edifice of international collaboration in mathematical research.

C.I.M.E. Director
Pietro ZECCA
Dipartimento di Energetica "S. Stecco"
Università di Firenze
Via S. Marta, 3
50139 Florence
Italy
e-mail: zecca@unifi.it

C.I.M.E. Secretary
Elvira MASCOLO
Dipartimento di Matematica
Università di Firenze
viale G.B. Morgagni 67/A
50134 Florence
Italy
e-mail: mascolo@math.unifi.it

For more information see CIME's homepage: http://www.cime.unifi.it

CIME's activity is supported by:

– Istituto Nazionale di Alta Mathematica "F. Severi"
– Ministero degli Affari Esteri - Direzione Generale per la Promozione e la Cooperazione - Ufficio V
– Ministero dell'Istruzione, dell'Università e della Ricerca

Grégoire Allaire · Anton Arnold
Pierre Degond · Thomas Y. Hou

Quantum Transport

Modelling, Analysis and Asymptotics -
Lectures given at the
C.I.M.E. Summer School
held in Cetraro, Italy
September 11–16, 2006

Editors:
Naoufel Ben Abdallah
Giovanni Frosali

 Springer

FONDAZIONE
CIME
ROBERTO CONTI

Authors and Editors

Grégoire Allaire
Centre de Mathématiques Appliquées
Ecole Polytechnique
Route de Saclay
91128 Palaiseau Cedex
France
gregoire.allaire@polytechnique.fr

Anton Arnold
Institute for Analysis and Scientific
Computing
Vienna University of Technology
Wiedner Hauptstraße 8
1040, Wien
Austria
anton.arnold@tuwien.ac.at

Naoufel Ben Abdallah
Institut de Mathématiques
Université de Toulouse
118 route de Narbonne
31062 Toulouse Cedex 9
France
naoufel@math.univ-toulouse.fr

Pierre Degond
Institut de Mathématiques
Université de Toulouse
118 route de Narbonne
31062 Toulouse Cedex 9
France
pierre.degond@math.univ-toulouse.fr

Thomas Yizhao Hou
Applied and Computational Math. 217-50
California Institute of Technology
1200 California Blvd
Pasadena, CA 91125
USA
hou@acm.caltech.edu

Giovanni Frosali
Dipartimento di Matematica Applicata
Università di Firenze
Via S.Marta 3
50139, Firenze
Italy
giovanni.frosali@unifi.it

ISBN: 978-3-540-79573-5   e-ISBN: 978-3-540-79574-2
DOI: 10.1007/978-3-540-79574-2

Lecture Notes in Mathematics ISSN print edition: 0075-8434
                              ISSN electronic edition: 1617-9692

Library of Congress Control Number: 2008927357

Mathematics Subject Classification (2000): 35Q40, 82C70, 76Y05, 35J10, 76M12, 76M50

© 2008 Springer-Verlag Berlin Heidelberg
This work is subject to copyright. All rights are reserved, whether the whole or part of the material is
concerned, specifically the rights of translation, reprinting, reuse of illustrations, recitation, broadcasting,
reproduction on microfilm or in any other way, and storage in data banks. Duplication of this publication
or parts thereof is permitted only under the provisions of the German Copyright Law of September 9,
1965, in its current version, and permission for use must always be obtained from Springer. Violations
are liable to prosecution under the German Copyright Law.

The use of general descriptive names, registered names, trademarks, etc. in this publication does not
imply, even in the absence of a specific statement, that such names are exempt from the relevant protective
laws and regulations and therefore free for general use.

Cover design: WMXDesign GmbH

Printed on acid-free paper

9 8 7 6 5 4 3 2 1

springer.com

This book is dedicated to the memory of Frédéric Poupaud.

Preface

Downscaling of semiconductor devices, which is now reaching the nanometer scale, makes it mandatory for us to understand the quantum phenomena involved in charge transport. Indeed, for nanoscale devices, the quantum nature of electrons cannot be neglected. In fact, it underlies the operation of an increasing number of devices. Unlike classical transport, the intuition of the physicist and the engineer is becoming insufficient for predicting the nature of device operation in the quantum context—the need for sufficiently accurate and numerically tractable models represents an outstanding challenge in which applied mathematics can play an important role.

The CIME Session "Quantum Transport: Modelling, Analysis and Asymptotics", which took place in Cetraro (Cosenza), Italy, from September 11 to September 16, 2006, was intended both to present an overview of up-to-date mathematical problems in this field and to provide the audience with techniques borrowed from other fields of application.

It was attended by about 50 scientists and researchers, coming from different countries. The list of participants is included at the end of this book.

The school was structured into four courses:

- **Grégoire Allaire** (École Polytechnique, Palaiseau, France) *Periodic Homogeneization and Effective Mass Theorems for the Schrödinger Equation.*
- **Anton Arnold** (Technische Universität, Vienna) *Mathematical Properties of Quantum Evolution Equations.*
- **Pierre Degond** (Université Paul Sabatier and CNRS, Toulouse, France) *Quantum Hydrodynamic and Diffusion Models Derived from the Entropy Principle.*
- **Thomas Yizhao Hou** (Caltech, Los Angeles, USA) *Multiscale Computations for Flow and Transport in Heterogeneous Media.*

This book contains the texts of the four series of lectures presented at the Summer School. Here follows a brief description of the subjects of these courses.

The first course, titled *Periodic Homogeneization and Effective Mass Theorems for the Schrödinger Equation,* was given by Professor Grégoire Allaire, a renowned specialist in homegeneization theory, and introduced the audience to the theory of homogeneization, a powerful tool for mathematically analyzing the multiscale aspects which are encountered in mathematical physics. First, the heuristic method of two-scale asymptotic expansions is discussed, then an entire section is devoted to the rigorous aspects and the main theoretical results that are at the root of the two-scale convergence. Such a method can be applied to the homogeneization of partial differential equations with periodically oscillating coefficients, like the model problem of diffusion in a periodic medium. These tools are then used to derive rigorously the so-called effective mass approximation which justifies the averaging of the crystal lattice effect on the transport of electrons in solids.

The course *Mathematical Properties of Quantum Evolution Equations,* given by Professor Anton Arnold, was aimed at introducing the basic mathematical properties of quantum evolution equations, such as the Schrödinger equation, the von Neumann equation and the Wigner formalism, and also dealt with the modelling of electron injection from reservoirs into semiconductor nano-devices, and the mathematical and numerical analysis of open boundary conditions for such equations. For the Schrödinger–Poisson analysis Strichartz inequalities are presented. In the density matrix formalism, both closed and open quantum systems can be treated. Their evolution is discussed in the space of trace class operators and energy subspaces, employing Lieb–Thirring inequalities. For the analysis of the Wigner–Poisson–Fokker–Planck system, quantum kinetic dispersion estimates are derived for Wigner–Poisson systems, inspired by the Vlasov–Poisson case. In this course, standard dispersion inequalities for the Schrödinger equation and novel stability properties of discrete artificial boundary conditions obtained by Professor Arnold are shown.

The course *Quantum Hydrodynamic and Diffusion Models Derived from the Entropy Principle,* given by Professor Pierre Degond, presents a novel methodology for the derivation of quantum macroscopic equations from kinetic-type models which is based on a deep understanding of the analogy between classical and quantum dynamics. The entropy minimization strategy of Levermore is considered in the quantum context, and is shown to produce diffusion or hydrodynamic-type equations in which the quantum features appear in particular in the nonlocal character of the relationship between the macroscopic variables such as the particle density and the entropic ones such as the chemical potential. The ad hoc corrections of classical fluid equations, which are commonly used in engineering simulations, are clarified and corrected using this approach. They appear as expansions of the fully quantum models in powers of the Planck constant. A whole field of difficult mathematical problems is open for researchers.

The last course was dedicated to numerical issues. In this course, titled *Multiscale Computations for Flow and Transport in Heterogeneous Media* and given by Professor Thomas Hou, a leader in this field, multiscale finite element methods are presented, analysed and illustrated in various physical situations. The author presents an exhaustive review of the more important recent advances in developing multiscale finite element methods for flow and transport in strongly heterogeneous porous media. The applications targeted by this course are in domains other than quantum transport, but the methodology will be of great interest to researchers involved in quantum transport modelling. Indeed, one of the main features of the quantum transport problem is the existence of different scales: macroscopic scales which are related to electrostatic forces and microscopic scales connected to oscillations of single wave functions and at which interference effects take place. Standard numerical methods need mesh sizes at microscopic scales, thus leading to unnecessarily high numerical cost. The goal of the course is to present methods that capture the small-scale effect on the large scales, but do not require resolving all the small-scale features. The course presented novel numerical multiscale methods which allow the use of coarse meshes while succeeding in resolving the microscopic scales.

The school succeeded in introducing the basic mathematical techniques to analyze quantum transport, to present new models dealing with collisions in the quantum context and to enlarge the knowledge in the domain to multiscale techniques borrowed from other fields of applied mathematics. The choice of the courses and the speakers was suggested by the demand of exhaustively presenting the state of art of quantum transport modelling, proposing both the kinetic and the hydrodynamic models, as well as the analytical and numerical aspects of the more significant problems.

During the course, the Thursday afternoon was dedicated to a poster session, where some of the young participants had the opportunity to show their more recent research and to discuss it with all the participants. The contributors to the poster session were D. Finco, E. Kalligiannaki, O. Maj, C. Manzini, O. Morandi, J.P. Milišic, C. Negulescu, G. Panati, T. Ryabukha, M. Schulte, and V.O. Shtyk.

The directors of this course thank the members of the CIME Scientific Committee for their invitation to organize it, and the Director, Professor Pietro Zecca, and the Secretary, Professor Elvira Mascolo, for their continuous help in the organization.

A special thanks has to be addressed to the lecturers for their good presentations, which stimulated scientific discussions. A thanks has to be addressed also to the attending people, constantly animated with genuine interest. The presence of some participants specialist in quantum transport topics favoured interactions among the students, and we are grateful for the attention paid by a precise and careful audience.

We also like to thank MIP Toulouse for its financial support of the French participants.

Finally, we thank the Director and all the staff of Hotel San Michele in Cetraro for their warm hospitality, which greatly contributed to the friendly atmosphere among all the participants.

Toulouse and Firenze, March 2007   *Naoufel Ben Abdallah*
                                                                Giovanni Frosali

Contents

Periodic Homogenization and Effective Mass Theorems for the Schrödinger Equation

Grégoire Allaire

Abstract The goal of this course is to give an introduction to periodic homogenization theory with an emphasis on applications to Schrödinger equation. We review the formal method of two-scale asymptotic expansions, then discuss the rigorous two-scale convergence method as well as the Bloch wave decomposition. Eventually these tools will be apply to the Schrödinger equation with a periodic potential perturbed by a small macroscopic potential. The notion of effective mass for the one electron model in solid state physics will be derived. Localization effects will also be considered

1 Introduction

This set of lecture notes is an elementary introduction to periodic homogenization, two-scale convergence, Bloch waves, and their application to the Schrödinger equation with a periodic potential. In particular, the notion of effective mass which is central in solid state physics will be rigorously derived here. The main mathematical results of homogenization in these lecture notes were originally published in [ACPSV04], [AP06], [AP05], [AV06]. Other mathematical works devoted to effective mass in solid state physics include [BLP78] (see Sect. 4 of Chap. 4), [Bec99], [PR96] and [Spa06], not to mention the many contributions in physics (see the books [Mye90], [Que98]).

Mathematically, homogenization can be defined as a theory for averaging partial differential equations and defining both effective properties and macroscopic models. Although this question of averaging and finding effective properties is very old in physics or mechanics, the mathematical theory

Grégoire Allaire

Centre de Mathématiques Appliquées, Ecole Polytechnique, Route de Saclay, 91128 Palaiseau Cedex, France, e-mail: gregoire.allaire@polytechnique.fr

N. Ben Abdallah, G. Frosali (eds.), *Quantum Transport.*
Lecture Notes in Mathematics 1946.
© Springer-Verlag Berlin Heidelberg 2008

of homogenization is quite recent, going back to the 1970s. The most general framework is known as the H-convergence, or G-convergence, introduced by Spagnolo [Spa68], [Spa76], and generalized by Tartar and Murat [MT78], [Tar78]. Although homogenization is not restricted to periodic problems, we shall focus here on periodic homogenization which is simpler and enough for the asymptotic analysis of periodic structures. There are many textbooks on periodic homogenization, see, e.g., [BP90], [BLP78], [CD99], [JKO94], [San80].

We discuss the theory of Bloch waves, a generalization of Fourier analysis devoted to the study of partial differential equations with periodic coefficients [BLP78], [CPV95], [Gel50], [Kuc93], [OK64], [RS78], [Wil78]. There is an intimate connection between homogenization and Bloch waves, which is by now well established [AC98], [BLP78], [COV02], [CPV95], [CV97], [MB91], [Sev82], but that we shall not explore in details. Rather we content ourselves in introducing Bloch waves as initial data in the Schrödinger equation.

Notations. The unit cube is denoted by $Y = (0,1)^N$. When Y is supplemented with periodic boundary conditions we identify it with the flat unit torus, denoted by $\mathbb{T}^N$. For any function $\phi(x,y)$ defined on $\mathbb{R}^N \times \mathbb{T}^N$, we denote by ϕ^ϵ the periodically oscillating function $\phi(x, \frac{x}{\epsilon})$.

2 Asymptotic Expansions in Periodic Homogenization

This section is devoted to an elementary introduction to periodic homogenization using the heuristic method of two-scale asymptotic expansions. A rigorous mathematical justification will be given in the next section on two-scale convergence.

We consider a model problem of conductivity in a periodic medium (for example, an heterogeneous domain made of periodic inclusions in a background matrix, see Fig. 1). The periodic domain is Ω, a bounded open set in $\mathbb{R}^N$ (with $N \geq 1$ the space dimension). Its period is denoted by ϵ, a positive number which is assumed to be very small in comparison with the size of the domain. The conductivity in Ω varies periodically with period ϵ in each direction. It is a matrix $A(y)$, where $y = x/\epsilon \in Y$ is the fast periodic (or microscopic) variable, while $x \in \Omega$ is the slow (or macroscopic) variable. The matrix A is not necessarily symmetric and is a bounded periodic function of y, with period Y (it may be discontinuous in y to model the discontinuity of conductivities between the inclusions and the matrix). It is positive definite, i.e., there exists a positive constant $\alpha > 0$ such that, for any vector $\xi \in \mathbb{R}^N$ and at any point $y \in Y$,

$$\alpha|\xi|^2 \leq \sum_{i,j=1}^{N} A_{ij}(y)\xi_i\xi_j.$$

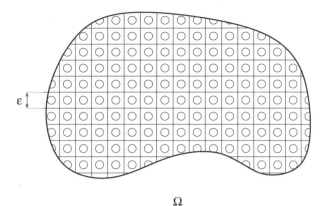

$$\Omega$$

Fig. 1 A periodic domain

Denoting by $f(x)$ the source term, and enforcing a Dirichlet boundary condition (for simplicity), the model problem of conductivity reads

$$\begin{cases} -\mathrm{div}\left(A\left(\dfrac{x}{\epsilon}\right)\nabla u_\epsilon\right) = f \text{ in } \Omega \\ u_\epsilon = 0 \qquad\qquad\qquad \text{ on } \partial\Omega, \end{cases} \tag{1}$$

where $u_\epsilon(x)$ is the unknown electrostatic potential or temperature.

The domain Ω, with its conductivity $A\left(\frac{x}{\epsilon}\right)$, is highly heterogeneous with periodic heterogeneities of lengthscale ϵ. Usually one does not need all the tiny details of the variations of u_ϵ, but rather some global or averaged behavior. In other words, it is enough to replace the heterogeneities by some effective macroscopic conductivity in Ω. From a numerical point of view, solving (1) by any method requires too much computational effort if ϵ is small since the number of elements or degrees of freedom should be at least of the order of the number of periods or inclusions, i.e., of the order of $1/\epsilon^N$. Again, it is preferable to average or homogenize the properties of Ω and compute an approximation of u_ϵ on a coarse mesh. Averaging the solution of (1) and finding the effective properties of Ω is precisely the goal of homogenization.

Although the true physical problem may involve a single value of the parameter ϵ, say ϵ_0, the mathematical theory of homogenization first embed the "real" problem for $\epsilon = \epsilon_0$ in a sequence of similar problems for which the lengthscale ϵ, becoming increasingly small, goes to zero. Then, an asymptotic analysis is performed as ϵ tends to zero, and the conductivity tensor of the limit problem is said to be the *effective* or *homogenized* conductivity. In this section this asymptotic analysis is carried out by means of the heuristic method of two-scale asymptotic expansions. We briefly present it and refer to the classical books [BP90], [BLP78], and [San80] for more details. A mathematical justification of its results will be given in the next section.

The method of two-scale asymptotic expansions is quite simple and start with the following postulate: the solution u_ϵ of (1) is written as a power series in ϵ

$$u_\epsilon(x) = \sum_{i=0}^{+\infty} \epsilon^i u_i\left(x, \frac{x}{\epsilon}\right),\tag{2}$$

where each term $u_i(x, y)$ is a function of both variables x and y, periodic in y with period $Y = (0,1)^N$ (u_i is called a Y-periodic function with respect to y). This series is plugged into the equation, and the following derivation rule is used:

$$\nabla\left(u_i\left(x, \frac{x}{\epsilon}\right)\right) = \left(\epsilon^{-1}\nabla_y u_i + \nabla_x u_i\right)\left(x, \frac{x}{\epsilon}\right),\tag{3}$$

where ∇_x and ∇_y denote the partial derivative with respect to the first and second variable of $u_i(x, y)$. For example, one has

$$\nabla u_\epsilon(x) = \epsilon^{-1}\nabla_y u_0\left(x, \frac{x}{\epsilon}\right) + \sum_{i=0}^{+\infty} \epsilon^i\left(\nabla_y u_{i+1} + \nabla_x u_i\right)\left(x, \frac{x}{\epsilon}\right).$$

Equation (1) becomes a series in ϵ

$$-\epsilon^{-2}\left[\operatorname{div}_y A\nabla_y u_0\right]\left(x, \frac{x}{\epsilon}\right)$$

$$-\epsilon^{-1}\left[\operatorname{div}_y A(\nabla_x u_0 + \nabla_y u_1) + \operatorname{div}_x A\nabla_y u_0\right]\left(x, \frac{x}{\epsilon}\right)$$

$$-\epsilon^0\left[\operatorname{div}_x A(\nabla_x u_0 + \nabla_y u_1) + \operatorname{div}_y A(\nabla_x u_1 + \nabla_y u_2)\right]\left(x, \frac{x}{\epsilon}\right)\tag{4}$$

$$-\sum_{i=1}^{+\infty} \epsilon^i\left[\operatorname{div}_x A(\nabla_x u_i + \nabla_y u_{i+1}) + \operatorname{div}_y A(\nabla_x u_{i+1} + \nabla_y u_{i+2})\right]\left(x, \frac{x}{\epsilon}\right)$$

$$= f(x).$$

Identifying each coefficient of the ϵ-series (4) as an individual equation yields a cascade of equations. It turns out that the three first equations are enough for our purpose. The ϵ^{-2} equation is

$$-\operatorname{div}_y A(y)\nabla_y u_0(x, y) = 0 \quad \text{in } \mathbb{T}^N,$$

which is nothing else than an equation in the unit torus $\mathbb{T}^N$ (equivalently in the unit cell Y with periodic boundary condition). In this equation, y is the variable, and x plays the role of a parameter. By virtue of Lemma 2.1 below, there exists a unique solution of this equation up to a constant, i.e., a function of x independent of y (since x is just a parameter). This implies that u_0 is a function that does not depend on y, i.e., there exists a function $u(x)$ such that

$$u_0(x, y) \equiv u(x).$$

Since $\nabla_y u_0 = 0$, the ϵ^{-1} equation is

$$-\text{div}_y A(y)\nabla_y u_1(x,y) = \text{div}_y A(y)\nabla_x u(x) \quad \text{in } \mathbb{T}^N, \tag{5}$$

which is an equation for the unknown u_1. Again, Lemma 2.1 implies that (5) admits a unique solution up to a constant. Equation (5) allows one to compute u_1 in terms of u, and it is easily seen that $u_1(x,y)$ depends linearly on the first derivative $\nabla_x u(x)$

$$u_1(x,y) = \sum_{i=1}^{N} \frac{\partial u}{\partial x_i}(x)w_i(y), \tag{6}$$

where w_i is the unique solution in $H^1(\mathbb{T}^N)/\mathbb{R}$ of the so-called cell problem

$$-\text{div}_y A(y)\,(e_i + \nabla_y w_i(y)) = 0 \quad \text{in } \mathbb{T}^N, \tag{7}$$

with $(e_i)_{1\leq i\leq N}$ the canonical basis of $\mathbb{R}^N$. A physical interpretation of w_i is that it is the local variation of potential or temperature created by an averaged (or macroscopic) gradient e_i. Remark that $u_1(x,y)$ is merely defined up to the addition of a function $\tilde{u}_1(x)$ (depending only on x), but this does not matter since only its gradient $\nabla_y u_1(x,y)$ is used in the sequel.

Finally, the ϵ^0 equation is

$$\begin{aligned}-\text{div}_y A(y)\nabla_y u_2(x,y) &= \text{div}_y A(y)\nabla_x u_1 \\ &+ \text{div}_x A(y)\,(\nabla_y u_1 + \nabla_x u) + f(x) \quad \text{in } \mathbb{T}^N,\end{aligned} \tag{8}$$

which is an equation for the unknown u_2 in the periodic unit cell Y. According to Lemma 2.1, (8) admits a unique solution if a compatibility condition is satisfied (the so-called Fredholm alternative): the average over Y of the right-hand side of (8) must vanish, i.e.,

$$\int_Y [\text{div}_y A(y)\nabla_x u_1 + \text{div}_x A(y)\,(\nabla_y u_1 + \nabla_x u) + f(x)]\,dy = 0,$$

which simplifies to

$$-\text{div}_x \left(\int_Y A(y)\,(\nabla_y u_1 + \nabla_x u)\,dy \right) = f(x) \quad \text{in } \Omega. \tag{9}$$

Inserting (6) in (9), we obtain the homogenized equation for u that we supplement with a Dirichlet boundary condition on $\partial\Omega$,

$$\begin{cases} -\text{div}_x A^* \nabla_x u(x) = f(x) \text{ in } \Omega \\ u = 0 \qquad\qquad\qquad\quad \text{on } \partial\Omega. \end{cases} \tag{10}$$

The homogenized conductivity A^* is defined by its entries

$$A^*_{ij} = \int_Y \left[(A(y)\nabla_y w_i) \cdot e_j + A_{ij}(y) \right] dy = \int_Y A(y) \left(e_i + \nabla_y w_i \right) \cdot \left(e_j + \nabla w_j \right) dy. \tag{11}$$

The constant tensor A^* describes the effective or homogenized properties of the heterogeneous material $A\left(\frac{x}{\epsilon}\right)$. Note that A^* does not depend on the choice of domain Ω, source term f, or boundary condition on $\partial\Omega$.

Remark 2.1. This method of two-scale asymptotic expansions is unfortunately not rigorous from a mathematical point of view. In other words, it yields heuristically the homogenized equation, but it does not yield a correct proof of the homogenization process. The reason is that the ansatz (2) is usually not correct after the two first terms. For example, it does not include possible boundary layers in the vicinity of $\partial\Omega$ (for details, see, e.g., [Lio81]). Nevertheless, it is possible to rigorously justify the above homogenization process, in particular by the method of two-scale convergence as explained in the next section.

Lemma 2.1. *Let $f(y) \in L^2(\mathbb{T}^N)$. There exists a unique (periodic) solution in $H^1(\mathbb{T}^N)/\mathbb{R}$ of*

$$-\operatorname{div} A(y)\nabla w(y) = f \quad in \ \mathbb{T}^N, \tag{12}$$

if and only if $\int_{\mathbb{T}^N} f(y) dy = 0$ (this is called the Fredholm alternative).

Remark 2.2. Recall that the unit torus $\mathbb{T}^N$ is identified with the unit cell $Y = (0,1)^N$ supplemented with periodic boundary conditions. Therefore, the formulation of (12) in the unit torus automatically includes the periodicity of the solution.

Proof. Due to the periodic boundary condition, for any function $w \in H^1(\mathbb{T}^N)$, a simple integration by parts yields

$$\int_Y \operatorname{div}_y A(y)\nabla_y w(y) dy = \int_{\partial Y} [A(y)\nabla_y w(y)] \cdot n\, ds = 0,$$

because the normal n takes opposite values on opposite faces of Y. Therefore, $\int_Y f(y) dy = 0$ is a necessary condition of existence of solutions for (12). Defining the quotient space $H^1(\mathbb{T}^N)/\mathbb{R}$ of functions defined in $H^1(\mathbb{T}^N)$ up to a constant, it is easily seen that $\|\nabla w\|_{L^2(\mathbb{T}^N)^N}$ is a norm for this quotient space. We check the assumptions of the Lax–Milgram lemma on the variational formulation of (12). Clearly, $\int_{\mathbb{T}^N} A(y)\nabla w \cdot \nabla\phi\, dy$ is a coercive continuous bilinear form on $H^1(\mathbb{T}^N)/\mathbb{R}$. Furthermore, if $\int_{\mathbb{T}^N} f(y) dy = 0$, one finds

$$\int_{\mathbb{T}^N} f(y)w(y) dy = \int_{\mathbb{T}^N} f(y) \left(w(y) - \int_{\mathbb{T}^N} w(y) dy \right) dy,$$

which is a continuous linear form on $H^1(\mathbb{T}^N)/\mathbb{R}$ thanks to the following Poincaré–Wirtinger inequality

$$\left\| w(y) - \int_{\mathbb{T}^N} w(y)dy \right\|_{L^2(\mathbb{T}^N)} \leq C\|\nabla w\|_{L^2(\mathbb{T}^N)^N}.$$

This proves that there exists a unique solution $w \in H^1(\mathbb{T}^N)/\mathbb{R}$ of (12) if $\int_{\mathbb{T}^N} f(y)dy = 0$. $\qquad\Box$

3 Two-Scale Convergence

On the contrary of many homogenization methods, like the Γ-, G-, and H-convergence [Dal93], [JKO94], [Tar00], the two-scale convergence method is devoted only to periodic homogenization problems. It is therefore a less general method, but it is also more efficient and simple in the context of periodic homogenization. Two-scale convergence has been introduced by Nguetseng [Ngu89] and Allaire [All92] and is exposed in a self-content fashion below. This section is devoted to the main theoretical results which are at the root of this method. Section 4 is a detailed application of the method to the simple model problem of the previous section.

Let us introduce some notations. We denote by $C^\infty(\mathbb{T}^N)$ the space of infinitely differentiable functions in $\mathbb{T}^N$ (or equivalently of Y-periodic functions in $\mathbb{R}^N$), and by $C(\mathbb{T}^N)$ the Banach space of continuous functions in $\mathbb{T}^N$. For an open set Ω of $\mathbb{R}^N$, $C_c^\infty(\Omega; C^\infty(\mathbb{T}^N))$ denotes the space of infinitely smooth and compactly supported functions in Ω with values in the space $C^\infty(\mathbb{T}^N)$.

Definition 3.1. A sequence of functions u_ϵ in $L^2(\Omega)$ is said to two-scale converge to a limit $u_0(x, y)$ belonging to $L^2(\Omega \times Y)$ if, for any function $\varphi(x, y)$ in $C_c^\infty(\Omega; C^\infty(\mathbb{T}^N))$, it satisfies

$$\lim_{\epsilon \to 0} \int_\Omega u_\epsilon(x)\varphi\left(x, \frac{x}{\epsilon}\right) dx = \int_\Omega \int_Y u_0(x, y)\varphi(x, y)dxdy.$$

Here are a few basic examples of two-scale convergence:

1. Any sequence u_ϵ which converges strongly in $L^2(\Omega)$ to a limit $u(x)$, two-scale converges to the same limit $u(x)$.
2. For any smooth function $u_0(x, y)$, being Y-periodic in y, the associated sequence $u_\epsilon(x) = u_0\left(x, \frac{x}{\epsilon}\right)$ two-scale converges to $u_0(x, y)$.
3. For the same smooth and Y-periodic function $u_0(x, y)$ the sequence defined by $v_\epsilon(x) = u_0(x, \frac{x}{\epsilon^2})$ has the same two-scale limit and weak-L^2 limit, namely $\int_Y u_0(x, y)dy$ (this is a consequence of the difference of orders in the speed of oscillations for v_ϵ and the test functions $\varphi\left(x, \frac{x}{\epsilon}\right)$). Clearly the two-scale limit captures only the oscillations which are in resonance with those of the test functions $\varphi\left(x, \frac{x}{\epsilon}\right)$.

The notion of "two-scale convergence" makes sense because of the next compactness theorem.

Theorem 3.1. *From each bounded sequence u_ϵ in $L^2(\Omega)$ one can extract a subsequence, and there exists a limit $u_0(x,y) \in L^2(\Omega \times Y)$ such that this subsequence two-scale converges to u_0.*

To simplify the proof of Theorem 3.1 we assume that Ω is bounded although it is not necessary. We first need the following elementary lemma (the proof of which is left to the reader).

Lemma 3.1. *Let $B = C(\bar{\Omega} \times \mathbb{T}^N)$ be the space of continuous functions $\varphi(x,y)$ on $\bar{\Omega} \times \mathbb{T}^N$ (which are thus Y-periodic in y). Then, B is a separable Banach space (i.e., it contains a dense countable family), is dense in $L^2(\Omega \times Y)$, and there exists a constant C such that, for any $\varphi \in B$,*

$$\int_\Omega |\varphi\left(x, \frac{x}{\epsilon}\right)|^2 dx \le C\|\varphi\|_B^2,$$

and

$$\lim_{\epsilon \to 0} \int_\Omega |\varphi\left(x, \frac{x}{\epsilon}\right)|^2 dx = \int_\Omega \int_Y |\varphi(x,y)|^2 dx dy.$$

Proof of Theorem 3.1. By Schwarz inequality, we have

$$\left|\int_\Omega u_\epsilon(x)\varphi\left(x, \frac{x}{\epsilon}\right) dx\right| \le C \left|\int_\Omega |\varphi\left(x, \frac{x}{\epsilon}\right)|^2 dx\right|^{\frac{1}{2}} \le C\|\varphi\|_B. \qquad (13)$$

This implies that the left-hand side of (13) is a continuous linear form on B which can be identified to a duality product $\langle \mu_\epsilon, \varphi \rangle_{B',B}$ for some bounded sequence of measures μ_ϵ. Since B is separable, one can extract a subsequence and there exists a limit μ_0 such μ_ϵ converges to μ_0 in the weak * topology of B' (the dual of B). On the other hand, Lemma 3.1 allows us to pass to the limit in the middle term of (13). Combining these two results, yields

$$|\langle \mu_0, \varphi \rangle_{B',B}| \le C \left|\int_\Omega \int_Y |\varphi(x,y)|^2 dx dy\right|^{\frac{1}{2}}. \qquad (14)$$

Equation (14) shows that μ_0 is actually a continuous form on $L^2(\Omega \times Y)$, by density of B in this space. Thus, there exists $u_0(x,y) \in L^2(\Omega \times Y)$ such that

$$\langle \mu_0, \varphi \rangle_{B',B} = \int_\Omega \int_Y u_0(x,y)\varphi(x,y)dx dy,$$

which concludes the proof of Theorem 3.1.   □

The next theorem shows that more information is contained in a two-scale limit than in a weak-L^2 limit; some of the oscillations of a sequence are contained in its two-scale limit. When all of them are captured by the two-scale limit (condition (16) below), one can even obtain a strong convergence (a corrector result in the vocabulary of homogenization).

Theorem 3.2. *Let u_ϵ be a sequence of functions in $L^2(\Omega)$ which two-scale converges to a limit $u_0(x,y) \in L^2(\Omega \times Y)$.*

1. Then, u_ϵ converges weakly in $L^2(\Omega)$ to $u(x) = \int_Y u_0(x,y)dy$ and

$$\lim_{\epsilon \to 0} \|u_\epsilon\|_{L^2(\Omega)} \geq \|u_0\|_{L^2(\Omega \times Y)} \geq \|u\|_{L^2(\Omega)}. \tag{15}$$

2. Assume further that $u_0(x,y)$ is smooth and that

$$\lim_{\epsilon \to 0} \|u_\epsilon\|_{L^2(\Omega)} = \|u_0\|_{L^2(\Omega \times Y)}. \tag{16}$$

Then, we have

$$\left\| u_\epsilon(x) - u_0\left(x, \frac{x}{\epsilon}\right) \right\|_{L^2(\Omega)} \to 0. \tag{17}$$

Proof. By taking test functions depending only on x in Definition 3.1, the weak convergence in $L^2(\Omega)$ of the sequence u_ϵ is established. Then, for a smooth test function $\varphi_n(x,y)$, developing the inequality

$$\int_\Omega |u_\epsilon(x) - \varphi_n\left(x, \frac{x}{\epsilon}\right)|^2 dx \geq 0,$$

and passing to the limit first $\epsilon \to 0$ and second $\varphi_n \to u_0$ yields formula (15). Furthermore, under assumption (16), a similar argument shows that

$$\lim_{\epsilon \to 0} \int_\Omega |u_\epsilon(x) - \varphi\left(x, \frac{x}{\epsilon}\right)|^2 dx = \int_\Omega \int_Y |u_0(x,y) - \varphi(x,y)|^2 dx dy.$$

If u_0 is smooth enough to be a test function φ, it yields (17).   □

Remark 3.1. The smoothness assumption on u_0 in the second part of Theorem 3.2 is needed to ensure the measurability of $u_0\left(x, \frac{x}{\epsilon}\right)$ (which otherwise is not guaranteed for a function of $L^2(\Omega \times Y)$). One can further check that any function in $L^2(\Omega \times Y)$ is attained as a two-scale limit (see Lemma 1.13 in [All92]), which implies that two-scale limits have no extra regularity.

Theorem 3.3. *Let u_ϵ be a bounded sequence in $H^1(\Omega)$. Then, up to a subsequence, u_ϵ two-scale converges to a limit $u(x) \in H^1(\Omega)$, and ∇u_ϵ two-scale converges to $\nabla_x u(x) + \nabla_y u_1(x,y)$, with $u_1(x,y) \in L^2(\Omega; H^1(\mathbb{T}^N)/\mathbb{R})$.*

Proof. Since u_ϵ (resp. ∇u_ϵ) is bounded in $L^2(\Omega)$ (resp. $L^2(\Omega)^N$), up to a subsequence, it two-scale converges to a limit $u_0(x,y) \in L^2(\Omega \times Y)$ (resp. $\xi_0(x,y) \in L^2(\Omega \times Y)^N$). Thus for any $\psi(x,y) \in C_c^\infty\left(\Omega; C^\infty(\mathbb{T}^N)^N\right)$, we have

$$\lim_{\epsilon \to 0} \int_\Omega \nabla u_\epsilon(x) \cdot \psi\left(x, \frac{x}{\epsilon}\right) dx = \int_\Omega \int_Y \xi_0(x,y) \cdot \psi(x,y) dx dy. \tag{18}$$

Integrating by parts the left-hand side of (18) gives

$$\epsilon \int_\Omega \nabla u_\epsilon(x) \cdot \psi \left(x, \frac{x}{\epsilon} \right) dx = - \int_\Omega u_\epsilon(x) \left(\operatorname{div}_y \psi + \epsilon \operatorname{div}_x \psi \right) \left(x, \frac{x}{\epsilon} \right) dx.$$

Passing to the limit yields

$$0 = - \int_\Omega \int_Y u_0(x,y) \operatorname{div}_y \psi(x,y) dx dy. \tag{19}$$

Another integration by parts in (19) shows that $u_0(x,y)$ does not depend on y, i.e., there exists $u(x) \in L^2(\Omega)$, such that $u_0(x,y) \equiv u(x)$. Next, in (18) we choose a function ψ such that $\operatorname{div}_y \psi(x,y) = 0$. Integrating by parts we obtain

$$\int_\Omega \int_Y u(x) \operatorname{div}_x \psi(x,y) dx dy = \lim_{\epsilon \to 0} \int_\Omega u_\epsilon(x) \operatorname{div}_x \psi \left(x, \frac{x}{\epsilon} \right) dx$$
$$= - \int_\Omega \int_Y \xi_0(x,y) \cdot \psi(x,y) dx dy. \tag{20}$$

If ψ does not depend on y, (20) proves that $u(x)$ belongs to $H^1(\Omega)$. Furthermore, we deduce from (20) that

$$\int_\Omega \int_Y (\xi_0(x,y) - \nabla u(x)) \cdot \psi(x,y) dx dy = 0 \tag{21}$$

for any function $\psi(x,y) \in C_c^\infty \left(\Omega; C^\infty(\mathbb{T}^N)^N \right)$ with $\operatorname{div}_y \psi(x,y) = 0$. Recall that the orthogonal of divergence-free functions are exactly the gradients (this well-known result can be very easily proved in the present context by means of Fourier analysis in $\mathbb{T}^N$). Thus, there exists a unique function $u_1(x,y)$ in $L^2(\Omega; H^1(\mathbb{T}^N)/\mathbb{R})$ such that $\xi_0(x,y) = \nabla u(x) + \nabla_y u_1(x,y)$. $\quad\square$

There are many generalizations of Theorem 3.3 which gives the precise form of the two-scale limit of a sequence of functions for which we know some extra estimates on part of their derivatives. In the sequel we shall need the following example of a generalization of Theorem 3.3, the proof of which may be found in [All92].

Theorem 3.4. *Let u_ϵ be a bounded sequence in $L^2(\Omega)$ such that $\epsilon \nabla u_\epsilon$ is also bounded in $L^2(\Omega)^N$. Then, there exists a two-scale limit $u_0(x,y) \in L^2(\Omega; H^1(\mathbb{T}^N)/\mathbb{R})$ such that, up to a subsequence, u_ϵ two-scale converges to $u_0(x,y)$, and $\epsilon \nabla u_\epsilon$ to $\nabla_y u_0(x,y)$.*

Remark 3.2. It is well known that, in general, non-linear functionals are not continuous with respect to the weak topologies of $L^p(\Omega)$ spaces $(1 \leq p \leq +\infty)$. Unfortunately, the same is true with the two-scale convergence which is also a weak-type convergence. As for the usual weak $L^p(\Omega)$

topology, we can merely establish a lower semi-continuity result for convex functionals of the same type than inequality (15) on the norm of the two-scale limit. For more details we refer to Sect. 3 in [All92].

4 Application to Homogenization

This section is devoted to the application of two-scale convergence to the homogenization of partial differential equations with periodically oscillating coefficients. For simplicity we focus on the model problem of diffusion in a periodic medium, as in Sect. 2. Of course, the principles of the two-scale convergence method are valid in many other cases with some changes, including non-linear (monotone or convex) problems.

We consider the following model problem of diffusion

$$\begin{cases} -\text{div}\left(A\left(x, \frac{x}{\epsilon}\right)\nabla u_\epsilon\right) = f \text{ in } \Omega \\ u_\epsilon = 0 \qquad\qquad\qquad \text{on } \partial\Omega. \end{cases} \quad (22)$$

We assume that Ω is a bounded open set in $\mathbb{R}^N$, and that the tensor of diffusion $A(x, y)$, not necessarily symmetric, belongs to $C(\bar{\Omega}; L^\infty(\mathbb{T}^N))^{N^2}$ and satisfies everywhere in $\Omega \times \mathbb{T}^N$

$$\alpha|\xi|^2 \le \sum_{i,j=1}^{N} A_{i,j}(x,y)\xi_i\xi_j \le \beta|\xi|^2,$$

for any vector $\xi \in \mathbb{R}^N$, α and β being two constants such that $0 < \alpha \le \beta$. For a source term $f(x)$ in $L^2(\Omega)$, (22) admits a unique solution u_ϵ in $H_0^1(\Omega)$ by application of Lax–Milgram lemma. Moreover, u_ϵ satisfies the following a priori estimate

$$\|u_\epsilon\|_{H_0^1(\Omega)} \le C\|f\|_{L^2(\Omega)} \quad (23)$$

where C is a positive constant which does not depend on ϵ or f.

We now describe the so-called "two-scale convergence method" for homogenizing problem (22). In a *first step*, we deduce from the a priori estimate (23) the precise form of the two-scale limit of the sequence u_ϵ. By application of Theorem 3.3, there exist two functions, $u(x) \in H_0^1(\Omega)$ and $u_1(x, y) \in L^2(\Omega; H^1(\mathbb{T}^N)/\mathbb{R})$, such that, up to a subsequence, u_ϵ two-scale converges to $u(x)$, and ∇u_ϵ two-scale converges to $\nabla_x u(x) + \nabla_y u_1(x, y)$. In view of these limits, u_ϵ is expected to behave as $u(x) + \epsilon u_1\left(x, \frac{x}{\epsilon}\right)$.

Thus, in a *second step*, we multiply (22) by a test function similar to the limit of u_ϵ, namely $\varphi(x) + \epsilon\varphi_1\left(x, \frac{x}{\epsilon}\right)$, where $\varphi(x) \in C_c^\infty(\Omega)$ and $\varphi_1(x,y) \in C_c^\infty(\Omega; C^\infty(\mathbb{T}^N))$. This yields

$$\int_{\Omega} A\left(x, \frac{x}{\epsilon}\right) \nabla u_{\epsilon}(x) \cdot \left(\nabla \varphi(x) + \nabla_y \varphi_1\left(x, \frac{x}{\epsilon}\right) + \epsilon \nabla_x \varphi_1\left(x, \frac{x}{\epsilon}\right)\right) dx \quad (24)$$

$$= \int_{\Omega} f(x) \left(\varphi(x) + \epsilon \varphi_1\left(x, \frac{x}{\epsilon}\right)\right) dx.$$

Regarding $A^t\left(x, \frac{x}{\epsilon}\right)\left(\nabla \varphi(x) + \nabla_y \varphi_1\left(x, \frac{x}{\epsilon}\right)\right)$ as a test function for the two-scale convergence (see Definition 3.1), we pass to the two-scale limit in (24) for the sequence ∇u_{ϵ}. Although this test function is not necessarily very smooth, as required by Definition 3.1, it belongs at least to $C\left(\bar{\Omega}; L^2(\mathbb{T}^N)\right)$ which can be shown to be enough for the two-scale convergence Theorem 3.1 to hold (see [All92] for details). Thus, the two-scale limit of (24) is

$$\int_{\Omega} \int_Y A(x, y) \left(\nabla u(x) + \nabla_y u_1(x, y)\right) \cdot \left(\nabla \varphi(x) + \nabla_y \varphi_1(x, y)\right) dx dy$$

$$= \int_{\Omega} f(x) \varphi(x) dx. \quad (25)$$

In a *third step*, we read off a variational formulation for (u, u_1) in (25). Remark that (25) holds true for any (φ, φ_1) in the Hilbert space $H_0^1(\Omega) \times L^2\left(\Omega; H^1(\mathbb{T}^N)/\mathbb{R}\right)$ by density of smooth functions in this space. Endowing it with the norm $\left(\|\nabla u\|_{L^2(\Omega)} + \|\nabla_y u_1\|_{L^2(\Omega \times \mathbb{T}^N)}\right)$, the assumptions of the Lax–Milgram lemma are easily checked for the variational formulation (25). The main point is the coercivity of the bilinear form defined by the left-hand side of (25): the coercivity of A yields

$$\int_{\Omega} \int_Y A(x, y) \left(\nabla \varphi(x) + \nabla_y \varphi_1(x, y)\right) \cdot \left(\nabla \varphi(x) + \nabla_y \varphi_1(x, y)\right) dx dy \geq$$

$$\alpha \int_{\Omega} \int_Y |\nabla \varphi + \nabla_y \varphi_1|^2 dx dy = \alpha \int_{\Omega} |\nabla \varphi|^2 dx + \alpha \int_{\Omega} \int_Y |\nabla_y \varphi_1|^2 dx dy.$$

By application of the Lax–Milgram lemma, we conclude that there exists a unique solution (u, u_1) of the variational formulation (25) in $H_0^1(\Omega) \times L^2\left(\Omega; H^1(\mathbb{T}^N)/\mathbb{R}\right)$. Consequently, the entire sequences u_{ϵ} and ∇u_{ϵ} converge to $u(x)$ and $\nabla u(x) + \nabla_y u_1(x, y)$. An easy integration by parts shows that (25) is a variational formulation associated to the following system of equations, the so-called "two-scale homogenized problem",

$$\begin{cases} -\text{div}_y \left(A(x, y) \left(\nabla u(x) + \nabla_y u_1(x, y)\right)\right) = 0 & \text{in } \Omega \times \mathbb{T}^N \\ -\text{div}_x \left(\int_Y A(x, y) \left(\nabla u(x) + \nabla_y u_1(x, y)\right) dy\right) = f(x) & \text{in } \Omega \\ u = 0 & \text{on } \partial\Omega. \end{cases} \quad (26)$$

At this point, the homogenization process could be considered as achieved since the entire sequence of solutions u_{ϵ} converges to the solution of a well-posed limit problem, namely the two-scale homogenized problem (26). However, it is usually preferable, from a physical or numerical point of view, to eliminate the microscopic variable y. In other words, we want to extract and decouple the usual homogenized and cell equations from (26).

Thus, in a *fourth (and optional) step*, the y variable and the u_1 unknown are eliminated from (26). As in the previous section we obtain u_1 in terms of the gradient of u relationship

$$u_1(x, y) = \sum_{i=1}^{N} \frac{\partial u}{\partial x_i}(x) w_i(x, y), \tag{27}$$

where $w_i(x, y)$ is defined, at each point $x \in \Omega$, as the unique solution in $H^1(\mathbb{T}^N)/\mathbb{R}$ of the cell problem

$$-\text{div}_y \left(A(x, y) \left(e_i + \nabla_y w_i(x, y) \right) \right) = 0 \quad \text{in } \mathbb{T}^N. \tag{28}$$

Then, plugging formula (27) in (26) yields the homogenized problem

$$\begin{cases} -\text{div}_x \left(A^*(x) \nabla u(x) \right) = f(x) \text{ in } \Omega \\ u = 0 \qquad\qquad\qquad\qquad\ \text{ on } \partial\Omega, \end{cases} \tag{29}$$

where the homogenized diffusion tensor is given by its entries

$$A_{ij}^*(x) = \int_Y A(x, y) \left(e_i + \nabla_y w_i(x, y) \right) \cdot \left(e_j + \nabla_y w_j(x, y) \right) dy. \tag{30}$$

Of course, all the above formulas coincide with the previous ones, obtained by using asymptotic expansions in Sect. 2.

Due to the simple form of our model problem the two equations of (26) can be decoupled in a microscopic and a macroscopic equation, (28) and (29) respectively, but we emphasize that it is not always possible, and sometimes it leads to very complicate forms of the homogenized equation.

Remark 4.1. It is possible to obtain so-called "corrector" results which give strong (or pointwise) convergences instead of just weak ones by adding some extra information stemming from the local equations. Typically for the above example, introducing u_1 defined by (27), we can prove

$$\left(u_\epsilon(x) - u(x) - \epsilon u_1 \left(x, \frac{x}{\epsilon} \right) \right) \to 0 \text{ in } H_0^1(\Omega) \text{ strongly} \tag{31}$$

under mild regularity assumptions on $u_1(x, y)$. We refer to [All92] for a proof.

5 Bloch Waves

The method of Bloch waves [Blo28], or the Bloch transform, is a generalization of Fourier transform that leaves invariant periodic functions. It is also known as Floquet theory [Flo83]. For a modern treatment of this topic see [BLP78], [CPV95], [Gel50], [Kuc93], [OK64], [RS78], [Wil78]. There are two steps in the Bloch decomposition: the first one is general (as is the Fourier transform), while the second one is specialized to a given self-adjoint partial differential equation.

The first step of the Bloch transform is the following result.

Theorem 5.1. *For any function $u(y) \in L^2(\mathbb{R}^N)$ there exists a unique function $\hat{u}(y, \theta) \in L^2(Y \times Y)$ such that*

$$u(y) = \int_Y \hat{u}(y, \theta)e^{2i\pi\theta \cdot y}d\theta. \tag{32}$$

The function $y \to \hat{u}(y, \theta)$ is Y-periodic while the function $\theta \to e^{2i\pi\theta \cdot y}\hat{u}(y, \theta)$ is Y-periodic. Furthermore, the linear map $\mathcal{B}$, called the Bloch transform and defined by $\mathcal{B}u = \hat{u}$, is an isometry from $L^2(\mathbb{R}^N)$ into $L^2(Y \times Y)$, i.e., Parseval formula holds for any $u, v \in L^2(\mathbb{R}^N)$

$$\int_{\mathbb{R}^N} u(y)\overline{v(y)}\, dy = \int_Y \int_Y \hat{u}(y, \theta)\overline{\hat{v}(y, \theta)}\, dy\, d\theta. \tag{33}$$

Remark 5.1. In Theorem 5.1 θ plays the role of a dual variable (like for the Fourier transform). It is often called Bloch parameter, reduced frequency (since its range is restricted to Y), or quasi momentum (in solid state physics). The physical interpretation of (32) is that any function in $L^2(\mathbb{R}^N)$ is a superposition (more precisely an integral with respect to θ) of the product of a periodic function, $y \to \hat{u}(y, \theta)$ and of a plane wave with wave number θ.

Proof. Let $u(y)$ be a smooth compactly supported function in $\mathbb{R}^N$. We define the function $\hat{u}$ by

$$\hat{u}(y, \theta) = \sum_{k \in \mathbb{Z}^N} u(y + k)e^{-2i\pi\theta \cdot (y+k)}.$$

This sum is well defined because it has a finite number of terms since u has compact support. It is also clearly a Y-periodic function of y. On the other hand, for $j \in \mathbb{Z}^N$, since $e^{-2i\pi j \cdot k} = 1$, we have

$$\hat{u}(y, \theta + j) = e^{-2i\pi j \cdot y}\sum_{k \in \mathbb{Z}^N} u(y + k)e^{-2i\pi\theta \cdot (y+k)} = e^{-2i\pi j \cdot y}\hat{u}(y, \theta).$$

Thus, $\theta \to e^{2i\pi\theta \cdot y}\hat{u}(y, \theta)$ is Y-periodic. Next, we compute

$$\int_Y \hat{u}(y, \theta)e^{2i\pi\theta \cdot y}d\theta = \sum_{k \in \mathbb{Z}^N} u(y + k)\int_Y e^{-2i\pi\theta \cdot k}d\theta = u(y)$$

since all integrals vanish except for $k = 0$. Therefore (32) is proved for smooth compactly supported functions. A similar argument works also for (33) which, in particular, shows that the Bloch transform $\mathcal{B}$ is a linear map, well defined on $C_c^\infty(\mathbb{R}^N)$ and bounded on $L^2(\mathbb{R}^N)$. Since $C_c^\infty(\mathbb{R}^N)$ is dense in $L^2(\mathbb{R}^N)$, $\mathcal{B}$ can be extended by continuity and (32), (33) hold true in $L^2(\mathbb{R}^N)$.   □

We now give a lemma (the proof of which is left to the reader as a simple exercise) showing in which sense the Bloch transform leaves invariant the periodic functions.

Lemma 5.1. Let $a(y) \in L^\infty(\mathbb{T}^N)$ be a periodic function. For any $u(y) \in L^2(\mathbb{R}^N)$, we have

$$\mathcal{B}(au) = a\mathcal{B}(u) \equiv a(y)\hat{u}(y, \theta).$$

We now turn to the second step of the Bloch transform which relies on the choice of a self-adjoint partial differential equation. For simplicity, we consider the following second-order symmetric p.d.e.

$$-\operatorname{div}_y\left(A(y)\nabla_y u\right) + c(y)u = f \quad \text{in } \mathbb{R}^N, \tag{34}$$

where the right-hand side f belongs to $L^2(\mathbb{R}^N)$. We assume that the coefficients $A(y)$ and $c(y)$ are real measurable bounded periodic functions, i.e., their entries belong to $L^\infty(\mathbb{T}^N)$. The tensor A is symmetric and uniformly coercive, i.e., there exists $\nu > 0$ such that for a.e. $y \in \mathbb{T}^N$

$$A(y)\xi \cdot \xi \geq \nu|\xi|^2 \text{ for any } \xi \in \mathbb{R}^N.$$

Furthermore, we assume that c is uniformly positive, i.e., there exists $c_0 > 0$ such that for a.e. $y \in \mathbb{T}^N$

$$c(y) \geq c_0 > 0.$$

The variational formulation of (34) is to find $u \in H^1(\mathbb{R}^N)$ such that, for any $\phi \in H^1(\mathbb{R}^N)$,

$$\int_{\mathbb{R}^N} \left(A(y)\nabla_y u \cdot \overline{\nabla_y \phi} + c(y)u\overline{\phi}\right) dy = \int_{\mathbb{R}^N} f\overline{\phi}\, dy. \tag{35}$$

Thanks to our assumptions on the coefficients, a simple application of the Lax–Milgram lemma yields the existence and uniqueness of a solution of (35) and thus of (34).

A first interesting application of Theorem 5.1 is to simplify (35) which is posed on the whole space and to reduce it to a family of problems posed on the simpler compact set $\mathbb{T}^N$. To show this we need the following simple lemma, the proof of which is left to the reader as a simple exercise.

Lemma 5.2. Let $u(y) \in H^1(\mathbb{R}^N)$. The Bloch transform of its gradient is

$$\mathcal{B}(\nabla_y u) = (\nabla_y + 2i\pi\theta)\mathcal{B}(u) \equiv \nabla_y \hat{u}(y, \theta) + 2i\pi\theta\hat{u}(y, \theta).$$

Proposition 5.1. *The p.d.e. (34) is equivalent to the family of p.d.e.'s, indexed by $\theta \in Y$,*

$$-(\mathrm{div}_y + 2i\pi\theta)\Big(A(y)(\nabla_y + 2i\pi\theta)\mathcal{B}u\Big) + c(y)\mathcal{B}u = \mathcal{B}f \quad in\ \mathbb{T}^N, \qquad (36)$$

which admits a unique solution $y \to (\mathcal{B}u)(y,\theta) \in H^1(\mathbb{T}^N)$ for any $\theta \in Y$.

Proof. We apply Lemmas 5.1 and 5.2 to (35) and obtain

$$\int_{\mathbb{T}^N}\int_Y \Big(A(\nabla_y+2i\pi\theta)\mathcal{B}u \cdot \overline{(\nabla_y+2i\pi\theta)\mathcal{B}\phi} + c\mathcal{B}u\overline{\mathcal{B}\phi}\Big)dy\,d\theta = \int_{\mathbb{T}^N}\int_Y \mathcal{B}f\overline{\mathcal{B}\phi}\,dy\,d\theta$$

which is just the variational formulation of (36) integrated with respect to θ (a mere parameter since there is no derivatives with respect to θ). This yields (36). □

One can still go further in the simplification of (36), and thus of (34), by using the Hilbertian basis of eigenfunctions of (36). Indeed, the Green operator for (36) is now compact (on the contrary of that for (34)). More precisely, for a given $\theta \in Y$, let us consider the Green operator $\mathcal{G}_\theta$ defined by

$$\begin{cases} L^2(\mathbb{T}^N) \to L^2(\mathbb{T}^N) \\ g(y) \quad \to \mathcal{G}_\theta g(y) = v(y) \end{cases} \qquad (37)$$

where $v \in H^1(\mathbb{T}^N)$ is the unique solution of

$$-(\mathrm{div}_y + 2i\pi\theta)\Big(A(y)(\nabla_y + 2i\pi\theta)v\Big) + c(y)v = g \quad in\ \mathbb{T}^N.$$

Denoting by $\langle,\rangle$ the complex Hilbertian product on $L^2(\mathbb{T}^N)$, the reader can easily check that

$$\langle \mathcal{G}_\theta g_1, g_2\rangle = \int_{\mathbb{T}^N}\Big(A(y)(\nabla_y+2i\pi\theta)v_1\cdot\overline{(\nabla_y+2i\pi\theta)v_2} + c(y)v_1\overline{v_2}\Big)dy = \langle g_1, \mathcal{G}_\theta g_2\rangle.$$

Clearly, $\mathcal{G}_\theta$ is a self-adjoint compact complex-valued linear operator acting on $L^2(\mathbb{T}^N)$. As such its inverse admits a countable sequence of real increasing eigenvalues $(\lambda_n)_{n\geq 1}$ (repeated with their multiplicity) and normalized eigenfunctions $(\psi_n)_{n\geq 1}$ with $\|\psi_n\|_{L^2(\mathbb{T}^N)} = 1$. The eigenvalues and eigenfunctions depend on the Bloch frequency $\theta \in Y$. In other words, the eigenvalues and eigenfunctions satisfy the so-called Bloch (or shifted) spectral cell equation

$$-(\mathrm{div}_y + 2i\pi\theta)\Big(A(y)(\nabla_y + 2i\pi\theta)\psi_n\Big) + c(y)\psi_n = \lambda_n(\theta)\psi_n \quad in\ \mathbb{T}^N. \quad (38)$$

The second step of the Bloch transform is the following result.

Theorem 5.2. *For any function $u(y) \in L^2(\mathbb{R}^N)$ there exists a unique countable family of functions $\hat{u}_n(\theta) \in L^2(Y)$, $n \geq 1$, such that*

$$u(y) = \sum_{n \geq 1} \int_Y \hat{u}_n(\theta) \psi_n(y, \theta) e^{2i\pi\theta \cdot y} d\theta. \tag{39}$$

Furthermore, the linear map $\mathcal{B}$, called the Bloch transform and defined by $\mathcal{B}u = (\hat{u}_n)_{n \geq 1}$, is an isometry from $L^2(\mathbb{R}^N)$ into $\ell_2\left(L^2(Y)\right)$, i.e., Parseval formula holds for any $u, v \in L^2(\mathbb{R}^N)$

$$\int_{\mathbb{R}^N} u(y)\overline{v(y)} \, dy = \sum_{n \geq 1} \int_Y \hat{u}_n(\theta)\overline{\hat{v}_n(\theta)} \, d\theta. \tag{40}$$

Proof. With the notations of Theorem 5.1 we decompose each $\hat{u}(y, \theta)$ on the corresponding eigenbasis

$$\hat{u}(y, \theta) = \sum_{n \geq 1} \hat{u}_n(\theta)\psi_n(y, \theta) \quad \text{with } \hat{u}_n(\theta) = \int_{\mathbb{T}^N} \hat{u}(y, \theta)\overline{\psi_n(y, \theta)}dy.$$

Commuting the sum with respect to n and the integral with respect to θ is a standard Fubini type result. There is a subtle point about the measurability, with respect to θ, of the eigenfunctions $\psi_n(y, \theta)$. A special choice of their normalization (they are defined up to multiplication by a unit complex function of θ) allows to state a measurable selection result (for details, see [Wil78]). $\square$

Remark 5.2. As a matter fact, our assumption on the coefficient $c(y)$ to be uniformly positive is not necessary for Theorem 5.2 to hold true. Indeed, if $c(y)$ just belongs to $L^\infty(\mathbb{T}^N)$, it is bounded from below and adding a large positive constant makes it positive, does not change the eigenfunctions and simply shifts the entire spectrum.

We come back to our model p.d.e. (34). Applying the Bloch transform of Theorem 5.2 to the right-hand side and solution of (34) we obtain an explicit algebraic formula for the solution

$$\hat{u}_n(\theta) = \frac{\hat{f}_n(\theta)}{\lambda_n(\theta)} \quad \forall n \geq 1, \forall \theta \in Y,$$

which is a generalization of a similar formula, using Fourier transform, for a constant coefficient p.d.e..

Definition. According to the context of Theorem 5.1 or of Theorem 5.2, a Bloch wave is either a function of the type $\psi(y)e^{2i\pi\theta \cdot y}$ where ψ is any periodic function defined on $\mathbb{T}^N$, or is precisely $\psi_n(y, \theta)e^{2i\pi\theta \cdot y}$ where ψ_n is an eigenfunction of (38).

A crucial point in the study of Bloch waves is to know the regularity of the eigenvalues $\lambda_n(\theta)$ and eigenfunctions $\psi_n(y, \theta)$ with respect to θ. Remark that the coefficients of (36) are polynomials of degree 2 in θ, so that the Green operator $\mathcal{G}_\theta$ is analytic with respect to θ. However it is well known that all eigenvalues and eigenfunctions are not as smooth, and some caution is in order [Kat66]. Nevertheless, if an eigenvalue $\lambda_n(\theta)$ is simple at the value $\theta = \theta^n$, then it remains simple in a small neighborhood of θ^n and it is a classical matter to prove that the n-th eigencouple of (38) is analytic in this neighborhood of θ^n [Kat66].

Remark 5.3. In one space dimension $N = 1$ it is well-known that all eigenvalues $\lambda_n(\theta)$ are simple, except possibly for $\theta = 0$ or $\theta = \pm 1/2$ when there is no gap below or above the n-th band (the so-called co-existence case, see [MW66]). In higher dimensions, $\lambda_n(\theta)$ has no reason to be simple although there are some results of generic simplicity in similar contexts, see [Alb75]. A multiple eigenvalue corresponds to the occurrence of "crossing" for smooth branches of eigenvalues, viewed as functions of θ.

In the sequel, we shall consider an energy level $n \geq 1$ and a Bloch parameter $\theta^n \in Y$ such that the eigenvalue

$$\lambda_n(\theta^n) \text{ is a simple eigenvalue.} \tag{41}$$

Under assumption (41) the n-th eigencouple of (38) is smooth in a neighborhood of θ^n [Kat66]. Introducing the operator $\mathbb{A}_n(\theta)$ defined on $L^2(\mathbb{T}^N)$ by

$$\mathbb{A}_n(\theta)\psi = -(\operatorname{div}_y + 2i\pi\theta)\Big(A(y)(\nabla_y + 2i\pi\theta)\psi\Big) + c(y)\psi - \lambda_n(\theta)\psi, \tag{42}$$

it is easy to differentiate (38). Denoting by $(e_k)_{1 \leq k \leq N}$ the canonical basis of $\mathbb{R}^N$ and by $(\theta_k)_{1 \leq k \leq N}$ the components of θ, the first derivative satisfies

$$\mathbb{A}_n(\theta)\frac{\partial \psi_n}{\partial \theta_k} = 2i\pi e_k A(\nabla_y + 2i\pi\theta)\psi_n + (\operatorname{div}_y + 2i\pi\theta)(A2i\pi e_k \psi_n) + \frac{\partial \lambda_n}{\partial \theta_k}(\theta)\psi_n, \tag{43}$$

and the second derivative is

$$\begin{aligned}
\mathbb{A}_n(\theta)\frac{\partial^2 \psi_n}{\partial \theta_k \partial \theta_l} &= 2i\pi e_k A(\nabla_y + 2i\pi\theta)\frac{\partial \psi_n}{\partial \theta_l} + (\operatorname{div}_y + 2i\pi\theta)\left(A2i\pi e_k \frac{\partial \psi_n}{\partial \theta_l}\right) \\
&\quad + 2i\pi e_l A(\nabla_y + 2i\pi\theta)\frac{\partial \psi_n}{\partial \theta_k} + (\operatorname{div}_y + 2i\pi\theta)\left(A2i\pi e_l \frac{\partial \psi_n}{\partial \theta_k}\right) \\
&\quad + \frac{\partial \lambda_n}{\partial \theta_k}(\theta)\frac{\partial \psi_n}{\partial \theta_l} + \frac{\partial \lambda_n}{\partial \theta_l}(\theta)\frac{\partial \psi_n}{\partial \theta_k} \\
&\quad - 4\pi^2 e_k A e_l \psi_n - 4\pi^2 e_l A e_k \psi_n + \frac{\partial^2 \lambda_n}{\partial \theta_l \partial \theta_k}(\theta)\psi_n
\end{aligned} \tag{44}$$

There exists a unique solution of (43), up to the addition of a multiple of ψ_n. Indeed, since there necessarily exists a partial derivative of ψ_n with respect to θ_k, the right-hand side of (43) satisfies the required compatibility condition or Fredholm alternative (i.e., it is orthogonal to ψ_n)

$$\int_{\mathbb{T}^N} \mathbb{A}_n(\theta^n) \frac{\partial \psi_n}{\partial \theta_k} \overline{\psi}_n \, dy = 0 \tag{45}$$

which yields a formula for $\nabla_\theta \lambda_n(\theta^n)$ in terms of ψ_n. On the same token, there exists a unique solution of (44), up to the addition of a multiple of ψ_n. The compatibility condition of (44) is

$$\int_{\mathbb{T}^N} \mathbb{A}_n(\theta^n) \frac{\partial^2 \psi_n}{\partial \theta_k \partial \theta_l} \overline{\psi}_n \, dy = 0 \tag{46}$$

which yields a formula for the Hessian matrix $\nabla_\theta \nabla_\theta \lambda_n(\theta^n)$.

6 Schrödinger Equation in Periodic Media

We study the homogenization of the following Schrödinger equation

$$\begin{cases} i\dfrac{\partial u_\epsilon}{\partial t} - \text{div}\left(A\left(\dfrac{x}{\epsilon}\right)\nabla u_\epsilon\right) + \left(\epsilon^{-2}c\left(\dfrac{x}{\epsilon}\right) + d\left(x, \dfrac{x}{\epsilon}\right)\right) u_\epsilon = 0 & \text{in } \mathbb{R}^N \times \mathbb{R}^+ \\ u_\epsilon(t = 0, x) = u_\epsilon^0(x) & \text{in } \mathbb{R}^N, \end{cases}$$

$$\tag{47}$$

where the unknown function u_ϵ is complex-valued. The coefficients $A_{ij}(y)$ and $c(y)$ are assumed to be real, measurable, bounded, periodic functions, i.e., belong to $L^\infty(\mathbb{T}^N)$, the tensor $A(y)$ is symmetric uniformly coercive, while $d(x,y)$ is real, measurable and bounded with respect to x, and periodic continuous with respect to y, i.e., belongs to $L^\infty(\mathbb{R}^N; C(\mathbb{T}^N))$. Remark that $c(y)$ and $d(x,y)$ do not satisfy any positivity assumption. Of course, the "usual" Schrödinger equation corresponds to the choice $A(y) \equiv Id$. Other choices may be interpreted as a periodic metric. The scaling of (47) is a "parabolic" scaling, typical of homogenization (see, e.g., [ACPSV04], or Chap. 4 in [BLP78]), but is different from the "semi-classical" scaling which is discussed in Sect. 7 below. In particular, in (47) we consider much larger times than in the semi-classical setting.

If the initial data u_ϵ^0 belongs to $H^1(\mathbb{R}^N)$, there exists a unique solution of the Schrödinger equation (47) in $C\left(\mathbb{R}^+; H^1(\mathbb{R}^N)\right)$ which satisfies the following a priori estimate.

Lemma 6.1. *There exists a constant $C > 0$ that does not depend on ϵ such that the solution of (47) satisfies*

$$\|u_\epsilon\|_{L^\infty(\mathbb{R}^+; L^2(\mathbb{R}^N))} = \|u_\epsilon^0\|_{L^2(\mathbb{R}^N)},$$

$$\epsilon \|\nabla u_\epsilon\|_{L^\infty(\mathbb{R}^+; L^2(\mathbb{R}^N)^N)} \le C\left(\|u_\epsilon^0\|_{L^2(\mathbb{R}^N)} + \epsilon \|\nabla u_\epsilon^0\|_{L^2(\mathbb{R}^N)^N}\right). \tag{48}$$

Proof. We multiply (47) by $\overline{u_\epsilon}$ and take the imaginary part to obtain

$$\frac{d}{dt} \int_{\mathbb{R}^N} |u_\epsilon(t,x)|^2 dx = 0.$$

Next we multiply (47) by $\frac{\partial \overline{u_\epsilon}}{\partial t}$ and we take the real part to get

$$\frac{d}{dt} \int_{\mathbb{R}^N} \left(\epsilon^2 A \left(\frac{x}{\epsilon} \right) \nabla u_\epsilon \cdot \nabla \overline{u_\epsilon} + \left(c \left(\frac{x}{\epsilon} \right) + \epsilon^2 d \left(x, \frac{x}{\epsilon} \right) \right) |u_\epsilon|^2 \right) dx = 0.$$

This yields the required a priori estimates.   □

We are interested in initial data that are Bloch wave packets of the type

$$u_\epsilon^0(x) = \psi_n \left(\frac{x}{\epsilon}, \theta^n \right) e^{2i\pi \frac{\theta^n \cdot x}{\epsilon}} v^0(x) \tag{49}$$

with ψ_n a Bloch eigenfunction defined by (38). Such initial data are important in solid state physics where (47) is the so-called one-electron model in a periodic crystal (see, e.g., [Mye90], [Ped97], [Que98]). The goal of homogenizing (47) is to rigorously derived the homogenized coefficients for the initial data (49) which are called *effective mass*.

7 Semiclassical Analysis and WKB Ansatz

The scaling of the Schrödinger equation (47) that we shall study in the next sections is called "parabolic scaling". It is different from the "semi-classical scaling", which is usual in semi-classical analysis (see, e.g., [Bus87], [DGR02], [Ger91], [GMNP97], [GMS91], [GRT88], [PST03], [PR96]), and which reads

$$\frac{i}{\epsilon} \frac{\partial u_\epsilon}{\partial t} - \text{div} \left(A \left(x, \frac{x}{\epsilon} \right) \nabla u_\epsilon \right) + \epsilon^{-2} c \left(x, \frac{x}{\epsilon} \right) u_\epsilon = 0 \quad \text{in } \mathbb{R}^N \times \mathbb{R}^+. \tag{50}$$

Note the ϵ^{-1} factor in front of the time derivative which is absent in (47) and means that shorter times are considered in (50). We also have introduced a macroscopic modulation in x of the coefficients which is possible in the semi-classical context. Therefore, the Bloch spectral cell problem (38) now depends on x as a parameter. In other words, (38) is replaced by

$$-(\text{div}_y + 2i\pi\theta)(A(x,y)(\nabla_y + 2i\pi\theta)\psi_n) + c(x,y)\psi_n = \lambda_n(x,\theta)\psi_n \text{ in } \mathbb{T}^N. \tag{51}$$

The goal of this section is to recall the famous WKB method (following the names of Wentzel, Kramers, Brillouin, see, e.g., [BLP78]) which is at the basis of geometric optics and relies on a different type of asymptotic expansions than those studied in Sect. 2. The starting point of the WKB method is the following postulated ansatz for the solution of (50)

$$u_\epsilon(x) = e^{2i\pi \frac{S(t,x)}{\epsilon}} \sum_{i=0}^{+\infty} \epsilon^i v_i\left(t, x, \frac{x}{\epsilon}\right), \tag{52}$$

where each term $v_i(t, x, y)$ is a Y-periodic function with respect to y. Note that the ansatz (52) is different from the previous one (2) because of the highly oscillating phase in front of the series. The function $S(t, x)$ is called the phase, while the series plays the role of an amplitude. The first derivatives of (52) are

$$\epsilon \frac{\partial u_\epsilon}{\partial t} = e^{2i\pi \frac{S(t,x)}{\epsilon}} \left(2i\pi v_0 \frac{\partial S}{\partial t} + \sum_{i=0}^{+\infty} \epsilon^i \left(\frac{\partial v_i}{\partial t} + v_{i+1} \frac{\partial S}{\partial t}\right)\right),$$

$$\epsilon \nabla u_\epsilon = e^{2i\pi \frac{S(t,x)}{\epsilon}} \left((\nabla_y + 2i\pi \nabla S)v_0 + \sum_{i=0}^{+\infty} \epsilon^i \left(\nabla_x v_i + (\nabla_y + 2i\pi \nabla S)v_{i+1}\right)\right).$$

We compute also the spatial second derivatives, plug the ansatz in (50) and deduce a cascade of equations. Inspired by (42), let us introduce the operator $\mathbb{A}(x, \theta)$ defined on $L^2(\mathbb{T}^N)$ by

$$\mathbb{A}(x, \theta)\psi = -(\mathrm{div}_y + 2i\pi\theta)\Big(A(x, y)(\nabla_y + 2i\pi\theta)\psi\Big) + c(x, y)\psi. \tag{53}$$

The ϵ^{-2} equation is

$$\mathbb{A}(x, \nabla S)v_0 = 2\pi \frac{\partial S}{\partial t} v_0 \quad \text{in} \quad \mathbb{T}^N, \tag{54}$$

which is a spectral problem, similar to (38), y being the space variable and (t, x) being fixed parameters. If the initial data is a Bloch wave packet of the type (49), i.e., concentrated on the n-th band, then we deduce from (54) that

$$2\pi \frac{\partial S}{\partial t}(t, x) = \lambda_n\Big(x, \nabla S(t, x)\Big), \tag{55}$$

and by simplicity of the eigenvalue

$$v_0(t, x, y) = u(t, x)\psi_n(x, y, \nabla S(t, x)).$$

Equation (55) is an eikonal equation (or Hamilton–Jacobi equation) for the phase S which has to be supplemented by some initial condition $S(0, x) = S^0(x)$.

The ϵ^{-1} equation is

$$\mathbb{A}(x, \nabla S)v_1 = \lambda_n(x, \nabla S)v_1 + f \quad \text{in} \quad \mathbb{T}^N, \tag{56}$$

with a right-hand side

$$f = -i\frac{\partial v_0}{\partial t} + \Big(\mathrm{div}_y + 2i\pi\nabla S\Big)(A\nabla_x v_0) + \mathrm{div}_x\Big(A(\nabla_y + 2i\pi\nabla S)v_0\Big).$$

To solve (56) for v_1, the Fredholm alternative requires that, for any (t, x),

$$\int_{\mathbb{T}^N} f(t, x, y)\overline{\psi_n}(x, y, \nabla S)\, dy = 0. \tag{57}$$

In (57) we replace $v_0(t, x, y)$ by its value $u(t, x)\psi_n(x, y, \nabla S)$ and we obtain an homogenized transport equation. Indeed, from the Fredholm alternative (45) for $\nabla_\theta \psi_n$, solution of (43), we deduce

$$-\nabla_x u \cdot \nabla_\theta \lambda_n = 2i\pi \int_{\mathbb{T}^N} \Big((\mathrm{div}_y + 2i\pi\theta)\,(A\nabla_x u\psi_n) + A(\nabla_y + 2i\pi\theta)\psi_n \cdot \nabla_x u\Big)dy.$$

Introducing the so-called group velocity

$$\mathcal{V} = \frac{\nabla_\theta \lambda_n(x, \nabla S)}{2\pi}, \tag{58}$$

(57) is equivalent to the following transport equation

$$\frac{\partial u}{\partial t} - \mathcal{V} \cdot \nabla_x u + b^* u = 0 \tag{59}$$

where $b^*(t, x)$ is some complicated coefficient.

Applying the method of characteristics to the eikonal equation (55) leads to the following Hamiltonian system in the phase space $(x, \theta) \in \mathbb{R}^N \times \mathbb{T}^N$

$$\begin{cases} \dot{x} = \dfrac{-1}{2\pi}\nabla_\theta \lambda_n(x, \theta), \\ \dot{\theta} = \nabla_x \lambda_n(x, \theta), \end{cases} \tag{60}$$

which therefore describes the dynamic of the semi-classical limit (see Chap. 4 in [BLP78] for details). At first order, the ansatz (52) yields an approximate solution of (50) of the type

$$u_\epsilon(t, x) \approx e^{2i\pi \frac{S(t, x)}{\epsilon}} \psi_n\Big(x, \frac{x}{\epsilon}, \nabla S(t, x)\Big) u(t, x, \nabla S(t, x)) \tag{61}$$

where the phase S is a solution of the eikonal equation (55) and the amplitude u is a solution of the transport equation (59).

As exposed, the WKB method is purely heuristic. The difficulty for justifying the approximation (61) is that, although (59) is a simple and well-posed linear transport equation in the phase space, the eikonal equation (55) is non-linear and does not admit smooth solution after a finite time (even for infinitely smooth initial data). In particular, one can apply the method of characteristics to (55) for building its smooth solutions, but this method breaks down at the occurrence of so-called caustics which correspond to the crossing of characteristic lines. Beyond caustics there are no more smooth solutions and the ansatz (52) breaks down (one can introduce viscosity solution of (55) but their regularity is not enough for our purpose). There is a possibility of rigorously justifying the WKB method (globally in time) by using semi-classical or Wigner measures [Ger91], [GMNP97], [MNP94]. Note that the WKB method is useful for computations too [BP06].

There is a special case for which (55) and (59) can be solved explicitly and globally in time: periodic coefficients and monochromatic initial data. Let us assume that

$$A(x, y) \equiv A(y) \quad \text{and} \quad c(x, y) \equiv c(y),$$

and consider an initial data given by (49), i.e., the initial phase is

$$S^0(x) = \theta \cdot x.$$

Then the explicit solution of the eikonal equation (55) is

$$S(t, x) = \theta \cdot x + 2\pi \lambda_n(\theta) \, t$$

Furthermore $\mathcal{V}$ is constant, $b^* = 0$ and $\nabla_x \lambda_n(\theta) = 0$, thus the explicit solution of (59) is

$$u(t, x) = u_0(x + \mathcal{V}t)$$

In such a simpler case we can find a better "long time" ansatz of the solution as explained in the next sections.

8 Homogenization Without Drift

In this section we assume that the initial data (49) is such that, for the energy level n and Bloch parameter θ^n, the eigenvalue $\lambda_n(\theta^n)$ satisfies

$$\begin{cases} (i) \ \lambda_n(\theta^n) \text{ is a simple eigenvalue,} \\ (ii) \ \theta^n \text{ is a critical point of } \lambda_n(\theta) \text{ i.e., } \nabla_\theta \lambda_n(\theta^n) = 0. \end{cases} \quad (62)$$

Physically, it implies that the particle modeled by the limit wave function does not experience any drift since the group velocity, defined by (58), is zero. This assumption of simplicity has two important consequences. First, if $\lambda_n(\theta^n)$ is simple, then it is infinitely differentiable in a vicinity of θ^n. Second, if $\lambda_n(\theta^n)$ is simple, then the limit problem is going to be a single effective Schrödinger equation. In Sect. 10 we make another assumption of a multiple eigenvalue with smooth branches. Then the homogenized limit is a system of several coupled Schrödinger equations (as many as the multiplicity).

Remark 8.1. Concerning the existence of critical points of $\lambda_n(\theta)$, it is easily checked that for the first band or energy level $n = 1$ assumption (62) is always satisfied with $\theta^1 = 0$ which is a minimum point of λ_1 (see, e.g., [BLP78],

[CV97]). In full generality, there may be or not a critical point of $\lambda_n(\theta)$. For example, in the case of constant coefficients, $\lambda_n(\theta)$ has no critical points for $n > 1$. However, in $N = 1$ space dimension it is well known (see, e.g., [MW66], [RS78]) that the top and the bottom of Bloch bands are attained alternatively for $\theta^n = 0$ or $\theta^n = \pm 1/2$, and that the corresponding eigenvalue $\lambda_n(\theta^n)$ is simple if it bounds a gap in the spectrum. Therefore, the maximum point θ^n below a gap, or the minimum point θ^n above a gap, do satisfy assumption (62).

Under assumption (62), i.e., $\nabla_\theta \lambda_n(\theta^n) = 0$, (43) and (44) simplify for $\theta = \theta^n$ and we find

$$\frac{\partial \psi_n}{\partial \theta_k} = 2i\pi \zeta_k, \quad \frac{\partial^2 \psi_n}{\partial \theta_k \partial \theta_l} = -4\pi^2 \chi_{kl}, \tag{63}$$

where ζ_k is the solution in $\mathbb{T}^N$ of

$$\mathbb{A}_n(\theta^n)\zeta_k = e_k A(y)(\nabla_y + 2i\pi\theta^n)\psi_n + (\mathrm{div}_y + 2i\pi\theta^n)(A(y)e_k\psi_n), \tag{64}$$

and χ_{kl} is the solution in $\mathbb{T}^N$ of

$$\begin{aligned}
\mathbb{A}_n(\theta^n)\chi_{kl} &= e_k A(y)(\nabla_y + 2i\pi\theta^n)\zeta_l + (\mathrm{div}_y + 2i\pi\theta^n)(A(y)e_k\zeta_l) \\
&+ e_l A(y)(\nabla_y + 2i\pi\theta^n)\zeta_k + (\mathrm{div}_y + 2i\pi\theta^n)(A(y)e_l\zeta_k) \\
&+ e_k A(y)e_l\psi_n + e_l A(y)e_k\psi_n - \frac{1}{4\pi^2}\frac{\partial^2\lambda_n}{\partial\theta_l\partial\theta_k}(\theta^n)\psi_n.
\end{aligned} \tag{65}$$

We obtain the following homogenized problem.

Theorem 8.1 ([AP05]). *Assume (62) and that the initial data $u_\epsilon^0 \in H^1(\mathbb{R}^N)$ is of the form*

$$u_\epsilon^0(x) = \psi_n\left(\frac{x}{\epsilon}, \theta^n\right) e^{2i\pi \frac{\theta^n \cdot x}{\epsilon}} v^0(x), \tag{66}$$

with $v^0 \in H^1(\mathbb{R}^N)$. The solution of (47) can be written as

$$u_\epsilon(t,x) = e^{i\frac{\lambda_n(\theta^n)t}{\epsilon^2}} e^{2i\pi \frac{\theta^n \cdot x}{\epsilon}} \psi_n\left(\frac{x}{\epsilon}, \theta^n\right) v(t,x) + r_\epsilon(t,x), \tag{67}$$

where r_ϵ is a small remainder term, i.e.,

$$\lim_{\epsilon \to 0} \int_{\mathbb{R}^N} |r_\epsilon(t,x)|^2 \, dx = 0, \tag{68}$$

uniformly on compact time intervals in $\mathbb{R}^+$, and $v \in C\left(\mathbb{R}^+; L^2(\mathbb{R}^N)\right)$ is the unique solution of the homogenized Schrödinger equation

$$\begin{cases} i\dfrac{\partial v}{\partial t} - \operatorname{div}\left(A_n^* \nabla v\right) + d_n^*(x)\, v = 0 & in\ \mathbb{R}^N \times \mathbb{R}^+ \\ v(t=0,x) = v^0(x) & in\ \mathbb{R}^N, \end{cases} \tag{69}$$

with $A_n^* = \frac{1}{8\pi^2}\nabla_\theta \nabla_\theta \lambda_n(\theta^n)$ and $d_n^*(x) = \int_{\mathbb{T}^N} d(x,y)|\psi_n(y)|^2\, dy$.

In the context of quantum mechanics or solid state physics Theorem 8.1 is called an effective mass theorem [Mye90], [Ped97], [Que98]. More precisely, the inverse tensor $(A_n^*)^{-1}$ is the effective mass of an electron in the n-th band of a periodic crystal (characterized by the periodic metric $A(y)$ and the periodic potential $c(y)$). Since we did not assume that θ^n was a minimum point, the tensor $A_n^* = \frac{1}{8\pi^2}\nabla_\theta \nabla_\theta \lambda_n(\theta^n)$ can be neither definite nor positive, which is quite surprising for a notion of mass (but this fact is well understood in solid state physics).

We recognize in the phase factor in the right-hand side of (67) the WKB phase which has been computed in the periodic case for monochromatic initial data in the previous section.

Remark 8.2. Theorem 8.1 is due to [AP05]. Other mathematical references on effective mass theorems include Sect. 4 of Chap. 4 in [BLP78] (with a different semi-classical scaling), [Bec99], [PR96] (with a different method of semi-classical measures) and [Spa06] (for nonlinear Schrödinger equations).

Remark 8.3. Theorem 8.1 does not fit into the framework of G- or H-convergence (see, e.g., [MT78], [Spa76]). Indeed these classical theories of homogenization state that the homogenized coefficients are independent of the initial data, which is not the case here. There is no contradiction in our result since H-convergence does not apply because we lack a uniform a priori estimate in $L^\infty(\mathbb{R}^+; H^1(\mathbb{R}^N))$ for the sequence of solutions u_ϵ, as required by H-convergence.

Remark 8.4. Assumption (66) can be weakened for proving variants of Theorem 8.1. For example, it still holds true if we merely assume that $u_\epsilon^0(x)e^{-2i\pi\frac{\theta^n \cdot x}{\epsilon}}$ two-scale converges strongly, or even weakly, to $\psi_n(y,\theta^n)v^0(x)$, provided that we use strong or weak two-scale convergence to state the result.

Remark 8.5. It is possible to obtain the homogenized problem (69) by using the formal method of two-scale asymptotic expansions with a WKB ansatz (52). More precisely, we postulate that the solution u_ϵ of (47) can be written as

$$u_\epsilon(t,x) = e^{i\frac{\lambda_n(\theta^n)t}{\epsilon^2}} e^{2i\pi\frac{\theta^n \cdot x}{\epsilon}} \sum_{i=0}^{+\infty} \epsilon^i u_i\left(t, x, \frac{x}{\epsilon}\right). \tag{70}$$

The phase factor in front of the series is already the result of the WKB method in Sect. 7. Plugging (70) into (47) we deduce that

$$u_0(x,y) = \psi_n(y,\theta^n)\, v(t,x), \quad u_1(x,y) = \frac{1}{2i\pi}\sum_{k=1}^{N}\frac{\partial v}{\partial x_k}(t,x)\frac{\partial\psi_n}{\partial\theta_k}(y,\theta^n)$$

and

$$u_2(x, y) = \frac{-1}{4\pi^2} \sum_{k,l=1}^{N} \frac{\partial^2 v}{\partial x_k \partial x_l}(t, x) \frac{\partial^2 \psi_n}{\partial \theta_k \partial \theta_l}(y, \theta^n).$$

The compatibility condition or Fredholm alternative for solving in u_2 yields the homogenized problem (69).

Proof of Theorem 8.1. This proof is based on ideas of [ACPSV04]. Define a sequence v_ϵ by

$$v_\epsilon(t, x) = u_\epsilon(t, x) e^{-i\frac{\lambda_n(\theta^n)t}{\epsilon^2}} e^{-2i\pi \frac{\theta^n \cdot x}{\epsilon}}.$$

Since $|v_\epsilon| = |u_\epsilon|$, by the a priori estimates of Lemma 6.1 we have

$$\|v_\epsilon\|_{L^\infty(\mathbb{R}^+;L^2(\mathbb{R}^N))} + \epsilon \|\nabla v_\epsilon\|_{L^\infty(\mathbb{R}^+ \times \mathbb{R}^N)} \le C.$$

Introducing a finite time $0 < T < +\infty$, and applying the compactness of two-scale convergence (see Theorem 3.1), up to a subsequence, there exists a limit $v^*(t, x, y) \in L^2\left((0, T) \times \mathbb{R}^N; H^1(\mathbb{T}^N)\right)$ such that v_ϵ and $\epsilon \nabla v_\epsilon$ two-scale converge to v^* and $\nabla_y v^*$, respectively. Similarly, by definition of the initial data, $v_\epsilon(0, x)$ two-scale converges to $\psi_n(y, \theta^n) v^0(x)$.

First step. We multiply (47) by the complex conjugate of

$$\epsilon^2 \phi(t, x, \frac{x}{\epsilon}) e^{i\frac{\lambda_n(\theta^n)t}{\epsilon^2}} e^{2i\pi \frac{\theta^n \cdot x}{\epsilon}}$$

where $\phi(t, x, y)$ is a smooth test function defined on $[0, T) \times \mathbb{R}^N \times \mathbb{T}^N$, with compact support in $[0, T) \times \mathbb{R}^N$. Integrating by parts yields

$$i\epsilon^2 \int_{\mathbb{R}^N} u_\epsilon^0 \overline{\phi}^\epsilon e^{-2i\pi \frac{\theta^n \cdot x}{\epsilon}} dx - i\epsilon^2 \int_0^T \int_{\mathbb{R}^N} v_\epsilon \frac{\partial \overline{\phi}^\epsilon}{\partial t} dt \, dx$$

$$+ \int_0^T \int_{\mathbb{R}^N} A^\epsilon (\epsilon \nabla + 2i\pi\theta^n) v_\epsilon \cdot (\epsilon \nabla - 2i\pi\theta^n) \overline{\phi}^\epsilon \, dt \, dx$$

$$+ \int_0^T \int_{\mathbb{R}^N} (c^\epsilon - \lambda_n(\theta^n) + \epsilon^2 d^\epsilon) v_\epsilon \overline{\phi}^\epsilon \, dt \, dx \qquad = 0.$$

Passing to the two-scale limit yields the variational formulation of

$$-(\operatorname{div}_y + 2i\pi\theta^n)\left(A(y)(\nabla_y + 2i\pi\theta^n)v^*\right) + c(y)v^* = \lambda_n(\theta^n)v^* \quad \text{in } \mathbb{T}^N.$$

By the simplicity of $\lambda_n(\theta^n)$, this implies that there exists a scalar function $v(t, x) \in L^2\left((0, T) \times \mathbb{R}^N\right)$ such that

$$v^*(t, x, y) = v(t, x)\psi_n(y, \theta^n). \tag{71}$$

Second step. We multiply (47) by the complex conjugate of

$$
\Psi_\epsilon = e^{i\frac{\lambda_n(\theta^n)t}{\epsilon^2}} e^{2i\pi\frac{\theta^n\cdot x}{\epsilon}}\left(\psi_n(\frac{x}{\epsilon},\theta^n)\phi(t,x) + \epsilon\sum_{k=1}^{N}\frac{\partial\phi}{\partial x_k}(t,x)\zeta_k(\frac{x}{\epsilon})\right) \quad (72)
$$

where $\phi(t,x)$ is a smooth test function with compact support in $[0,T)\times\mathbb{R}^N$, and $\zeta_k(y)$ is the solution of (64). As usual in periodic homogenization, the choice (72) of the test function is dictated by the formal two-scale asymptotic expansion (70). More precisely, the purpose of the corrector ζ_k is to compensate by its second derivatives the first derivatives of ψ_n. Since ζ_k is proportional to $\partial\psi_n/\partial\theta_k$, the rule of thumb is that derivatives with respect to x correspond to derivatives with respect to θ. After some algebra we found that

$$
\int_{\mathbb{R}^N} A^\epsilon \nabla u_\epsilon \cdot \nabla\overline{\Psi}_\epsilon dx = \int_{\mathbb{R}^N} A^\epsilon(\nabla + 2i\pi\frac{\theta^n}{\epsilon})(\overline{\phi}v_\epsilon)\cdot(\nabla - 2i\pi\frac{\theta^n}{\epsilon})\overline{\psi}_n^\epsilon
$$
$$
+\epsilon\int_{\mathbb{R}^N} A^\epsilon(\nabla + 2i\pi\frac{\theta^n}{\epsilon})(\frac{\partial\overline{\phi}}{\partial x_k}v_\epsilon)\cdot(\nabla - 2i\pi\frac{\theta^n}{\epsilon})\overline{\zeta}_k^\epsilon
$$
$$
-\int_{\mathbb{R}^N} A^\epsilon e_k\frac{\partial\overline{\phi}}{\partial x_k}v_\epsilon\cdot(\nabla - 2i\pi\frac{\theta^n}{\epsilon})\overline{\psi}_n^\epsilon
$$
$$
+\int_{\mathbb{R}^N} A^\epsilon(\nabla + 2i\pi\frac{\theta^n}{\epsilon})(\frac{\partial\overline{\phi}}{\partial x_k}v_\epsilon)\cdot e_k\overline{\psi}_n^\epsilon
$$
$$
-\int_{\mathbb{R}^N} A^\epsilon v_\epsilon\nabla\frac{\partial\overline{\phi}}{\partial x_k}\cdot e_k\overline{\psi}_n^\epsilon
$$
$$
-\int_{\mathbb{R}^N} A^\epsilon v_\epsilon\nabla\frac{\partial\overline{\phi}}{\partial x_k}\cdot(\epsilon\nabla - 2i\pi\theta^n)\overline{\zeta}_k^\epsilon
$$
$$
+\int_{\mathbb{R}^N} A^\epsilon\overline{\zeta}_k^\epsilon(\epsilon\nabla + 2i\pi\theta^n)v_\epsilon\cdot\nabla\frac{\partial\overline{\phi}}{\partial x_k}
$$
$$
(73)
$$

Now, for any smooth compactly supported test function Φ, we deduce from the definition of ψ_n that

$$
\int_{\mathbb{R}^N} A^\epsilon(\nabla + 2i\pi\frac{\theta^n}{\epsilon})\psi_n^\epsilon\cdot(\nabla - 2i\pi\frac{\theta^n}{\epsilon})\overline{\Phi} + \frac{1}{\epsilon^2}\int_{\mathbb{R}^N}(c^\epsilon - \lambda_n(\theta^n))\psi_n^\epsilon\overline{\Phi} = 0, \quad (74)
$$

and from the definition of ζ_k

$$
\int_{\mathbb{R}^N} A^\epsilon(\nabla + 2i\pi\frac{\theta^n}{\epsilon})\zeta_k^\epsilon\cdot(\nabla - 2i\pi\frac{\theta^n}{\epsilon})\overline{\Phi} + \frac{1}{\epsilon^2}\int_{\mathbb{R}^N}(c^\epsilon - \lambda_n(\theta^n))\zeta_k^\epsilon\overline{\Phi} =
$$
$$
\epsilon^{-1}\int_{\mathbb{R}^N} A^\epsilon(\nabla + 2i\pi\frac{\theta^n}{\epsilon})\psi_n^\epsilon\cdot e_k\overline{\Phi} - \epsilon^{-1}\int_{\mathbb{R}^N} A^\epsilon e_k\psi_n^\epsilon\cdot(\nabla - 2i\pi\frac{\theta^n}{\epsilon})\overline{\Phi}.
$$
$$
(75)
$$

Combining (73) with the other terms of the variational formulation of (47), we easily check that the first line of its right-hand side cancels out because of (74) with $\Phi = \overline{\phi}v_\epsilon$, and the next three lines cancel out because of (75) with $\Phi = \frac{\partial \overline{\phi}}{\partial x_k}v_\epsilon$. On the other hand, we can pass to the limit in the three last terms of (73). Finally, (47) multiplied by $\overline{\Psi}_\epsilon$ yields after simplification

$$
i\int_{\mathbb{R}^N} u_\epsilon^0 \overline{\Psi}_\epsilon(t=0)dx - i\int_0^T \int_{\mathbb{R}^N} v_\epsilon\left(\overline{\psi}_n^\epsilon \frac{\partial \overline{\phi}}{\partial t} + \epsilon\frac{\partial^2 \overline{\phi}}{\partial x_k \partial t}\overline{\zeta}_k^\epsilon\right)dt\,dx
$$

$$
-\int_0^T \int_{\mathbb{R}^N} A^\epsilon v_\epsilon \nabla \frac{\partial \overline{\phi}}{\partial x_k} \cdot e_k \overline{\psi}_n^\epsilon dt\,dx
$$

$$
-\int_0^T \int_{\mathbb{R}^N} A^\epsilon v_\epsilon \nabla \frac{\partial \overline{\phi}}{\partial x_k} \cdot (\epsilon\nabla - 2i\pi\theta^n)\overline{\zeta}_k^\epsilon dt\,dx \qquad (76)
$$

$$
+\int_0^T \int_{\mathbb{R}^N} A^\epsilon \overline{\zeta}_k^\epsilon(\epsilon\nabla + 2i\pi\theta^n)v_\epsilon \cdot \nabla\frac{\partial \overline{\phi}}{\partial x_k}dt\,dx
$$

$$
+\int_0^T \int_{\mathbb{R}^N} d^\epsilon v_\epsilon \overline{\Psi}_\epsilon\,dt\,dx \qquad\qquad\qquad\qquad = 0.
$$

Passing to the two-scale limit in each term of (76) gives

$$
i\int_{\mathbb{R}^N} \int_{\mathbb{T}^N} \psi_n v^0 \overline{\psi}_n \overline{\phi}(t=0)\,dx\,dy - i\int_0^T \int_{\mathbb{R}^N}\int_{\mathbb{T}^N} \psi_n v \overline{\psi}_n \frac{\partial \overline{\phi}}{\partial t}dt\,dx\,dy
$$

$$
-\int_0^T \int_{\mathbb{R}^N}\int_{\mathbb{T}^N} A\psi_n v\nabla\frac{\partial \overline{\phi}}{\partial x_k} \cdot e_k\overline{\psi}_n dt\,dx\,dy
$$

$$
-\int_0^T \int_{\mathbb{R}^N}\int_{\mathbb{T}^N} A\psi_n v\nabla\frac{\partial \overline{\phi}}{\partial x_k} \cdot (\nabla_y - 2i\pi\theta^n)\overline{\zeta}_k dt\,dx\,dy
$$

$$
+\int_0^T \int_{\mathbb{R}^N}\int_{\mathbb{T}^N} A\overline{\zeta}_k(\nabla_y + 2i\pi\theta^n)\psi_n v \cdot \nabla\frac{\partial \overline{\phi}}{\partial x_k}dt\,dx\,dy
$$

$$
+\int_0^T \int_{\mathbb{R}^N}\int_{\mathbb{T}^N} d(x,y)\psi_n v\overline{\psi}_n \overline{\phi}\,dt\,dx\,dy \qquad\qquad = 0.
$$
$$
(77)
$$

Recalling the normalization $\int_{\mathbb{T}^N} |\psi_n|^2 dy = 1$, and introducing

$$
2\,(A_n^*)_{jk} = \int_{\mathbb{T}^N} \Big(A\psi_n e_j \cdot e_k\overline{\psi}_n + A\psi_n e_k \cdot e_j\overline{\psi}_n
$$

$$
+ A\psi_n e_j \cdot (\nabla_y - 2i\pi\theta^n)\overline{\zeta}_k + A\psi_n e_k \cdot (\nabla_y - 2i\pi\theta^n)\overline{\zeta}_j \qquad (78)
$$

$$
- A\overline{\zeta}_k(\nabla_y + 2i\pi\theta^n)\psi_n \cdot e_j - A\overline{\zeta}_j(\nabla_y + 2i\pi\theta^n)\psi_n \cdot e_k\Big)dy,
$$

and $d_n^*(x) = \int_{\mathbb{T}^N} d(x, y)|\psi_n(y)|^2\, dy$, (77) is equivalent to

$$i \int_{\mathbb{R}^N} v^0 \overline{\phi} dx - i \int_0^T \int_{\mathbb{R}^N} v \frac{\partial \overline{\phi}}{\partial t} dt\, dx - \int_0^T \int_{\mathbb{R}^N} A^* v \cdot \nabla \nabla \overline{\phi} dt\, dx$$

$$+ \int_0^T \int_{\mathbb{R}^N} d^*(x) v \overline{\phi} dt\, dx = 0$$

which is a very weak form of the homogenized equation (69). The compatibility condition (46) of (65) for the second derivative of ψ_n yields that the matrix A_n^*, defined by (78), is indeed equal to $\frac{1}{8\pi^2} \nabla_\theta \nabla_\theta \lambda_n(\theta^n)$, and thus is symmetric. Although, the tensor A_n^* is possibly non-coercive, the homogenized problem (69) is well posed. Indeed, by using semi-group theory (see, e.g., [Bre73] or Chap. X in [RS78], [CH98]), there exists a unique solution in $C(\mathbb{R}^+; L^2(\mathbb{R}^N))$, although it may not belong to $L^\infty(\mathbb{R}^+; H^1(\mathbb{R}^N))$. By uniqueness of the solution of the homogenized problem (69), we deduce that the entire sequence v_ϵ two-scale converges weakly to $\psi_n(y, \theta^n) v(t, x)$.

It remains to prove the strong two-scale convergence of v_ϵ. By Lemma 6.1 we have

$$\|v_\epsilon(t)\|_{L^2(\mathbb{R}^N)} = \|u_\epsilon(t)\|_{L^2(\mathbb{R}^N)} = \|u_\epsilon^0\|_{L^2(\mathbb{R}^N)} \to \|\psi_n v^0\|_{L^2(\mathbb{R}^N \times \mathbb{T}^N)} = \|v^0\|_{L^2(\mathbb{R}^N)}$$

by the normalization condition of ψ_n. From the conservation of energy of the homogenized equation (69) we have

$$\|v(t)\|_{L^2(\mathbb{R}^N)} = \|v^0\|_{L^2(\mathbb{R}^N)},$$

and thus we deduce the strong convergence (68) from Theorem 3.2. □

Remark 8.6. By changing the main assumption on the Bloch spectrum it is possible to obtain a fourth order homogenized equation instead of the usual Schrödinger equation. Specifically, if we consider

$$i\epsilon^2 \frac{\partial u_\epsilon}{\partial t} - \operatorname{div}\left(A\left(\frac{x}{\epsilon}\right) \nabla u_\epsilon\right) + \left(\epsilon^{-2} c\left(\frac{x}{\epsilon}\right) + \epsilon^2 d\left(x, \frac{x}{\epsilon}\right)\right) u_\epsilon = 0, \qquad (79)$$

and if we make the following assumption, instead of (62),

$$\begin{cases} (i)\ \lambda_n(\theta^n) \text{ is a simple eigenvalue,} \\ (ii)\ \nabla_\theta \lambda_n(\theta^n) = 0, \nabla_\theta \nabla_\theta \lambda_n(\theta^n) = 0, \nabla_\theta \nabla_\theta \nabla_\theta \lambda_n(\theta^n) = 0, \end{cases} \qquad (80)$$

then, for the same type of initial data (66), we can prove that the solution of (79) can be written as

$$u_\epsilon(t, x) = e^{i \frac{\lambda_n(\theta^n)t}{\epsilon^4}} e^{2i\pi \frac{\theta^n \cdot x}{\epsilon}} \psi_n\left(\frac{x}{\epsilon}, \theta^n\right) v(t, x) + r_\epsilon(t, x), \qquad (81)$$

where r_ϵ converges strongly to 0 in $L_{loc}^\infty\left(\mathbb{R}^+; L^2(\mathbb{R}^N)\right)$ and $v \in C\left(\mathbb{R}^+; L^2(\mathbb{R}^N)\right)$ is the solution of the fourth-order homogenized problem

$$\begin{cases} i\dfrac{\partial v}{\partial t} + \operatorname{div} \operatorname{div} \left(A_n^* \nabla \nabla v \right) + d_n^*(x)\, v = 0 & \text{in } \mathbb{R}^N \times \mathbb{R}^+ \\ v(t = 0, x) = v^0(x) & \text{in } \mathbb{R}^N, \end{cases} \tag{82}$$

with $A_n^* = \frac{1}{(2\pi)^4 4!} \nabla_\theta \nabla_\theta \nabla_\theta \nabla_\theta \lambda_n(\theta^n)$ and $d_n^*(x) = \int_{\mathbb{T}^N} d(x, y) |\psi_n(y)|^2 \, dy$.

Remark that the time scaling in (79) is not the same than that in (47): this means that we are looking for an asymptotic for longer time of order ϵ^{-2} in (79), compared to (47).

9 Generalization with Drift

The Schrödinger equation (47) can still be homogenized when θ^n is not a critical point of $\lambda_n(\theta)$. In other words we generalize Theorem 8.1 by weakening assumption (62) that we now replace by

$$\lambda_n(\theta^n) \text{ is a simple eigenvalue.} \tag{83}$$

This yields a large drift in the homogenized problem associated to the group velocity (58), as predicted for shorter times by the WKB method of Sect. 7. Because of this large drift we need to replace the standard two-scale convergence by the following generalization which was introduced in [MP05]. As usual T denotes some final time, $0 \leq T \leq +\infty$.

Theorem 9.1 ([MP05]). *Let $\mathcal{V} \in \mathbb{R}^N$ be a given drift velocity. Let $(u_\epsilon)_{\epsilon > 0}$ be a uniformly bounded sequence in $L^2((0, T) \times \mathbb{R}^N)$. There exists a subsequence, still denoted by ϵ, and a limit function $u_0(t, x, y) \in L^2((0, T) \times \mathbb{R}^N \times \mathbb{T}^N)$ such that u_ϵ two-scale converges with drift weakly to u_0 in the sense that*

$$\lim_{\epsilon \to 0} \int_0^T \int_{\mathbb{R}^N} u_\epsilon(t, x) \phi\left(t, x + \frac{\mathcal{V}}{\epsilon} t, \frac{x}{\epsilon}\right) dt\, dx =$$

$$\int_0^T \int_{\mathbb{R}^N} \int_{\mathbb{T}^N} u_0(t, x, y) \phi(t, x, y)\, dt\, dx\, dy \tag{84}$$

for all functions $\phi(t, x, y) \in L^2\left((0, T) \times \mathbb{R}^N; C(\mathbb{T}^N)\right)$.

Recall that, $\mathbb{T}^N$ being the unit torus, the test function ϕ in (84) is $(0, 1)^N$-periodic with respect to the y variable. Remark that Theorem 9.1 does not reduce to the usual definition of two-scale convergence upon the change of variable $z = x + \frac{\mathcal{V}}{\epsilon} t$ because there is no drift in the fast variable $y = \frac{x}{\epsilon}$. The proof of Theorem 9.1 is similar to that of Theorem 3.1, except that it relies on the following simple lemma.

Lemma 9.1. Let $\phi(t, x, y) \in L^2\left((0, T) \times \mathbb{R}^N; C(\mathbb{T}^N)\right)$. Then

$$\lim_{\epsilon \to 0} \int_0^T \int_{\mathbb{R}^N} \left| \phi\left(t, x + \frac{\mathcal{V}}{\epsilon}t, \frac{x}{\epsilon}\right) \right|^2 dt\, dx = \int_0^T \int_{\mathbb{R}^N} \int_{\mathbb{T}^N} |\phi(t, x, y)|^2 dt\, dx\, dy.$$

Proof. We introduce a partition of $\mathbb{T}^N$ in small cubes $(Y_i)_{1 \leq i \leq n^N}$ of size $1/n$ and approximate $\phi(t, x, y)$ by

$$\sum_{i=1}^{n^N} \phi(t, x, y_i)\chi_i(y)$$

where χ_i is the characteristic function of Y_i and y_i is its center. By a standard approximation argument (see Sect. 5 of [All92]), since $y \to \phi(t, x, y)$ is continuous, it is enough to prove the lemma for $\phi(t, x, y_i)\chi_i(y)$. We make the change of variables $x' = x + \frac{\mathcal{V}}{\epsilon}t$ which yields

$$\int_0^T \int_{\mathbb{R}^N} \left| \phi\left(t, x + \frac{\mathcal{V}}{\epsilon}t, y_i\right) \chi_i\left(\frac{x}{\epsilon}\right) \right|^2 dt\, dx =$$
$$\int_0^T \int_{\mathbb{R}^N} \left| \phi(t, x', y_i) \chi_i\left(\frac{x'}{\epsilon} - \frac{\mathcal{V}}{\epsilon^2}t\right) \right|^2 dt\, dx'. \tag{85}$$

For each fixed time t the sequence $\chi_i^2\left(\frac{x'}{\epsilon} - \frac{\mathcal{V}}{\epsilon^2}\right)$ is periodically oscillating in x' and converges weakly to its average in $\mathbb{T}^N$. Therefore, we deduce that, as ϵ goes to 0, (85) converges to

$$\int_0^T \int_{\mathbb{R}^N} \int_{\mathbb{T}^N} |\phi(t, x', y_i) \chi_i(y)|^2 dt\, dx'\, dy$$

which is the desired result. $\square$

The next Proposition asserts a corrector-type result, the proof of which is a simple adaptation of Proposition 3.2.

Proposition 9.1. Let $(u_\epsilon)_{\epsilon > 0}$ be a sequence in $L^2((0, T) \times \mathbb{R}^N)$ which two-scale converges with drift to a limit $u_0(t, x, y) \in L^2((0, T) \times \mathbb{R}^N \times \mathbb{T}^N)$. It satisfies

$$\lim_{\epsilon \to 0} \|u_\epsilon\|_{L^2((0,T) \times \mathbb{R}^N)} \geq \|u_0\|_{L^2((0,T) \times \mathbb{R}^N \times \mathbb{T}^N)}.$$

Assume further that

$$\lim_{\epsilon \to 0} \|u_\epsilon\|_{L^2((0,T) \times \mathbb{R}^N)} = \|u_0\|_{L^2((0,T) \times \mathbb{R}^N \times \mathbb{T}^N)}.$$

Then, it is said to two-scale converges with drift strongly *and it satisfies*

$$\lim_{\epsilon \to 0} \int_0^T \int_{\mathbb{R}^N} \left| u_\epsilon(t,x) - u_0\left(t, x + \frac{\mathcal{V}}{\epsilon} t, \frac{x}{\epsilon}\right)\right|^2 dx \, dt = 0,$$

if $u_0(t,x,y)$ *is smooth, say* $u_0(t,x,y) \in L^2\left((0,T) \times \mathbb{R}^N; C(\mathbb{T}^N)\right)$.

Because of the large drift due to the group velocity (58) we have to make an assumption on the behavior at infinity of the macroscopic potential $d(x,y)$. There are many different possibilities but we choose the simplest one: we assume that

$$\lim_{|x| \to +\infty} d(x,y) = d^\infty(y) \quad \text{uniformly in } \mathbb{T}^N. \tag{86}$$

Under assumptions (83) and (86) the following generalization of Theorem 8.1 is obtained in [AP05].

Theorem 9.2 ([AP05]). *Assume that the initial data* $u_\epsilon^0 \in H^1(\mathbb{R}^N)$ *is of the form*

$$u_\epsilon^0(x) = \psi_n\left(\frac{x}{\epsilon}, \theta^n\right) e^{2i\pi \frac{\theta^n \cdot x}{\epsilon}} v^0(x), \tag{87}$$

with $v^0 \in H^1(\mathbb{R}^N)$. *The solution of (47) can be written as*

$$u_\epsilon(t,x) = e^{i\frac{\lambda_n(\theta^n)t}{\epsilon^2}} e^{2i\pi \frac{\theta^n \cdot x}{\epsilon}} \psi_n\left(\frac{x}{\epsilon}, \theta^n\right) v\left(t, x + \frac{\mathcal{V}}{\epsilon} t\right) + r_\epsilon(t,x), \tag{88}$$

where $\mathcal{V}$ *is the group velocity defined by (58) and* $r_\epsilon(t,x)$ *is a small remainder term, i.e., for any* $0 < T < +\infty$,

$$\lim_{\epsilon \to 0} \int_0^T \int_{\mathbb{R}^N} |r_\epsilon(t,x)|^2 \, dx \, dt = 0, \tag{89}$$

and $v \in C\left(\mathbb{R}^+; L^2(\mathbb{R}^N)\right)$ *is the unique solution of the Schrödinger homogenized problem*

$$\begin{cases} i\dfrac{\partial v}{\partial t} - \operatorname{div}\left(A_n^* \nabla v\right) + d_n^* \, v = 0 & \text{in } \mathbb{R}^N \times \mathbb{R}^+, \\ v(t=0, x) = v^0(x) & \text{in } \mathbb{R}^N, \end{cases} \tag{90}$$

with $A_n^* = \frac{1}{8\pi^2} \nabla_\theta \nabla_\theta \lambda_n(\theta^n)$ *and* $d_n^* = \int_{\mathbb{T}^N} d^\infty(y) |\psi_n(y)|^2 \, dy$.

Proof. The proof is similar to that of Theorem 8.1 since the a priori estimate of Lemma 6.1 still holds true. In a first step, by multiplying (47) by a test function

$$\epsilon^2 \phi\left(t, x + \frac{\mathcal{V}}{\epsilon} t, \frac{x}{\epsilon}\right) e^{i\frac{\lambda_n(\theta^n)t}{\epsilon^2}} e^{2i\pi \frac{\theta^n \cdot x}{\epsilon}},$$

where $\phi(t, x, y)$ is a smooth test function defined on $[0, T] \times \mathbb{R}^N \times \mathbb{T}^N$, with compact support in $[0, T] \times \mathbb{R}^N$, we prove that the sequence

$$v_\epsilon(t, x) = u_\epsilon(t, x) e^{-i\frac{\lambda_n(\theta^n)t}{\epsilon^2}} e^{-2i\pi\frac{\theta^n \cdot x}{\epsilon}}$$

two-scale converges with drift to a limit $\psi_n(y, \theta^n) v(t, x)$. Then, in a second step we multiply (47) by the complex conjugate of

$$\Psi_\epsilon = e^{i\frac{\lambda_n(\theta^n)t}{\epsilon^2}} e^{2i\pi\frac{\theta^n \cdot x}{\epsilon}} \left(\psi_n(\frac{x}{\epsilon}, \theta^n)\phi(t, x + \frac{\mathcal{V}}{\epsilon}t) \right.$$

$$\left. + \frac{\epsilon}{2i\pi}\sum_{k=1}^{N} \frac{\partial\phi}{\partial x_k}(t, x + \frac{\mathcal{V}}{\epsilon}t)\frac{\partial\psi_n}{\partial\theta_k}(\frac{x}{\epsilon}) \right).$$

Integrating by parts we perform a computation which is identical to that in the proof of Theorem 8.1 except that two new terms of order ϵ^{-1} arise and cancel out exactly, namely the term

$$-\frac{1}{2i\pi\epsilon}\frac{\partial\lambda_n}{\partial\theta_k}\int_0^T \int_{\mathbb{R}^N} v_\epsilon\overline{\psi}_n^\epsilon\frac{\partial\overline{\phi}}{\partial x_k}\, dt\, dx$$

which comes from the additional term in (43) satisfied by $\frac{\partial\psi_n}{\partial\theta_k}$ (additional with respect to (64) for ζ_k), and the same term with positive sign which arises in the integration by parts, with respect to time, of

$$\int_0^T \int_{\mathbb{R}^N} i\frac{\partial u_\epsilon}{\partial t}\overline{\psi}_n^\epsilon\overline{\phi}(t, x + \frac{\mathcal{V}}{\epsilon}t)dt\, dx.$$

The rest of the proof is as in Theorem 8.1, provided the usual two-scale convergence is replaced by the two-scale convergence with drift which relies on test functions having a large drift in the macroscopic variable. $\square$

10 Homogenized System of Equations

In this section we investigate the case of a Bloch eigenvalue which is not simple. Physically speaking it can be interpreted as a tangential crossing of modes. The semi-classical limit of modes crossing yields the so-called Landau–Zener formula, recently analyzed in [FG02], [FG03]. Our study is different in two respects. First our scaling is not that of semi-classical analysis. Second the crossing is tangential, i.e., the group velocities $\nabla_\theta\lambda_n(\theta)$ are assumed to be the same for each mode. To simplify the exposition we consider an eigenvalue of multiplicity two, but the argument works through for any multiplicity. We replace assumption (62) by the following one: for $n \geq 1$, we consider a Bloch parameter $\theta^n \in \mathbb{T}^N$ such that

$$
\begin{cases}
(i) \quad \lambda_n(\theta^n) = \lambda_{n+1}(\theta^n) \neq \lambda_k(\theta^n) \quad \forall k \neq n, n+1, \\
(ii) \quad \text{locally near } \theta^n, \ \lambda_n(\theta) \text{ and } \lambda_{n+1}(\theta) \text{ form two} \\
\qquad \text{smooth branches of eigenvalues with corresponding} \\
\qquad \text{smooth eigenfunctions } \psi_n(\theta) \text{ and } \psi_{n+1}(\theta), \\
(iii) \quad \nabla_\theta \lambda_n(\theta^n) = \nabla_\theta \lambda_{n+1}(\theta^n) = 0.
\end{cases}
\qquad (91)
$$

By a convenient abuse of language we still denote by $\lambda_n(\theta)$ and $\lambda_{n+1}(\theta)$ the two smooth (local) branches of eigenvalues passing through θ^n (this is equivalent to a pointwise relabeling of these two eigenvalues, not necessarily following the usual increasing order). In dimension $N = 1$ a double eigenvalue can only occur when there is no gap between two consecutive Bloch bands and assumption (91) is automatically satisfied [MW66]. However, in dimension $N > 1$ it is not even clear that, near a double eigenvalue, one can find two smooth branches because θ is a vector-valued parameter [Kat66]. Therefore, (91) is a very strong mathematical assumption which is physically not very relevant in dimension $N > 1$.

Theorem 10.1 ([AP05]). *Assume (91) and that the initial data $u_\epsilon^0 \in H^1(\mathbb{R}^N)$ is of the form*

$$
u_\epsilon^0(x) = \psi_n\left(\frac{x}{\epsilon}, \theta^n\right) e^{2i\pi \frac{\theta^n \cdot x}{\epsilon}} v_1^0(x) + \psi_{n+1}\left(\frac{x}{\epsilon}, \theta^n\right) e^{2i\pi \frac{\theta^n \cdot x}{\epsilon}} v_2^0(x), \qquad (92)
$$

with $v_1^0, v_2^0 \in H^1(\mathbb{R}^N)$. The solution of (47) can be written as

$$
\begin{aligned}
u_\epsilon(t, x) = e^{i\frac{\lambda_n(\theta^n)t}{\epsilon^2}} e^{2i\pi \frac{\theta^n \cdot x}{\epsilon}} \bigg(&\psi_n\left(\frac{x}{\epsilon}, \theta^n\right) v_1(t, x) \\
&+ \psi_{n+1}\left(\frac{x}{\epsilon}, \theta^n\right) v_2(t, x)\bigg) + r_\epsilon(t, x),
\end{aligned}
\qquad (93)
$$

with $\lim_{\epsilon \to 0} \int_{\mathbb{R}^N} |r_\epsilon(t, x)|^2 \, dx = 0$ uniformly on compact time intervals in $\mathbb{R}^+$, and $(v_1, v_2) \in C\left(\mathbb{R}^+; L^2(\mathbb{R}^N)^2\right)$ is the unique solution of the homogenized Schrödinger system of two equations

$$
\begin{cases}
i\dfrac{\partial v_1}{\partial t} - \operatorname{div}\left(A_n^* \nabla v_1\right) + d_{11}^*(x)\, v_1 + d_{12}^*(x)\, v_2 = 0 & \text{in } \mathbb{R}^N \times \mathbb{R}^+ \\
i\dfrac{\partial v_2}{\partial t} - \operatorname{div}\left(A_{n+1}^* \nabla v_2\right) + d_{21}^*(x)\, v_1 + d_{22}^*(x)\, v_2 = 0 & \text{in } \mathbb{R}^N \times \mathbb{R}^+ \\
(v_1, v_2)(t = 0, x) = (v_1^0, v_2^0)(x) & \text{in } \mathbb{R}^N,
\end{cases}
\qquad (94)
$$

with $A_n^* = \frac{1}{8\pi^2} \nabla_\theta \nabla_\theta \lambda_n(\theta^n)$, $A_{n+1}^* = \frac{1}{8\pi^2} \nabla_\theta \nabla_\theta \lambda_{n+1}(\theta^n)$ and

$$
\begin{pmatrix} d_{11}^*(x) & d_{12}^*(x) \\ d_{21}^*(x) & d_{22}^*(x) \end{pmatrix} = \int_{\mathbb{T}^N} d(x, y) \begin{pmatrix} \psi_n(y)\overline{\psi}_n(y) & \psi_n(y)\overline{\psi}_{n+1}(y) \\ \psi_{n+1}(y)\overline{\psi}_n(y) & \psi_{n+1}(y)\overline{\psi}_{n+1}(y) \end{pmatrix} dy.
$$

Remark 10.1. The main point in Theorem 10.1 is that the homogenized system is of dimension equal to the multiplicity of the eigenvalue $\lambda_n(\theta^n)$.

However, the homogenized system (94) is coupled only by zero-order terms since the diffusion operator is diagonal.

Remark 10.2. Of course, Theorem 10.1 can easily be generalized in the case of a common non-zero group velocity $\mathcal{V} = \nabla_\theta \lambda_n(\theta^n)/2\pi = \nabla_\theta \lambda_{n+1}(\theta^n)/2\pi \neq 0$. If assumption (iii) in (91) is not satisfied, i.e., if there are two different values of the drift velocity, $\nabla_\theta \lambda_n(\theta^n) \neq \nabla_\theta \lambda_{n+1}(\theta^n)$, then we obtain an uncoupled limit system, i.e., each branch of eigenfunctions yields a different homogenized Schrödinger equation.

Proof of Theorem 10.1. The a priori estimate of Lemma 6.1 still holds true for u_ϵ and also for the sequence v_ϵ defined by

$$v_\epsilon(t,x) = u_\epsilon(t,x) e^{-i\frac{\lambda_n(\theta^n)t}{\epsilon^2}} e^{-2i\pi\frac{\theta^n \cdot x}{\epsilon}}.$$

Applying Theorem 3.1, there exists a limit $v^*(t,x,y) \in L^2\left((0,T) \times \mathbb{R}^N; H^1(\mathbb{T}^N)\right)$ such that, up to a subsequence, v_ϵ and $\epsilon \nabla v_\epsilon$ two-scale converge to v^* and $\nabla_y v^*$, respectively.

First step. We multiply (47) by the complex conjugate of

$$\epsilon^2 \phi(t,x,\frac{x}{\epsilon}) e^{i\frac{\lambda_n(\theta^n)t}{\epsilon^2}} e^{2i\pi\frac{\theta^n \cdot x}{\epsilon}}$$

where $\phi(t,x,y)$ is a smooth test function defined on $[0,T) \times \mathbb{R}^N \times \mathbb{T}^N$, with compact support in $[0,T) \times \mathbb{R}^N$. Integrating by parts and passing to the two-scale limit yields the variational formulation of

$$-(\mathrm{div}_y + 2i\pi\theta)\Big(A(y)(\nabla_y + 2i\pi\theta)v^*\Big) + c(y)v^* = \lambda_n(\theta^n)v^* \quad \text{in } \mathbb{T}^N.$$

Since $\lambda_n(\theta^n) = \lambda_{n+1}(\theta^n)$ is of multiplicity 2, there exist two scalar functions $v_1(t,x), v_2(t,x) \in L^2\left((0,T) \times \mathbb{R}^N\right)$ such that

$$v^*(t,x,y) = v_1(t,x)\psi_n(y,\theta^n) + v_2(t,x)\psi_{n+1}(y,\theta^n). \tag{95}$$

Second step. We multiply (47) by the complex conjugate of

$$\Psi_\epsilon = e^{i\frac{\lambda_n(\theta^n)t}{\epsilon^2}} e^{2i\pi\frac{\theta^n \cdot x}{\epsilon}} \left(\psi_n(\frac{x}{\epsilon},\theta^n)\phi_1(t,x) + \psi_{n+1}(\frac{x}{\epsilon},\theta^n)\phi_2(t,x)\right.$$
$$\left. + \epsilon \sum_{k=1}^N \left(\frac{\partial\phi_1}{\partial x_k}(t,x)\zeta_k^1(\frac{x}{\epsilon}) + \frac{\partial\phi_2}{\partial x_k}(t,x)\zeta_k^2(\frac{x}{\epsilon})\right)\right)$$

where ϕ_1, ψ_2 are two smooth test functions with compact support in $[0,T) \times \mathbb{R}^N$, and $\zeta_k^1(y)$ is the solution of (64) with ψ_n in the right-hand side (respectively, $\zeta_k^2(y)$ with ψ_{n+1}). Note that at this point we strongly use the assumption on the smoothness of the eigenfunctions since $\zeta_k^1(y)$ (respectively, $\zeta_k^2(y)$) is defined as the partial derivative of ψ_n (respectively, ψ_{n+1}) with respect to θ_k. We integrate by parts and we pass to the two-scale limit using the same algebra as in the proof of Theorem 8.1. We also use the orthogonality property

$$\int_{\mathbb{T}^N} \psi_n \overline{\psi}_{n+1}\, dy = 0,$$

to obtain

$$i \int_{\mathbb{R}^N} \left(v_1^0 \overline{\phi}_1(0) + v_2^0 \overline{\phi}_2(0) \right) dx - i \int_0^T \int_{\mathbb{R}^N} \left(v_1 \frac{\partial \overline{\phi}_1}{\partial t} + v_2 \frac{\partial \overline{\phi}_2}{\partial t} \right) dt\, dx$$

$$- \int_0^T \int_{\mathbb{R}^N} \sum_{p,q=1}^2 A^*_{pq} v_p \cdot \nabla \nabla \overline{\phi}_q\, dt\, dx$$

$$+ \int_0^T \int_{\mathbb{R}^N} \int_{\mathbb{T}^N} d(\psi_n v_1 + \psi_{n+1} v_2)(\overline{\psi}_n \overline{\phi}_1 + \overline{\psi}_{n+1}\overline{\phi}_2)\, dt\, dx\, dy \qquad = 0,$$

$$\tag{96}$$

where $A^*_{11} = A^*_n$ and $A^*_{22} = A^*_{n+1}$, defined by (78), and A^*_{12} is defined by

$$2\left(A^*_{12}\right)_{jk} = \int_{\mathbb{T}^N} \Big(A\psi_n e_j \cdot e_k \overline{\psi}_{n+1} + A\psi_n e_k \cdot e_j \overline{\psi}_{n+1}$$

$$+ A\psi_n e_j \cdot (\nabla_y - 2i\pi\theta^n)\overline{\zeta}^2_k + A\psi_n e_k \cdot (\nabla_y - 2i\pi\theta^n)\overline{\zeta}^2_j \tag{97}$$

$$- A\overline{\zeta}^2_k(\nabla_y + 2i\pi\theta^n)\psi_n \cdot e_j - A\overline{\zeta}^2_j(\nabla_y + 2i\pi\theta^n)\psi_n \cdot e_k \Big) dy,$$

with a symmetric formula for A^*_{21}. Recall that $A^*_n = \frac{1}{8\pi^2} \nabla_\theta \nabla_\theta \lambda_n(\theta^n)$ because of the compatibility condition (46) of (65) for the second derivative of ψ_n, obtained by multiplying (65) by ψ_n. However, the same holds true if we multiply (65) by ψ_{n+1}

$$\int_{\mathbb{T}^N} \mathbb{A}_n(\theta^n) \chi_{kl} \overline{\psi}_{n+1}\, dy = \int_{\mathbb{T}^N} \chi_{kl} \overline{\mathbb{A}_n(\theta^n)\psi_{n+1}}\, dy = 0$$

because $\mathbb{A}_n(\theta^n)\psi_{n+1} = 0$. Therefore, we deduce that (97) is equivalent to

$$2\left(A^*_{12}\right)_{lk} = \frac{1}{4\pi^2} \frac{\partial^2 \lambda_n}{\partial \theta_l \partial \theta_k}(\theta^n) \int_{\mathbb{T}^N} \psi_n \overline{\psi}_{n+1}\, dy = 0$$

by orthogonality of ψ_n and ψ_{n+1}. Thus $A^*_{12} = A^*_{21} = 0$ and (96) is a weak formulation of the limit system (94) which is thus coupled only through the zero-order terms. It is easily seen that (94) is well-posed in $C\left(\mathbb{R}^+; L^2(\mathbb{R}^N)^2\right)$. The rest of the proof is as for Theorem 8.1. □

11 Localization

We now come back to the scaling of semi-classical analysis (as in Sect. 7) and consider the following Schrödinger equation

$$\begin{cases} \dfrac{i}{\epsilon}\dfrac{\partial u_\epsilon}{\partial t} - \operatorname{div}\left(A\left(x,\dfrac{x}{\epsilon}\right)\nabla u_\epsilon\right) + \dfrac{1}{\epsilon^2}c\left(x,\dfrac{x}{\epsilon}\right)u_\epsilon = 0 & \text{in } \mathbb{R}^N \times \mathbb{R}^+ \\ u_\epsilon(0,x) = u_\epsilon^0(x) & \text{in } \mathbb{R}^N. \end{cases} \tag{98}$$

The main difference with the previous sections is that the periodic coefficients are now macroscopically modulated. We have seen in Sect. 7 the WKB method which shows that, in some sense, the homogenized limit problem for (98) is a Liouville transport equation. For an initial data living in the n-th Bloch band (and under some technical assumptions on the Bloch spectral cell problem (51)) the semi-classical limit of (98) is given by the dynamic of the Hamiltonian system (60).

The results of this section are going to be completely different, featuring in particular a localization phenomenon. Our approach to (98) is different since we consider special initial data that are monochromatic, have zero group velocity and zero applied force. Namely the initial data is concentrating at a point (x^n, θ^n) of the phase space where $\nabla_\theta \lambda_n(x^n, \theta^n) = \nabla_x \lambda_n(x^n, \theta^n) = 0$. In such a case, the previous Hamiltonian system (60) degenerates (its solution is constant) and is unable to describe the precise dynamic of the wave function u_ϵ. We exhibit another limit problem which is again a Schrödinger equation with quadratic potential. In other words we build a sequence of approximate solutions of (98) which are the product of a Bloch wave and of the solution of an homogenized Schrödinger equation. Furthermore, if the full Hessian tensor of the Bloch eigenvalue $\lambda_n(x, \theta)$ is positive definite at (x^n, θ^n), we prove that all the eigenfunctions of an homogenized Schrödinger equation are exponentially decreasing at infinity. In other words, we exhibit a localization phenomenon for (98) since we build a sequence of approximate solutions that decay exponentially fast away from x^n. The root of this localization phenomenon is the macroscopic modulation (i.e., with respect to x) of the periodic coefficients which is similar in spirit to the randomness that causes Anderson's localization (see [CL90] and references therein).

Our main assumptions are that there exist $x^n \in \mathbb{R}^N$ and $\theta^n \in \mathbb{T}^N$ such that

$$\begin{aligned} &(i)\ x \to A(x,y), c(x,y) \text{ are } C^2 \text{ in a neighborhood of } x^n, \\ &(ii)\ \lambda_n(x^n, \theta^n) \text{ is a simple eigenvalue}, \\ &(iii)\ (x^n, \theta^n) \text{ is a critical point of } \lambda_n(x, \theta), i.e., \\ &\qquad \nabla_x \lambda_n(x^n, \theta^n) = \nabla_\theta \lambda_n(x^n, \theta^n) = 0. \end{aligned} \tag{99}$$

Notations. We introduce a new intermediate scale variable z, defined by

$$z := \sqrt{\epsilon}(y - y^n) \equiv \frac{x - x^n}{\sqrt{\epsilon}}.$$

Theorem 11.1 ([AP06]). *Under assumption (99) and for an initial data*

$$u_\epsilon^0(x) = \psi_n\left(x^n, \frac{x}{\epsilon}, \theta^n\right)e^{2i\pi\frac{\theta^n \cdot x}{\epsilon}}v^0\left(\frac{x - x^n}{\sqrt{\epsilon}}\right), \tag{100}$$

the solution of (98) can be written as

$$u_\epsilon(t,x) = e^{i\frac{\lambda_n(x^n,\theta^n)t}{\epsilon}} e^{2i\pi\frac{\theta^n\cdot x}{\epsilon}} v_\epsilon\left(t, \frac{x-x^n}{\sqrt{\epsilon}}\right), \tag{101}$$

where $v_\epsilon(t,z)$ two-scale converges strongly to $\psi_n(x^n, y, \theta^n)v(t,z)$, i.e.,

$$\lim_{\epsilon\to 0} \int_{\mathbb{R}^N} \left| v_\epsilon(t,z) - \psi_n\left(x^n, \frac{z}{\sqrt{\epsilon}} + \frac{x^n}{\epsilon}, \theta^n\right) v(t,z)\right|^2 dz = 0, \tag{102}$$

uniformly on compact time intervals in $\mathbb{R}^+$, and $v(t,z)$ is the unique solution of the homogenized Schrödinger equation

$$\begin{cases} i\dfrac{\partial v}{\partial t} - \mathrm{div}_z\left(A^*\nabla_z v\right) + \mathrm{div}_z(vB^*z) + c^*v + vD^*z\cdot z = 0 \ \text{in } \mathbb{R}^N\times\mathbb{R}^+ \\ v(0,z) = v^0(z) \qquad\qquad\qquad\qquad\qquad\qquad\qquad\qquad\quad \text{in } \mathbb{R}^N \end{cases} \tag{103}$$

where $A^* = \dfrac{1}{8\pi^2}\nabla_\theta\nabla_\theta\lambda_n(x^n,\theta^n)$, $B^* = \dfrac{1}{2i\pi}\nabla_\theta\nabla_x\lambda_n(x^n,\theta^n)$,

$D^* = \dfrac{1}{2}\nabla_x\nabla_x\lambda_n(x^n,\theta^n)$, *and c^* is given by*

$$\begin{aligned} c^* = \int_{\mathbb{T}^N} &\Big[A(\nabla_y + 2i\pi\theta^n)\psi_n \cdot \frac{\partial\bar{\psi}_n}{\partial x_k} e_k - A(\nabla_y - 2i\pi\theta^n)\frac{\partial\bar{\psi}_n}{\partial x_k} \cdot \psi_n e_k \\ &- \frac{\partial A}{\partial x_k}(\nabla_y - 2i\pi\theta^n)\bar{\psi}_n \cdot \psi_n e_k\Big](x^n, y)\, dy\,. \end{aligned} \tag{104}$$

Notice that even if the tensor A^* might be non-coercive, the homogenized problem (103) is well posed. Indeed the operator $\mathbb{A}^* : L^2(\mathbb{R}^N) \to L^2(\mathbb{R}^N)$ defined by

$$\mathbb{A}^*\phi = -\mathrm{div}\left(A^*\nabla\phi\right) + \mathrm{div}(\phi B^*z) + c^*\phi + \phi D^*z\cdot z \tag{105}$$

is self-adjoint by virtue of Proposition 11.1 below and therefore by using semi-group theory (see, e.g., [Bre73] or Chap. X in [RS78]), one can show that there exists a unique solution in $C(\mathbb{R}^+; L^2(\mathbb{R}^N))$, although it may not belong to $L^\infty(\mathbb{R}^+; H^1(\mathbb{R}^N))$. The next result establishes the conservation of the L^2-norm for the solution v of the homogenized equation (103) and the self-adjointness of the operator $\mathbb{A}^*$.

Proposition 11.1. *Let $v \in C(\mathbb{R}^+; L^2(\mathbb{R}^N))$ be a solution of (103). Then*

$$\|v(t,\cdot)\|_{L^2(\mathbb{R}^N)} = \|v^0\|_{L^2(\mathbb{R}^N)} \quad \forall t \in \mathbb{R}^+\,. \tag{106}$$

Moreover the operator $\mathbb{A}^*$ defined in (105) is self-adjoint.

Proof. We multiply (103) by $\bar{v}$ and take the imaginary part to obtain

$$\frac{1}{2}\frac{d}{dt}\int_{\mathbb{R}^N}|v|^2\,dz = \text{Im}\left(\int_{\mathbb{R}^N}vB^*z\cdot\nabla\bar{v} - c^*|v|^2\,dz\right). \tag{107}$$

After integration by parts the right-hand side of (107) is equal to

$$-\left(\frac{1}{2i}\text{tr}\,B^* + \text{Im}c^*\right)\int_{\mathbb{R}^N}|v|^2\,dz$$

and therefore (106) is proved as soon as we show that

$$\frac{1}{2i}\text{tr}\,B^* + \text{Im}c^* = 0. \tag{108}$$

Actually, (108) is a consequence of the Fredholm alternatives for the derivatives, with respect to x and θ, of the cell spectral equation (51) (for details, see [AP06]). In order to prove the self-adjointness of the operator $\mathbb{A}^*$, one first checks that $\mathbb{A}^*$ is symmetric, which easily follows by (108) and the fact that $\overline{B}^* = -B^*$, and then observes that up to addition of a multiple of the identity the operator $\mathbb{A}^*$ is monotone (see, e.g., [Bre83], Chap. VII). $\square$

In the next proposition we will denote by $\nabla\nabla\lambda_n$ the full Hessian matrix of the function $\lambda_n(x,\theta)$ evaluated at the point (x^n,θ^n), namely

$$\nabla\nabla\lambda_n = \begin{pmatrix} \nabla_x\nabla_x\lambda_n & \nabla_\theta\nabla_x\lambda_n \\ \nabla_\theta\nabla_x\lambda_n & \nabla_\theta\nabla_\theta\lambda_n \end{pmatrix}(x^n,\theta^n).$$

Proposition 11.2. *Assume that the matrix $\nabla\nabla\lambda_n$ is positive definite. Then there exists an orthonormal basis $\{\phi_n\}_{n\geq 1}$ of eigenfunctions of $\mathbb{A}^*$; moreover for each n there exists a real constant $\gamma_n > 0$ such that*

$$e^{\gamma_n|z|}\phi_n\,,\ e^{\gamma_n|z|}\nabla\phi_n \in L^2(\mathbb{R}^N). \tag{109}$$

Proof. Up to shifting the spectrum of the operator $\mathbb{A}^*$, we may assume that $\text{Re}(c^*) = 0$. In order to prove the existence of an orthonormal basis of eigenfunctions we introduce the inverse operator of $\mathbb{A}^*$, denoted by G^*

$$\begin{aligned} G^* : L^2(\mathbb{R}^N) &\to L^2(\mathbb{R}^N) \\ f &\to \phi \text{ unique solution in } H^1(\mathbb{R}^N) \text{ of} \\ &\quad \mathbb{A}^*\phi = f \quad \text{in } \mathbb{R}^N \end{aligned} \tag{110}$$

and we show that G^* is compact. Indeed multiplication of (110) by $\bar{\phi}$ yields

$$\int_{\mathbb{R}^N}[A^*\nabla\phi\cdot\nabla\bar{\phi} - iB^*\text{Im}(\phi z\cdot\nabla\bar{\phi}) + D^*z\cdot z|\phi|^2]\,dz = \int_{\mathbb{R}^N}f\bar{\phi}\,dz. \tag{111}$$

Upon defining the $2N$-dimensional vector-valued function Φ

$$\Phi := \begin{pmatrix} 2i\pi z\phi \\ \nabla\phi \end{pmatrix}$$

we rewrite (111) in agreement with this block notation

$$\int_{\mathbb{R}^N} \frac{1}{8\pi^2} \nabla\nabla\lambda_n \Phi \cdot \overline{\Phi}\, dz = \int_{\mathbb{R}^N} f\overline{\phi}\, dz\,.$$

By the positivity assumption on the matrix $\nabla\nabla\lambda_n$ it follows that there exists a positive constant c_0 such that

$$c_0 \left(||\nabla\phi||^2_{L^2(\mathbb{R}^N)} + ||z\phi||^2_{L^2(\mathbb{R}^N)} \right) \leq ||f||_{L^2(\mathbb{R}^N)} ||\phi||_{L^2(\mathbb{R}^N)}\,,$$

which implies by a standard argument

$$||\phi||^2_{L^2(\mathbb{R}^N)} + ||\nabla\phi||^2_{L^2(\mathbb{R}^N)} + ||z\phi||^2_{L^2(\mathbb{R}^N)} \leq C||f||^2_{L^2(\mathbb{R}^N)}\,,$$

from which we deduce the compactness of G^* in $L^2(\mathbb{R}^N)$-strong. Thus there exists an infinite countable number of eigenvalues for $\mathbb{A}^*$. The proof of the exponential decay (109) of the corresponding eigenfunctions is a classical matter (see [AP06] for details). □

Proof of Theorem 11.1. We content ourselves in giving a sketch of it. Without loss of generality we assume from now on that $x^n = 0$, which simplifies the change of variables, and we define $\lambda_n = \lambda_n(0, \theta^n)$. The main idea is to rescale the space variable by introducing $z = x/\sqrt{\epsilon}$, and to perform a Taylor expansion in the coefficients for z close to the origin. We define a sequence v_ϵ by

$$v_\epsilon(t, z) := e^{-i\frac{\lambda_n t}{\epsilon}} e^{-2i\pi \frac{\theta^n \cdot x}{\epsilon}} u_\epsilon(t, x)\,. \tag{112}$$

The a priori estimates of Lemma 6.1 are still valid here and imply that $v_\epsilon(t, z)$ satisfies

$$||v_\epsilon||_{L^\infty(\mathbb{R}^+; L^2(\mathbb{R}^N))} + \sqrt{\epsilon}||\nabla v_\epsilon||_{L^\infty(\mathbb{R}^+; L^2(\mathbb{R}^N))} \leq C\,.$$

We apply the compactness of two-scale convergence (see Theorem 3.1) with test functions oscillating periodically in z with period $\sqrt{\epsilon}$ (instead of ϵ as before). Therefore, up to a subsequence, there exists a limit $v^*(t, z, y) \in L^2\left((0, T) \times \mathbb{R}^N; H^1(\mathbb{T}^N)\right)$ such that v_ϵ and $\sqrt{\epsilon}\nabla v_\epsilon$ two-scale converge to v^* and $\nabla_y v^*$, respectively. Similarly, by definition of the initial data, $v_\epsilon(0, z)$ two-scale converges to $\psi_n(y)v^0(z)$.

Although v_ϵ is the unknown which will pass to the limit in the sequel, it is simpler to write an equation for another function, namely

$$w_\epsilon(t, z) := e^{2i\pi \frac{\theta^n \cdot z}{\sqrt{\epsilon}}} v_\epsilon(t, z) = e^{-i\frac{\lambda_n t}{\epsilon}} u_\epsilon(t, x)\,. \tag{113}$$

Upon this change of unknown and of variable, it can be checked that w_ϵ solves the following equation

$$\begin{cases} i\dfrac{\partial w_\epsilon}{\partial t} - \operatorname{div}\left(A\left(\sqrt{\epsilon}z, z/\sqrt{\epsilon}\right)\nabla w_\epsilon\right) + \dfrac{1}{\epsilon}\left(c(\sqrt{\epsilon}z, z/\sqrt{\epsilon}) - \lambda_n\right)w_\epsilon = 0 \\[2mm] w_\epsilon(0,z) = u_\epsilon^0(\sqrt{\epsilon}z) \end{cases} \tag{114}$$

where the differential operators div and ∇ act with respect to the new variable z.

First step. As usual we multiply (114) by the complex conjugate of

$$\epsilon\phi\left(t, z, \frac{z}{\sqrt{\epsilon}}\right)e^{2i\pi\frac{\theta^n\cdot z}{\sqrt{\epsilon}}}$$

where $\phi(t,z,y)$ is a smooth test function defined on $\mathbb{R}^+ \times \mathbb{R}^N \times \mathbb{T}^N$, with compact support in $\mathbb{R}^+ \times \mathbb{R}^N$. Since this test function has compact support (fixed with respect to ϵ), the effect of the non-periodic variable in the coefficients is negligible for sufficiently small ϵ. Therefore we can replace the value of each coefficient at $(\sqrt{\epsilon}z, z/\sqrt{\epsilon})$ by its Taylor expansion of order two about the point $(0, z/\sqrt{\epsilon})$. Passing to the two-scale limit we get the variational formulation of

$$-(\operatorname{div}_y + 2i\pi\theta^n)\left(A(0,y)(\nabla_y + 2i\pi\theta^n)v^*\right) + c(0,y)v^* = \lambda_n v^* \quad \text{in } \mathbb{T}^N.$$

The simplicity of λ_n implies that there exists a scalar function $v(t,z) \in L^2\left(\mathbb{R}^+ \times \mathbb{R}^N\right)$ such that

$$v^*(t,z,y) = v(t,z)\psi_n(0,y,\theta^n). \tag{115}$$

Second step. We multiply (114) by the complex conjugate of

$$\Psi_\epsilon(t,z) = e^{2i\pi\theta^n\cdot\frac{z}{\sqrt{\epsilon}}}\left[\psi_n^\epsilon\phi(t,z) + \sqrt{\epsilon}\sum_{k=1}^N\left(\frac{1}{2i\pi}\frac{\partial\psi_n^\epsilon}{\partial\theta_k}\frac{\partial\phi}{\partial z_k}(t,z) + z_k\frac{\partial\psi_n^\epsilon}{\partial x_k}\phi(t,z)\right)\right], \tag{116}$$

where $\psi_n^\epsilon(z) = \psi_n(0, z/\sqrt{\epsilon}, \theta^n)$ and $\phi(t,z)$ is a smooth test function with compact support in $\mathbb{R}^+ \times \mathbb{R}^N$. Remark the new terms depending linearly on z in (116), new with respect to (72). Nevertheless a similar computation (see [AP06] for details) allows us to pass to the scale limit and obtain a weak formulation of (103).   □

Acknowledgements This work was partially supported by the MULTIMAT european network MRTN-CT-2004-505226 funded by the EEC. Most of the results presented here were obtained in collaboration with M. Palombaro, A. Piatnitski and M. Vanninathan.

References

[Alb75]   Albert, J.H.: Genericity of simple eigenvalues for elliptic pde's, Proc. A.M.S.
             48:413–418 (1975).
[All92]   Allaire, G.: Homogenization and two-scale convergence, SIAM J. Math. Anal.
             23(6):1482–1518 (1992).
[ACPSV04]   Allaire, G., Capdeboscq, Y., Piatnitski, A., Siess, V., Vanninathan, M.: Ho-
             mogenization of periodic systems with large potentials, Arch. Rat. Mech.
             Anal. **174**, pp.179–220 (2004).
[AC98]   Allaire, G., Conca, C.: Bloch wave homogenization and spectral asymptotic
             analysis, J. Math. Pures et Appli. **77**:153–208 (1998).
[AP06]   Allaire, G., Palombaro, M.: Localization for the Schrödinger equation in a
             locally periodic medium, SIAM J. Math. Anal. **38**, pp.127–142 (2006).
[AP05]   Allaire, G., Piatnistki, A.: Homogenization of the Schrödinger equation and
             effective mass theorems, Comm. Math Phys. **258**, pp.1–22 (2005).
[AV06]   Allaire, G., Vanninathan, M.: Homogenization of the Schrödinger equation
             with a time oscillating potential, DCDS series B, **6**, pp.1–16 (2006).
[BP90]   Bakhvalov, N., Panasenko, G.: Homogenization : averaging processes in pe-
             riodic media, Mathematics and its applications, vol.36, Kluwer Academic
             Publishers, Dordrecht (1990).
[Bec99]   Bechouche, Ph.: Semi-classical limits in a crystal with a Coulombian self-
             consistent potential: effective mass theorems, Asymptot. Anal. **19**, no. 2,
             pp.95–116 (1999).
[BP06]   Ben Abdallah, N., Pinaud, O.: Multiscale simulation of transport in an open
             quantum system: resonances and WKB interpolation, J. Comput. Phys. **213**,
             no. 1, pp.288–310 (2006).
[BLP78]   Bensoussan, A., Lions, J.-L., Papanicolaou, G.: Asymptotic analysis for pe-
             riodic structures, North-Holland, Amsterdam (1978).
[Blo28]   Bloch, F.: Uber die Quantenmechanik der Electronen in Kristallgittern, Z.
             Phys. **52**, pp.555–600 (1928).
[Bre73]   Brézis, H.: Opérateurs maximaux monotones et semi-groupes de contractions
             dans les espaces de Hilbert, North-Holland, Amsterdam (1973).
[Bre83]   Brézis, H.: Analyse fonctionelle, Masson, Paris (1983).
[Bus87]   Buslaev, V.: Semiclassical approximation for equations with periodic coeffi-
             cients, Russ. Math. Surv. **42**, pp.97–125 (1987).
[CL90]   Carmona, R., Lacroix, J.: Spectral Theory of Random Schrödinger Operators,
             Birkhäuser, Boston (1990).
[CH98]   Cazenave, Th., Haraux, A.: An introduction to semilinear evolution equa-
             tions, Oxford Lecture Series in Mathematics and its Applications, **13**, Oxford
             University Press, New York (1998).
[CD99]   Cioranescu, D., Donato, P.: An introduction to homogenization, Oxford Lec-
             ture Series in Mathematics and Applications **17**, Oxford (1999).
[COV02]   Conca, C., Orive, R., Vanninathan, M.: Bloch approximation in homogeniza-
             tion and applications, SIAM J. Math. Anal. **33**:1166–1198 (2002).
[CPV95]   Conca, C., Planchard, J., Vanninathan, M.: Fluids and periodic structures,
             RMA **38**, J. Wiley & Masson, Paris (1995).
[CV97]   Conca, C., Vanninathan, M.: Homogenization of periodic structures via Bloch
             decomposition, SIAM J. Appl. Math. **57**:1639–1659 (1997).
[Dal93]   Dal Maso, G.: An Introduction to Γ-Convergence, Progress in Nonlinear Dif-
             ferential Equations and their Applications, **8**, Birkhäuser, Boston (1993).
[DGR02]   Dimassi, M., Guillot, J.-C., Ralston, J.: Semiclassical asymptotics in magnetic
             Bloch bands, J. Phys. A **35**, no. 35, 7597–7605 (2002).
[FG02]   Fermanian-Kammerer, C., Gérard, P.: Mesures semi-classiques et croisement
             de modes, Bull. Soc. Math. France **130**, pp.123–168 (2002).

[FG03]   Fermanian-Kammerer, C., Gérard, P.: A Landau-Zener formula for non-
            degenerated involutive codimension 3 crossings, Ann. Henri Poincaré **4**,
            pp.513–552 (2003).
[Flo83]   Floquet, G.: Sur les équations différentielles linéaires à coefficients pério-
            diques, Ann. Ecole Norm. Sér. 2 **12**, pp.47–89 (1883).
[Gel50]   Gelfand, I.M.: Expansion in series of eigenfunctions of an equation with pe-
            riodic coefficients, Dokl. Akad. Nauk. SSSR **73**, pp.1117–1120 (1950).
[Ger91]   Gérard, P.: *Mesures semi-classiques et ondes de Bloch*, Séminaire sur les
            équations aux Dérivées Partielles, 1990–1991, Exp. No. XVI, 19 pp., École
            Polytech., Palaiseau (1991).
[GMNP97]   Gérard, P., Markowich, P., Mauser, N., Poupaud, F.: *Homogenization limits
            and Wigner transforms*, Comm. Pure Appl. Math. **50**, no. 4, 323–379 (1997).
[GMS91]   Gérard, C., Martinez, A., Sjöstrand, J.: *A mathematical approach to the ef-
            fective Hamiltonian in perturbed periodic problems*, Comm. Math. Phys. **142**,
            no. 2, 217–244 (1991).
[GRT88]   Guillot, J.-C., Ralston, J., Trubowitz, E.: *Semi-classical methods in solid
            state physics*, Comm. Math. Phys. **116**, 401–415 (1988).
[JKO94]   Jikov, V.V., Kozlov, S.M., Oleinik., O.A.: *Homogenization of Differential
            Operators and Integral Functionals.* Springer Verlag, (1994).
[Kat66]   Kato, T.: *Perturbation theory for linear operators*, Springer-Verlag, Berlin
            (1966).
[Kuc93]   Kuchment, P.: *Floquet theory for partial differential equations*, Operator The-
            ory: Advances and Applications, **60**, Birkhäuser Verlag, Basel (1993).
[Lio81]   Lions, J.-L.: *Some methods in the mathematical analysis of systems and their
            control*, Science Press, Beijing, Gordon and Breach, New York (1981).
[MW66]   Magnus, W., Winkler, S.: *Hill's equation*, Interscience Tracts in Pure and
            Applied Mathematics, No. 20, Interscience Publishers John Wiley & Sons,
            New York-London-Sydney (1966).
[MNP94]   Markowich, P., Mauser, N., Poupaud, P.: *A Wigner-function approach to
            (semi)classical limits: electrons in a periodic potential*, J. Math. Phys. **35**,
            no. 3, pp.1066–1094 (1994).
[MP05]   Marušić-Paloka, E., Piatnitski, A.: *Homogenization of a nonlinear
            convection-diffusion equation with rapidly oscillating coefficients and strong
            convection*, J. London Math. Soc. (2) **72**, no. 2, 391–409 (2005).
[MB91]   Morgan, R., Babuska, I.: *An approach for constructing families of equations
            for periodic media. I and II*, SIAM J. Math. Anal. **22**, pp.1–15, pp.16–33,
            (1991).
[MT78]   Murat, F., Tartar, L.: *H-convergence*, Séminaire d'Analyse Fonctionnelle
            et Numérique de l'Université d'Alger, mimeographed notes (1978). English
            translation in *Topics in the mathematical modelling of composite materials*,
            A. Cherkaev, R. Kohn, Editors, Progress in Nonlinear Differential Equations
            and their Applications, **31**, Birkhaüser, Boston (1997).
[Mye90]   Myers, H.P.: *Introductory solid state physics*, Taylor & Francis, London
            (1990).
[Ngu89]   Nguetseng, G.: *A general convergence result for a functional related to the
            theory of homogenization*, SIAM J. Math. Anal. **20**(3), pp.608–623 (1989).
[OK64]   Odeh, F., Keller, J.: *Partial differential equations with periodic coefficients
            and Bloch waves in crystals*, J. Math. Phys. **5**, 11, pp.1499–1504 (1964).
[PST03]   Panati, G., Sohn, H., Teufel, S.: *Effective dynamics for Bloch electrons:
            Peierls substitution and beyond*, Comm. Math. Phys. **242**, pp.547–578 (2003).
[Ped97]   Pedersen, F.: *Simple derivation of the effective-mass equation using a
            multiple-scale technique*, Eur. J. Phys., **18**, pp.43–45 (1997).
[PR96]   Poupaud, F., Ringhofer, C.: *Semi-classical limits in a crystal with exterior
            potentials and effective mass theorems*, Comm. Partial Differential Equations,
            21, no. 11–12, pp.1897–1918 (1996).

[Que98]   Quéré, Y.: *Physics of materials,* Taylor & Francis (1998).
[RS78]   Reed, M., Simon, B.: *Methods of modern mathematical physics,* Academic
            Press, New York (1978).
[San80]   Sanchez-Palencia, E.: *Non homogeneous media and vibration theory,* Lecture
            notes in physics **127**, Springer Verlag (1980).
[Sev82]   Sevost'janova, E.V.: *An asymptotic expansion of the solution of a second
            order elliptic equation with periodic rapidly oscillating coefficients,* Math.
            USSR Sbornik, **43**, pp.181–198 (1982).
[Spa68]   Spagnolo, S.: *Sulla convergenza di soluzioni di equazione paraboliche ed el-
            litiche,* Ann. Sc. Norm. Sup. Pisa **22**, pp.577–597 (1968).
[Spa76]   Spagnolo, S.: *Convergence in energy for elliptic operators,* Numerical solu-
            tions of partial differential equations III Synspade 1975, B. Hubbard ed.,
            Academic Press New York (1976).
[Spa06]   Sparber, Ch.: *Effective mass theorems for nonlinear Schrödinger equations,*
            SIAM J. Appl. Math. **66**, no. 3, pp.820–842 (2006).
[Tar78]   Tartar, L.: *Quelques remarques sur l'homogénéisation,* Proc. of the Japan-
            France Seminar 1976 "Functional Analysis and Numerical Analysis", Japan
            Society for the Promotion of Sciences pp.469–482 (1978).
[Tar00]   Tartar, L.: *An introduction to the homogenization method in optimal design,*
            in Optimal shape design (Tróia, 1998), A. Cellina and A. Ornelas eds., Lecture
            Notes in Mathematics **1740**, pp.47–156, Springer, Berlin (2000).
[Wil78]   Wilcox, C.: *Theory of Bloch waves,* J. Anal. Math. **33**, pp.146–167 (1978).

Mathematical Properties of Quantum Evolution Equations

Anton Arnold

Abstract This chapter focuses on the mathematical analysis of nonlinear quantum transport equations that appear in the modeling of nano-scale semiconductor devices. We start with a brief introduction on quantum devices like the resonant tunneling diode and quantum waveguides. For the mathematical analysis of quantum evolution equations we shall mostly focus on whole space problems to avoid the technicalities due to boundary conditions. We shall discuss three different quantum descriptions: Schrödinger wave functions, density matrices, and Wigner functions. For the Schrödinger–Poisson analysis (in H^1 and L^2) we present Strichartz inequalities. As for density matrices, we discuss both closed and open quantum systems (in Lindblad form). Their evolution is analyzed in the space of trace class operators and energy subspaces, employing Lieb–Thirring-type inequalities. For the analysis of the Wigner–Poisson–Fokker–Planck system we shall first derive (quantum) kinetic dispersion estimates (for Vlasov–Poisson and Wigner–Poisson). The large-time behavior of the linear Wigner–Fokker–Planck equation is based on the (parabolic) entropy method. Finally, we discuss boundary value problems in the Wigner framework.

List of Abbreviations and Symbols

$\Re$	Real part of a complex number
$\Im$	Imaginary part of a complex number
$\bar{z}$	Complex conjugate of $z \in \mathbb{C}$
$\mathcal{F}, \mathcal{F}_x, \mathcal{F}_{x\to\xi}$	Fourier transform (w.r.t. the x variable; from the x to the ξ variable)

Anton Arnold

Institute for Analysis and Scientific Computing, Vienna University of Technology, Wiedner Hauptstraße 8, A-1040 Vienna, Austria, e-mail: anton.arnold@tuwien.ac.at

N. Ben Abdallah, G. Frosali (eds.), *Quantum Transport.*

Lecture Notes in Mathematics 1946.

© Springer-Verlag Berlin Heidelberg 2008

$\mathcal{F}^{-1}, \mathcal{F}^{-1}_{\xi \to x}$ Inverse Fourier transform (from the ξ to the x variable)

$\hat{\varphi}$ Fourier transform of the function φ, i.e.
$\hat{\varphi}(\xi) = (2\pi)^{-N/2} \int_{\mathbb{R}^N} \varphi(x) \mathrm{e}^{-ix \cdot \xi} \mathrm{d}x$

ℓ^p Sequence spaces, $1 \le p \le \infty$

$L^p(\Omega)$ Lebesgue spaces on the open set Ω, $1 \le p \le \infty$

$L^p_{loc}(\Omega)$ Spaces of locally p-integrable functions, $1 \le p \le \infty$

$L^p(\Omega, \mathrm{d}\mu)$ Lebesgue spaces on the open set Ω w.r.t. the measure $\mathrm{d}\mu$, $1 \le p \le \infty$

$H^k(\mathbb{R}^N)$ Sobolev spaces on $\mathbb{R}^N$ with differentiability index $k \in \mathbb{Z}$ (and integrability $p = 2$)

$W^{k,p}(\mathbb{R}; X)$ Sobolev spaces of functions on $\mathbb{R}$ with values in the Banach space X (differentiability index $k \in \mathbb{N}_0$, integrability $1 \le p \le \infty$)

$L^{q,p}$ $:= L^q(\mathbb{R}; L^p(\mathbb{R}^N))$, $1 \le q, p \le \infty$

$L^{q,p}_T$ $:= L^q((-T,T); L^p(\mathbb{R}^N))$, with some fixed $T > 0$

$L^1_x(L^q_v)$ $:= L^1(\mathbb{R}^N_x; L^q(\mathbb{R}^N_v))$, $1 \le q \le \infty$

$C(\mathbb{R}; X)$ Space of continuous functions on $\mathbb{R}$ with values in the Banach space X

$C^1(\mathbb{R}; X)$ Space of continuously differentiable functions on $\mathbb{R}$ with values in the normed space X

$C_0(\Omega)$ Continuous functions on Ω with compact support

$C_0^\infty(\Omega)$ Infinitely differentiable functions on Ω with compact support

$C_B^\infty(\Omega)$ Infinitely differentiable, bounded functions on Ω

$\mathcal{S}(\mathbb{R}^N)$ Schwartz space of rapidly decreasing functions on $\mathbb{R}^N$

$\mathcal{S}'(\mathbb{R}^N)$ Tempered distributions on $\mathbb{R}^N$

$\mathcal{B}(X, Y)$ Space of bounded operators from the Banach space X to the Banach space Y

$\mathcal{J}_1(\mathcal{H})$ Space of trace class operators on the Hilbert space $\mathcal{H}$

$\tilde{\mathcal{J}}_1(\mathcal{H})$ (Sub)space of self-adjoint operators trace class operators on the Hilbert space $\mathcal{H}$

$\mathcal{J}_2(\mathcal{H})$ Space of Hilbert–Schmidt operators on the Hilbert space $\mathcal{H}$

Tr   Operator trace

$\| \cdot \|_1$ Trace class norm on $\mathcal{J}_1(\mathcal{H})$

$\| \cdot \|_2$ Hilbert–Schmidt norm on $\mathcal{J}_2(\mathcal{H})$

x, y Position variables, $x, y \in \mathbb{R}^N$

v Velocity variable, $v \in \mathbb{R}^N$

p Momentum, $p \in \mathbb{R}^N$

t Time, $t \in \mathbb{R}$

$\hbar$ (Reduced) Planck constant

ε Permittivity

e (Positive) elementary charge

m Particle mass (electron mass, e.g.)

$V(x, t)$ (Electrostatic) potential, $V \in \mathbb{R}$

$\psi(x, t)$ (Schrödinger) wave function, $\psi \in \mathbb{C}$

$w(x, v, t)$	Wigner function, $w \in \mathbb{R}$
$n(x, t)$	(Spatial) position density, $n \geq 0$
$e_{kin}(x, t)$	Kinetic energy density, $e_{kin} \geq 0$
$j(x, t)$	Current density, $j \in \mathbb{R}$
$D(x)$	Doping profile, density of donor ions, $D \geq 0$
$\varrho(x, y, t)$	Density matrix function, $\varrho \in \mathbb{C}$
$\hat{\varrho}$	Density matrix operator, $\hat{\varrho} \in \mathcal{J}_1$
$\mathcal{D}(A)$	Domain of the operator A
$\bar{A}$	Closure of the operator A
A^*	Adjoint of the operator A
$\hookrightarrow$	Continuous embedding of normed spaces
$f *_x g$	(Partial) Convolution of the functions f and g w.r.t. the x variable

1 Quantum Transport Models for Semiconductor Nano-Devices

The modern computer and telecommunication industry relies heavily on the use of semiconductor devices like transistors. A very important fact of the success of these devices is their rapidly shrinking size. Presently, their characteristic size (channel lengths in transistors, e.g.) has been decreased to some deca-nanometers only. On such small length scales, quantum properties of electrons and atoms cannot be neglected any longer and there are two different consequences: On the one hand classical simulation models of "conventional" devices must then be modified as to include quantum corrections. This concerns devices like the *metal oxide semiconductor field-effect transistor* (MOSFET), which is the dominant building block of today's integrated circuits. On the other hand, more and more *intrinsic quantum devices* (like resonant tunneling diodes, resonant tunneling field-effect transistors, single-electron transistors, quantum dots, quantum waveguides) are devised and manufactured. Their main operational features depend on actively exploiting quantum mechanical effects like tunneling, spatial confinements, and quantized energy levels. Such quantum devices have already applications as high frequency oscillators, in laser diodes, and as memory devices. Moreover, quantum dots are very promising candidates for use in solid-state quantum computation. Compared to off-the-shelf MOSFETs, however, they are still used in rather experimental settings with niche applications. But their great technological and commercial potential drives a tremendous research interest in such nano-devices.

The development of novel semiconductor devices is usually supported by computer simulations to optimize the desired operating features. Now, in order to perform the numerical simulations for the electron flow through a device, mathematical equations (mostly partial differential or integro-differential equations) are needed. They should be both physically accurate and numerically solvable with low computational cost. Semiconductor engineers use various quantum mechanical frameworks in their quantum transport models: Schrödinger wave functions, Wigner functions, density matrices, and Green's functions. Schrödinger models are used to describe the purely ballistic transport of electrons and holes, and they are employed for simulations of quantum waveguides and nano-scale semiconductor heterostructures, e.g. As soon as scattering mechanisms (between electrons and phonons, or with crystal impurities, e.g.) become important, it is convenient to adopt the Wigner formalism or the equivalent density matrix formalism. For practical applications Wigner functions have also the advantage to allow for a rather simple, intuitive formulation of boundary conditions at device contacts or interfaces. As a drawback, the Wigner equation is posed in a high dimensional phase space which makes its numerical solution extremely costly. As a compromise, fluid-type models can provide a reasonable approximation. Although less accurate, they are often used.

In these lecture notes we shall give a survey on the analytical problems and properties associated with Schrödinger, Wigner, and density matrix models that are used in quantum transport applications. For a basic introduction (both physical and mathematical) to quantum mechanics we refer the reader to [Boh89], [LL85], [Deg], Sect. I.6 of [DL88], and [Tha05]. An introduction to quantum transport for semiconductor devices can be found in [Fre94], [Rin03], Sects. 1.4–1.6 of [MRS90], and [KN06].

Before starting the mathematical analysis of quantum evolution equations, we first present the transport models used for two prototype devices – quantum waveguides and resonant tunneling diodes.

1.1 Quantum Waveguide with Adjustable Cavity

In this subsection we discuss the mathematical models used in the simulations of *quantum waveguides*. These are novel electronic switches of nanoscale dimensions. They are made of several different layers of semiconductor materials such that the electron flow is confined to small channels or waveguides. Due to their sandwiched structure the relevant geometry for the electron current is essentially two dimensional. Figure 1 shows the example of a T-shaped *quantum interference transistor*. The actual structure can be realized as an etched layer of GaAs (gallium arsenide) between two layers of doped AlGaAs (aluminum gallium arsenide). Applying an external potential at the gate (i.e. above the shaded portion of the stub), the "allowed region" for the electrons, and hence the geometry (in particular the stub length) can be modified. This allows to control the current flow through such an electronic device. It makes it a switch, which resembles a transistor – but on a nano-scale. With respect

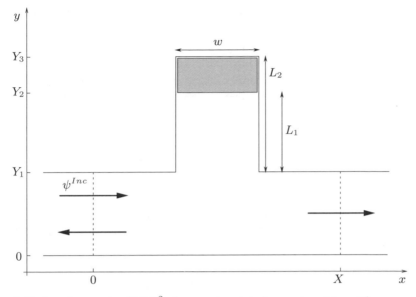

Fig. 1 T-shaped geometry $\Omega \subset \mathbb{R}^2$ of a quantum interference transistor with source and drain contacts to the left and right of the channel. Applying a gate voltage above the stub allows to modify the stub length from L_1 to L_2 and hence to switch the transistor between the on- and off-states. In numerical simulations, the domain Ω is artificially cut off at $x = 0$ and $x = X$ by adding transparent boundary conditions

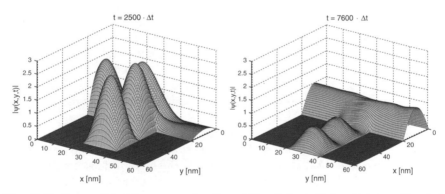

Fig. 2 Stationary Schrödinger wave functions $|\psi(x,y)|$ for a T-shaped waveguide. *Left*: short stub (i.e. $L_1 = 32$ nm) – "off state"; *Right*: long stub (i.e. $L_2 = 40.5$ nm) – "on state"

to small changes in the applied potential and the geometry, such a device shows sharp peaks in conductance that are due to the presence of trapped charges in the stub (see Fig. 2). It is expected that these novel devices will operate at low power and high speed.

The electron transport through a quantum waveguide can be modeled in good approximation by a two dimensional, time dependent Schrödinger–Poisson system for the (infinitely many) wave functions $\psi_\lambda(x,t)$, indexed by the energy variable $\lambda \in \Lambda \subset \mathbb{R}$. The (possibly time-dependent) spatial domain $\Omega \subset \mathbb{R}^2$ consists of (very long) leads and the active switching region (e.g. T-shaped as in Fig. 1). In typical applications electrons are constantly fed into the leads as a continuous superposition of plane waves $\psi_\lambda^{pw}(x,t)$, $\lambda \in \Lambda$. Theoretically, $\Lambda = \mathbb{R}$, but for practical simulations it is restricted to a finite interval, and ultimately discretized. The appropriate Schrödinger–Poisson system reads

$$ i\hbar \frac{\partial \psi_\lambda}{\partial t} = -\frac{\hbar^2}{2m_*} \Delta \psi_\lambda + eV(x,t)\psi_\lambda, \quad x \in \Omega, \ \lambda \in \Lambda, \ t > 0. \tag{1} $$

Here, $\hbar$ is the reduced Planck constant, m_* the effective electron mass in the semiconductor crystal lattice, and e denotes the (positive) elementary charge. The potential $V = V_e + V_{sc}$ consists of an external, applied potential V_c and the selfconsistent potential satisfying the Poisson equation with Dirichlet boundary conditions:

$$ -\varepsilon \Delta V_{sc}(x,t) = e\, n(x,t) = e \int_\Lambda |\psi_\lambda(x,t)|^2 g(\lambda)\, d\lambda, \quad x \in \Omega, \tag{2} $$

$$ V_s = 0, \quad \text{on } \partial\Omega. $$

Here, ε is the permittivity of the semiconductor material and n the spatial electron density. $g(\lambda)$ is a probability distribution, representing the statistics (Fermi–Dirac, e.g.) of the injected waves from both the left and right contact.

In this model we made the following simplifications: We considered only a *single band* and the Schrödinger equation is in the *effective mass approximation*. This means that the effect of the microscopic crystal lattice (yielding a highly oscillatory potential on the atomic length scale) is assumed to be homogenized, and this results in the (constant) effective mass m_*. In heterostructures, however, the effective mass might be space dependent, or even induce nonlocal effects.

This quantum waveguide is connected via leads to an electric circuit. Hence, it is an open system with current flowing through the device. As a consequence, the total electron mass inside the system does not stay constant in time. In typical applications the two leads or contact regions are much longer than suggested by Fig. 1. To reduce computational costs one is therefore obliged to reduce the simulation domain by introducing so-called *open* or *transparent boundary conditions* (TBCs), at $x = 0$ and $x = X$. The purpose of such TBCs is to cut-off the computational domain, but without changing the solution of the original equation. In the simplest case (i.e. a 1D approximation and $V \equiv 0$ in the leads) the TBC takes the form

$$ \frac{\partial}{\partial \eta}(\psi_\lambda - \psi_\lambda^{pw}) = -\sqrt{\frac{2m_*}{\hbar}}\, e^{-i\pi/4} \sqrt{\partial_t}\,(\psi_\lambda - \psi_\lambda^{pw}), \quad \text{for } \lambda \in \Lambda, \ x = 0 \text{ or } x = X, \tag{3} $$

where η denotes the unit outward normal vector at each interface. $\sqrt{\partial_t}$ is the fractional time derivative of order $\frac{1}{2}$, and it can be rewritten as a time-convolution of the boundary data with the kernel $t^{-3/2}$. For the derivation of the 2D-variant of such TBCs and the mathematical analysis of this coupled model (1)–(3) we refer to [BMP05, Arn01, AABES07] and to [LK90] for a stationary Schrödinger-TBC.

To close this subsection we present some simulations of the electron flow through the T-shaped waveguide from Fig. 1 with the dimensions $X = 60\,\text{nm}$, $Y_1 = 20\,\text{nm}$. These calculations are based on the linear Schrödinger equation for a single wave function with $V \equiv 0$ and the injection of a mono-energetic plane wave (i.e. $\Lambda = \{\lambda_0\}$) with $\lambda_0 = 130\,\text{meV}$ from the left lead. The corresponding function $\psi_{\lambda_0}^{pw}(t)$ then appears in formula (3) for the left TBC at $x = 0$. The simulation was based on a compact forth order finite difference scheme ("Numerov scheme") and a Crank–Nicolson discretization in time [AS07, SA07, AJ06].

There are two important device data for practitioners: the current–voltage (I–V) characteristics and the ratio between the on- and the (residual) off-current. This information can be obtained from computing the *stationary Schrödinger states* from the time independent analogue of (1)–(3). Moreover, a third important parameter is the *switching time* between these two stationary states. Depending on the size and shape of the stub, the electron current is either reflected ("off-state" of the device, see Fig. 2, left) or it can flow through the device ("on-state", see Fig. 2, right). In a numerical simulation, this device switching can be realized as follows. Starting from the stationary Schrödinger state shown in Fig. 2 left, we instantaneously extended the stub length from $L_1 = 32\,\text{nm}$ to $L_2 = 40.5\,\text{nm}$. This initiates an evolution of the wave function. After a transient phase of about 4 ps, the new steady state (cf. Fig. 2, right) is reached.

The (complex valued) Schrödinger wave function $\psi(x,t)$, obtained from (1) is rather an auxiliary quantity without intrinsic physical interpretation. Instead, one is rather interested in the following *macroscopic quantities*:

$$n(x,t) := |\psi(x,t)|^2 \ldots \text{ particle density,}$$

$$j(x,t) := \frac{\hbar}{m_*}\Im\left(\bar{\psi}\nabla\psi\right) \ldots \text{ (particle) current density,}$$

which satisfy the *continuity equation*:

$$n_t + \text{div}\,j = 0\,.$$

If we consider a finite, *closed system* (i.e. without inflow and outflow), mass is preserved in time. In this case one typically has $\psi \in L^2(\mathbb{R}^N; \mathbb{C})$, and frequently chooses the normalization $\int_{\mathbb{R}^N} |\psi(x)|^2\,dx = 1$. Then, $\|\psi(\cdot,t)\|_{L^2}^2$ is the (scaled) total mass of the system which is constant under the time evolution by the Schrödinger equation.

Table 1 Examples of the quantization rule

Classical quantity, $a(x, p)$	Quantization, operator A	Expectation value $\langle \psi, A\psi \rangle$
x ... position	x	$\int x \|\psi\|^2 \, \mathrm{d}x$
p ... momentum	$-\mathrm{i}\hbar\nabla_x$	$\mathrm{i}\hbar \int \psi\nabla\bar{\psi} \, \mathrm{d}x$
$\frac{\|p\|^2}{2m}$ ... kinetic energy	$-\frac{\hbar^2}{2m}\Delta_x$	$\frac{\hbar^2}{2m} \int \|\nabla\psi\|^2 \, \mathrm{d}x$
$V(x)$ ... potential (energy)	$V(x)$	$\int V(x)\|\psi\|^2 \, \mathrm{d}x$

As a final remark, we illustrate the relationship between physical variables for classical and quantum particles. Let x and p be, resp., the position and momentum of a classical particle with mass m, and let $a(x, p)$ denote a general physical observable. If that particle is described (at a fixed time) by a classical phase space distribution $f(x, p) \geq 0$ with $(x, p) \in \mathbb{R}^{2N}$ and the typical normalization $\|f\|_{L^1(\mathbb{R}^{2N})} = 1$, the corresponding expectation value of a is

$$\iint_{\mathbb{R}^{2N}} f(x, p)a(x, p) \, \mathrm{d}x\mathrm{d}p \,.$$

For a quantum particle the scalar function $a(x, p)$ is replaced by a (formally) self-adjoint operator A using the *quantization rules* $x \mapsto A := x$ and $p \mapsto A := -\hbar\nabla_x$. Here, $A = x$ denotes the multiplication operator by the variable x. The *expectation value* of the *observable* A in the quantum state ψ is computed as $\langle \psi, A\psi \rangle$ (cf. Table 1 of simple examples).

1.2 Resonant Tunneling Diode

A *resonant tunnel diode* (RTD) is a nano-device which uses quantum effects (tunneling and discrete energy levels) to yield an I–V curve with *negative differential resistance*. Even at room temperature a RTD is capable of generating a tera-Hertz wave, which explains its practical application for ultra high-speed oscillators and possibly for novel digital logic circuits. Presently, RTDs are on the verge of commercialization and extensively studied by engineers.

RTDs have a sandwiched structure of different semiconductor materials (*GaAs* and *AlGaAs*, e.g.) which form two barriers that are only a few nanometers thick (see Fig. 3). This double barrier structure gives rise to a single *quantum well* in its middle. The resulting barrier potential $V_{cont}(x)$ is sketched in Fig. 4. Charge carriers (such as electrons and holes) enter this well by tunneling through a barrier and they can only have particular discrete energy values inside the quantum well. When the energy of the incoming

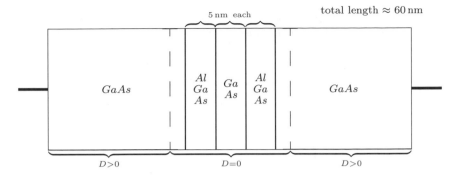

Fig. 3 Schematic 2D cut through a resonant tunneling diode. It consists of the two semi-conductor materials $GaAs$ and $AlGaAs$. Close to the two metallic contacts, the crystal lattice is highly doped (with Si, e.g.). $D(x) \geq 0$ is the spatial density of the implanted donor ions

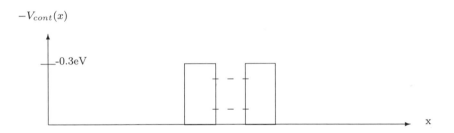

Fig. 4 Effective barrier potential (*contact potential*) V_{cont} for the electron transport induced by the semiconductor heterostructure

electrons coincides with one of these well-energies, their tunneling probability through a barrier rises significantly. Hence, the resulting tunneling current through the double barrier is very peaked at such *resonant energy levels*. In the contact regions of a RTD the crystal lattice is highly doped, i.e. there are donor ions (with Si, e.g.) intentionally implanted into the semiconductor material. Those ions cannot move and their concentration is described by the function $D(x) \geq 0$, the *doping profile*.

A popular and quite accurate simulation model for RTDs is based on Wigner functions [KKFR89, KN06]. Wigner functions are a phase space formulation of quantum mechanics that is equivalent to Schrödinger wave functions (cf. Sect. 5.1 for details). But they seem to be more practical for RTD-simulations, as it is easier to include scattering effects and to formulate (simple) boundary conditions. Their higher dimensionality, however, poses a serious numerical challenge.

The (real valued) Wigner function $w(x, v, t)$ describes the state of a quantum system at time t in the position–velocity phase space. In contrast to classical phase space probability distributions, w typically takes both positive and negative values. The main macroscopic quantities are obtained as follows:

$$n(x,t) := \int_{\mathbb{R}^N} w(x,v,t)\,\mathrm{d}v \geq 0 \ldots \text{ particle density}$$

$$j(x,t) := \int_{\mathbb{R}^N} v\,w(x,v,t)\,\mathrm{d}v \ldots \text{ (particle) current density}$$

$$e_{kin}(x,t) := \frac{m_*}{2} \int_{\mathbb{R}^N} |v|^2 w(x,v,t)\,\mathrm{d}v \geq 0 \ldots \text{ kinetic energy density}$$

Since the Wigner function takes also negative values, it is a-priori not clear why the macroscopic particle density and kinetic energy density should be non-negative, as indicated above. This physically important non-negativity is a consequence of the non-negativity of the density matrix (operator) that is associated with a Wigner function (see Sects. 4.2, 5.1 below).

In order to mathematically formulate a (Wigner function based) quantum transport model of a RTD, we make the following assumptions:

- Only one carrier species is considered: electrons (since the mobility of the holes is too small in such a device to contribute significantly to the charge transport).
- One-particle-like *mean field model* (Hartree approximation).
- Only one parabolic band (with effective mass m_*).
- Purely quantum mechanical transport.
- Ballistics dominates scattering effects (for device lengths up to the order of the electrons' mean free path).

Under the above assumptions, the *Wigner equation* describes the time evolution of the Wigner function w in a given, real valued (electrostatic) potential $V(x,t)$:

$$w_t + v \cdot \nabla_x w - e\,\Theta[V]w = 0, \quad x,v \in \mathbb{R}^N. \tag{4}$$

Here, $\Theta[V]$ is a *pseudo-differential operator* (typical abbreviation: "ΨDO"), defined via a multiplication operator for the v-Fourier transformed Wigner function $\mathcal{F}_v w$:

$$\Theta[V]w(x,v)$$
$$= \frac{\mathrm{i}}{\hbar}(2\pi)^{-N} \iint_{\mathbb{R}^{2N}} \left[V(x + \frac{\hbar\eta}{2m_*}) - V(x - \frac{\hbar\eta}{2m_*}) \right] w(x,\tilde{v})\mathrm{e}^{\mathrm{i}(v-\tilde{v})\cdot\eta}\,\mathrm{d}\tilde{v}\mathrm{d}\eta.$$

Under some regularity and decay assumptions on the potential V, it can be rewritten as convolution operator in v:

$$\Theta[V]w(x,v) = \alpha(x,v) *_v w(x,v)$$

$$\alpha(x,v) := \frac{2}{\hbar}(2\pi)^{-\frac{N}{2}}\left(\frac{2m_*}{\hbar}\right)^N \Im\left[e^{i\frac{2m_*}{\hbar}x\cdot v}\,(\mathcal{F}V)\left(\frac{2m_*}{\hbar}v\right)\right].$$

This convolution form illustrates the non-local effect of potentials in quantum mechanics. Indeed, a particle or wave packet already "feels" an upcoming potential barrier before actually hitting it. Such a "premature" reflection is clearly seen in numerical simulations based on Wigner functions.

For realistic device simulations, scattering (between electrons and impurities or with phonons, i.e. thermal vibrations of the crystal lattice) must be included in the model. Hence, the r.h.s. of (4) has to be augmented by some (at least simple) scattering term. For the 1D simulations of a RTD in [KKFR89] the following relaxation term was used as a phenomenological model for the electron–phonon interactions:

$$w_t + vw_x - e\,\Theta\left[V_{sc}(x,t) + V_{cont}(x)\right]w = \frac{w_{st} - w}{\tau(v)}, \tag{5}$$

$$0 < x < L,\ v \in \mathbb{R},\ t > 0.$$

Here, w_{st} is some appropriate steady state, and $\tau > 0$ denotes the relaxation time, which may be energy dependent. The spatial interval $(0,L)$ models the diode, and (5) is supplemented by some boundary conditions at the contact points $x = 0$, $x = L$. Motivated by the characteristic lines of the free transport equation $w_t + v \cdot \nabla_x w = 0$, the simplest choice is to prescribe the inflow, i.e. $w^+(0,v)$ for $v > 0$ and $w^-(L,v)$ for $v < 0$ (cf. Fig. 5). This procedure is inspired by classical kinetic theory. The statistical carrier distributions in the two contacts yield the prescribed boundary data $w^+(0,v)$, $w^-(L,v) \geq 0$.

In (5) the potential consists of two contributions: the (time independent) barrier potential $V_{cont}(x)$ and the *self-consistent potential* $V_{sc}(x,t)$, which is due to the mean field approximation. V_{sc} solves the (electrostatic) *Poisson equation*

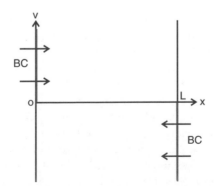

Fig. 5 Vertical slab of x–v-phase space $(0,L) \times \mathbb{R}$ for the 1D Wigner equation: inflow boundary conditions are prescribed at $x = 0$, $v > 0$ and at $x = L$, $v < 0$

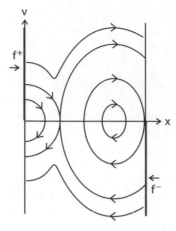

Fig. 6 I–V-characteristics of a RT-diode shows negative differential resistance: (*solid line*) experimental data, (*dashed line*) computed with a simple Schrödinger tunneling model. Reprinted figure with permission from [KKFR89]. Copyright (1989) by the American Physical Society

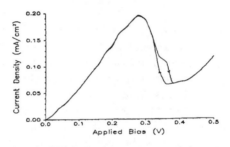

Fig. 7 I–V-characteristics of a RT-diode shows a hysteresis including two stable branches: numerical simulation based on a relaxation-time Wigner–Poisson model. Reprinted figure with permission from [KKFR89]. Copyright (1989) by the American Physical Society

$$\varepsilon\partial_x^2 V_{sc} = e\left(D(x) - n(x,t)\right), \quad 0 < x < L\,. \tag{6}$$

The non-linear *relaxation-time Wigner–Poisson* model (5)–(6) is used in [KKFR89] for numerical simulations of a RTD. Here, the main goal is to compute the I–V-characteristics and to verify the negative differential resistance of this device. Figures 6 and 7 compare the I–V-curve from experimental data with the numerical results.

For the (semi)classical semiconductor Boltzmann equation excellent models for the most important collisional mechanisms have been derived (cf. [MRS90]) and are incorporated into today's commercial simulation tools. In quantum kinetic theory, however, accurate and numerically usable collision models are much less developed. In contrast to classical kinetic theory, *quantum collision operators* are actually non-local in time (i.e. they include a time integral over the "past", cf. the Levinson equation [Lev70] as one possible model). However, since most of the existing numerical simulations involve only local in time approximations, we shall confine our discussion to such collision operators Q. The two most used models are the already mentioned relaxation time approximation

$$Qw := \frac{w_{st}(x, v) - w(x, v, t)}{\tau(v)}$$

and the *quantum Fokker–Planck model* with

$$Qw := \underbrace{D_{pp}\Delta_v w}_{\text{class. diffusion}} + \underbrace{2\gamma \operatorname{div}_v(vw)}_{\text{friction}} + \underbrace{D_{qq}\Delta_x w + 2D_{pq} \operatorname{div}_x(\nabla_v w)}_{\text{quantum diffusion}} \qquad (7)$$

(cf. [CL83, CEFM00] for a derivation). Both of these models are purely phenomenological, but quantum mechanically "correct" (if $\tau(v) = \tau_0 \geq 0$ or if the *Lindblad condition* (35) holds). And this is important for their mathematical analysis (cf. Sect. 6.2).

As a third option, the r.h.s. of the Wigner equation (4) is often replaced by a semiclassical Boltzmann scattering operator

$$Qw := \int_{\mathbb{R}^N} [S(v, v')w(x, v') - S(v', v)w(x, v)] \, dv' \,,$$

with the scattering rate $S(v, v')$ (for the electron–phonon interaction, e.g.). Such semiclassical Boltzmann operators give good simulation results [KN06], but they are quantum mechanically *not* "correct" (cf. Sect. 6.1). Hence, we shall not discuss their mathematical analysis.

The RTD-structure is also the key building block of another nano-device, the *resonant tunneling field-effect transistor* (see Fig. 8). This device with three contacts is currently in experimental stage but might become a major building block of logic circuits. It is promising to yield a simple integration of a tunneling diode with the conventional FET structure. Thus it advantageously combines the features of a regular transistor (gain, amplification) with a RTD (negative differential conductance at room temperature). Through an applied gate voltage one can adjust the barrier height, which allows the current peak, the peak-to-valley ratio, and the peak positions to be tuned.

The current from source to drain mainly flows along the channel, which has the same material structure as a RTD (see Fig. 9). Hence, this is also where the quantum effects mainly take place.

These lecture notes are motivated by applicable quantum transport models, which are typically nonlinear partial differential equations (PDEs) on a *bounded domain*, hence, initial boundary value problems (IBVPs). Nevertheless we shall focus the mathematical analysis of quantum evolution problems in the following sections mostly on whole space cases. This is motivated by the fact that much less mathematical analysis has been carried out for those IBVPs. Moreover, these quantum mechanical IBVPs often tend to be much

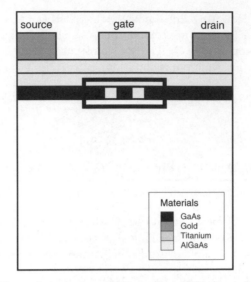

Fig. 8 Schematic 2D cut through a resonant tunneling–FET; dimensions: $0.2\,\mu\text{m}\times0.2\,\mu\text{m}$. The electron current and quantum effects mainly take place along the channel (*inside the box*)

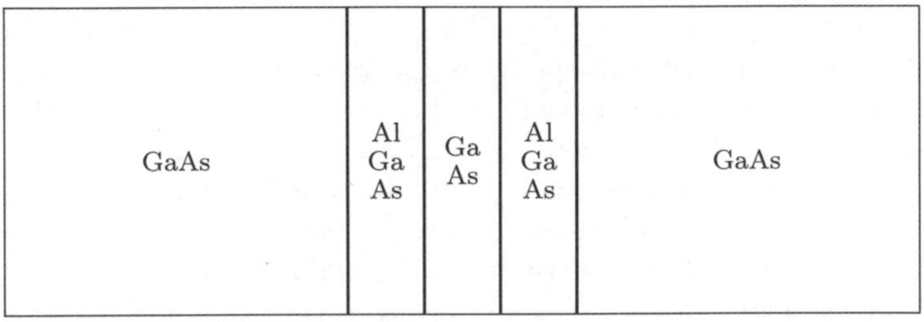

GaAs	Al Ga As	Ga As	Al Ga As	GaAs

Fig. 9 Enlargement of the central channel region from Fig. 8: There the channel of a is sandwiched structure of the two semiconductor materials *GaAs* and *AlGaAs*

"messier", both from a modeling point of view and mathematically. Consider, as an example, the presented relaxation-time Wigner–Poisson system (an IBVP) for a resonant tunneling diode. With those inflow boundary conditions, it seems impossible to guarantee that the Wigner function $w(t)$ will stay "positive", in the sense of corresponding to a positive density matrix. However, for the corresponding whole space model, this problem does not arise.

2 Linear Schrödinger Equation

In this section we collect some well-known analytical results for the Cauchy problem of the linear Schrödinger equation on $\mathbb{R}^N$. Most of this material is taken from Sect. 3 in [Caz96], and is will serve as our basic background for discussing the nonlinear Schrödinger–Poisson equation in Sect. 3. In these two section we assume that the equations are scaled such that the parameters satisfy $e = m_* = \hbar = 1$.

2.1 Free Schrödinger Group

We consider the *free Schrödinger equation* on $\mathbb{R}^N$ for the complex valued wave function $\psi = \psi(x, t)$:

$$i\psi_t = -\frac{1}{2}\Delta\psi, \quad t \in \mathbb{R}, \tag{8}$$

$$\psi(0) = \varphi.$$

The operator $A := \frac{1}{2}\Delta$ with the domain $\mathcal{D}(A) = H^2(R^N)$ is self-adjoint on $L^2(\mathbb{R}^N)$. By Stone's Theorem (cf. [Paz83]) iA hence generates a C_0-group of isometries on $L^2(\mathbb{R}^N)$:

$$T(t), \ t \in \mathbb{R}; \quad T(t)^* = T(-t)$$

$$T(t_1)T(t_2) = T(t_1+t_2), \quad T(0) = I$$

$$\lim_{t \to 0} T(t)\varphi = \varphi \quad \forall\varphi \in L^2(\mathbb{R}^N)$$

$$\lim_{t \to 0} \frac{T(t)\varphi - \varphi}{t} = iA\varphi \quad \forall\varphi \in \mathcal{D}(A)$$

The operator iA is call the *infinitesimal generator* of $T(t)$. This evolution group provides a solution to (8) in the following sense:

Proposition 2.1.

(a) Let $\varphi \in L^2(\mathbb{R}^N)$. Then $\psi(t) = T(t)\varphi$ is the unique solution of

$$\begin{cases} i\psi_t = -\frac{1}{2}\Delta\psi & \text{in } H^{-2}(\mathbb{R}^N), \quad \forall t \in \mathbb{R} \\ \psi \in C(\mathbb{R}; L^2(\mathbb{R}^N)) \cap C^1(\mathbb{R}; H^{-2}(\mathbb{R}^N)) \\ \psi(0) = \varphi \end{cases}$$

This mild solution satisfies mass conservation, i.e. $\|\psi(t)\|_{L^2} = \|\varphi\|_{L^2}$ $\forall t \in \mathbb{R}$ (since $T(t)$ is isometric).
(b) If $\varphi \in H^2(\mathbb{R}^N)$, the above solution is a classical solution with $\psi \in C(\mathbb{R}; H^2) \cap C^1(\mathbb{R}; L^2)$.

Lemma 2.1 (Representation of $T(t)$).

$$T(t)\varphi = K(t) * \varphi \quad \forall t \neq 0, \ \varphi \in \mathcal{S}(\mathbb{R}^N) \tag{9}$$

$$K(x,t) = (8\pi \mathrm{i}t)^{-\frac{N}{2}} e^{\frac{\mathrm{i}|x|^2}{8t}}$$

Proof. Define $\psi \in C(\mathbb{R}; \mathcal{S}(\mathbb{R}^N))$ by:

$$\hat{\psi}(\xi,t) := \underbrace{e^{-\frac{\mathrm{i}}{2}|\xi|^2 t}}_{=\hat{K}(\xi,t)} \hat{\varphi}(\xi), \quad \xi \in \mathbb{R}^N \tag{10}$$

$$\Rightarrow \quad \mathrm{i}\hat{\psi}_t = \frac{1}{2}|\xi|^2 \hat{\psi} \quad \text{on } \mathbb{R}_t \times \mathbb{R}^N$$

$$\hat{\psi}(0) = \hat{\varphi}(\xi)$$

$\square$

Note the (formal) similarity between $K(x,t)$, the Green's function of the Schrödinger equation and the heat kernel.

For more regular initial data, the regularity is propagated in time:

Remark 2.1.

(a) Let $\varphi \in H^s(\mathbb{R}^N)$, $s \in \mathbb{R}$. Then $\psi(t) = T(t)\varphi$ satisfies:

$$\psi \in \bigcap_{0 \leq j < \infty} C^j(\mathbb{R}; H^{s-2j}(\mathbb{R}^N)), \qquad \|\psi(t)\|_{H^s} = \|\varphi\|_{H^s}.$$

This follows from (10) with $\|\varphi\|_{H^s}^2 = \left\| (1+|\xi|^2)^{\frac{s}{2}} \hat{\varphi}(\xi) \right\|_{L^2}^2.$

(b)

$$T(t)\varphi = (8\pi \mathrm{i}t)^{-\frac{N}{2}} e^{\frac{\mathrm{i}|x|^2}{8t}} \int_{\mathbb{R}^N} e^{-\frac{\mathrm{i}x \cdot y}{4t}} e^{\frac{\mathrm{i}|y|^2}{8t}} \varphi(y) \, \mathrm{d}y, \quad t \neq 0. \tag{11}$$

I.e. $T(t)$ is a Fourier transform up to a rescaling and a multiplication by a function of modulus 1.

2.2 Smoothing Effects and Gain of Integrability in $\mathbb{R}^N$

We shall now discuss simple smoothing properties of the free Schrödinger group $T(t)$. On the one hand we can gain local integrability for $t \neq 0$. On the other hand, (11) shows that $T(t)$, $t \neq 0$ is almost a Fourier transform. And a Fourier transform maps nicely decaying functions into smooth functions. However, the regularity gain for $t \neq 0$ never appears directly on ψ, but it is always coupled to some spatial moments of ψ. This is caused by the multiplier $e^{\frac{\mathrm{i}|y|^2}{8t}}$ in (11):

Proposition 2.2. *Let the multi-index* $\alpha \in \mathbb{N}_0^N$, $\varphi \in \mathcal{S}'(\mathbb{R}^N)$ *with* $x^\alpha \varphi \in L^2(\mathbb{R}^N)$, *and let* $\psi(t) = T(t)\varphi \in C(\mathbb{R}; \mathcal{S}'(\mathbb{R}^N))$. *Then*

$$\partial_x^\alpha \left(e^{-\frac{i|x|^2}{8t}} \psi(t) \right) \in C(\mathbb{R}\backslash\{0\}; L^2(\mathbb{R}^N)),$$

$$(4|t|)^{|\alpha|} \left\| \partial_x^\alpha \left(e^{-\frac{i|x|^2}{8t}} \psi(t) \right) \right\|_{L^2} = \|x^\alpha \varphi\|_{L^2}, \quad t \in \mathbb{R}.$$

This follows directly from (11).

Example 2.1. Choose $|\alpha| = 1$:

$$\|(x + 4i\,t\nabla)\psi(t)\|_{L^2} = \text{const} = \|x\varphi\|_{L^2}, \quad t \in \mathbb{R}.$$

Now we consider the gain of local-in-x integrability:

Proposition 2.3. *Let* $2 \leq p \leq \infty$, $t \neq 0$. *Then* $T(t) \in \mathcal{B}(L^{p'}(\mathbb{R}^N), L^p(\mathbb{R}^N))$:

$$\|T(t)\varphi\|_{L^p} \leq (8\pi|t|)^{-N\left(\frac{1}{2} - \frac{1}{p}\right)} \|\varphi\|_{L^{p'}} \quad \forall \varphi \in L^{p'}(\mathbb{R}^N). \tag{12}$$

Here and in the sequel $p' = \frac{p}{p-1}$ is the Hölder conjugate of p.

Proof. Let $\varphi \in \mathcal{S}(\mathbb{R}^N)$:

$\|T(t)\varphi\|_{L^\infty} \leq (8\pi|t|)^{-\frac{N}{2}} \|\varphi\|_{L^1}$ follows from (9) by the Young inequality for convolutions,

$\|T(t)\varphi\|_{L^2} = \|\varphi\|_{L^2}$ since $T(t)$ is isometric.

The result then follows by interpolation (Riesz–Thorin Theorem, [RS75]) and the density of $\mathcal{S}(\mathbb{R}^N)$ in $L^{p'}(\mathbb{R}^N)$. $\square$

2.3 Potentials, Inhomogeneous Equation

Here, we first discuss homogeneous Schrödinger equations with bounded and relatively bounded potentials:

Proposition 2.4 (Bounded perturbations of generators, [Paz83]).
Let A be the infinitesimal generator of the C_0-semigroup $T(t)$ on the Banach space X with $\|T(t)\| \leq Me^{\omega t}$, and $B \in \mathcal{B}(X)$.
  Then $A + B$ generates a C_0-semigroup $S(t)$ on X with

$$\|S(t)\| \leq Me^{(\omega + M\|B\|)t}, \quad t \in \mathbb{R}.$$

Example 2.2. Schrödinger equation with bounded potential $V \in L^\infty(\mathbb{R}^N)$:

$$\begin{cases} i\psi_t = -\frac{1}{2}\Delta\psi + V\psi, & t \in \mathbb{R} \\ \psi(0) = \varphi \in L^2(\mathbb{R}^N) \end{cases}$$

has a unique mild solution. It is even a classical solution for $\varphi \in H^2(\mathbb{R}^N)$. The *Hamiltonian* of this equation is $H := -\frac{1}{2}\Delta + V$. It reveals conservation of the following energy:

$$\langle\psi(t), H\psi(t)\rangle = \frac{1}{2}\|\nabla\psi(t)\|_{L^2}^2 + \int_{\mathbb{R}^N} V n(t)\,dx = \text{const in } t. \qquad (13)$$

Next we perturb the free Hamiltonian by a special class of unbounded potentials:

Proposition 2.5 (Relatively bounded perturbations of generators, Kato–Rellich Th. [RS75]). *Let A be a self-adjoint operator on the Hilbert space X, and the operator B symmetric and A-bounded (i.e.*

$$\exists a, b \in \mathbb{R} : \|B\varphi\| < a\|A\varphi\| + b\|\varphi\| \quad \forall\varphi \in \mathcal{D}(A))$$

with $a < 1$. Then $A + B$ is self-adjoint on $\mathcal{D}(A)$.

Example 2.3. Hydrogen atom – motion of one electron in the attractive Coulomb potential of the fixed nucleus:

$$\begin{cases} i\psi_t = -\frac{1}{2}\Delta\psi - \frac{1}{|x|}\psi, & t \in \mathbb{R} \\ \psi(0) = \varphi \in L^2(\mathbb{R}^3) & (\text{or } \varphi \in H^2(\mathbb{R}^3)) \end{cases}$$

has a unique mild (or, resp., classical) solution. To prove this, we split the potential $V(x) = \frac{1}{|x|}$ into a short and long range potential:

$$V = V_1 + V_2 \quad \text{with}$$

$$V_1 \in L^2(\mathbb{R}^3), \; V_2 := \min\left(1, \frac{1}{|x|}\right) \in L^\infty(\mathbb{R}^3).$$

V_1 is Δ-bounded because of $\psi \in H^2(\mathbb{R}^3) \hookrightarrow L^\infty(\mathbb{R}^3)$ by a Sobolev embedding. Hence, Proposition 2.5 applies to V_1 and Proposition 2.4 applies to V_2.

Now we turn to inhomogeneous Schrödinger equations:

Proposition 2.6. *Let $\varphi \in L^2(\mathbb{R}^N)$, $f \in C([0,T]; L^2(\mathbb{R}^N))$.*

(a) Then $\exists!$ solution of

$$\begin{cases} i\psi_t + \frac{1}{2}\Delta\psi + f = 0 & \forall t \in [0,T] \\ \psi \in C([0,T]; L^2(\mathbb{R}^N)) \cap C^1([0,T]; H^{-2}(\mathbb{R}^N)) \\ \psi(0) = \varphi. \end{cases}$$

With $T(t)$ denoting the free Schrödinger group, this mild solution *satisfies*

$$\psi(t) = T(t)\varphi + i \int_0^t T(t-s)f(s)\,ds, \quad 0 \le t \le T. \tag{14}$$

(b) Let, additionally, $\varphi \in H^2(\mathbb{R}^N)$ and either $f \in W^{1,1}((0,T); L^2(\mathbb{R}^N))$ or $f \in L^1((0,T); H^2(\mathbb{R}^N))$. Then ψ is a (classical solution), satisfying $\psi \in C(\mathbb{R}; H^2) \cap C^1(\mathbb{R}; L^2)$.

2.4 Strichartz Estimates

The goal of this subsection is to derive combined space–time estimates for the inhomogeneous Schrödinger equation (14).

Definition 2.1. A pair of indices (q,p) is called *admissible* if

$$2 \le p < \frac{2N}{N-2} \quad \text{(or } 2 \le p \le \infty \text{ if } N = 1; \ 2 \le p < \infty \text{ if } N = 2),$$
$$\frac{2}{q} = N\left(\frac{1}{2} - \frac{1}{p}\right).$$

Notation:

$$L^{q,p} := L^q(\mathbb{R}_t; L^p(\mathbb{R}^N))$$
$$L_I^{q,p} := L^q(I; L^p(\mathbb{R}^N)) \quad \text{for any interval } I \subset \mathbb{R}$$

The following *Strichartz estimate* for the free and inhomogeneous Schrödinger equation describes a gain of local-in-x integrability. Since the following inequalities hold in a mixed space-time norm, this gain of integrability does not hold pointwise in time, but for almost all t:

Proposition 2.7. *Let (q,p), (a,b) be admissible pairs.*

(a) Let $\varphi \in L^2(\mathbb{R}^N)$. Then $T(t)\varphi \in L^{q,p} \cap C(\mathbb{R}; L^2(\mathbb{R}^N))$ with

$$\|T(\cdot)\varphi\|_{L^{q,p}} \le C(q)\|\varphi\|_{L^2}. \tag{15}$$

(b) Let $f \in L_I^{a',b'}$ and $t_0 \in \bar{I}$. Then it holds

$$\Lambda_f(t) := \int_{t_0}^t T(t-s)f(s)\,ds \in L_I^{q,p} \cap C(\bar{I}; L^2(\mathbb{R}^N))$$

with

$$\|\Lambda_f\|_{L_I^{q,p}} \le C(a,q)\|f\|_{L_I^{a',b'}}.$$

The constants $C(q)$ and $C(a,q)$ are independent of time.

Proof (of the inhomogeneous version for $(a,b) = (q,p)$. The general case depends on duality arguments, see Sect. 3.2 of [Caz96]).

Let $I = [0,T]$, $t_0 = 0$, $f \in C_0([0,T]; L^{p'})$; the result for general $f \in L^{q,p}$ then follows by density. Inequality (12) and $N(\frac{1}{2} - \frac{1}{p}) = \frac{2}{q}$ yield:

$$\|\Lambda_f(t)\|_{L^p} \le C(q) \int_0^t |t - s|^{-N(\frac{1}{2} - \frac{1}{p})} \|f(s)\|_{L^{p'}} \, ds$$

$$\le C(p) \int_0^T |t - s|^{-\frac{2}{q}} \|f(s)\|_{L^{p'}} \, ds.$$

With the weak Young inequality (cf. [RS75]) we conclude:

$$\|\Lambda_f\|_{L_I^{q,p}} \le C(q)\|f\|_{L_I^{q',p'}}.$$

$\square$

Remark 2.2.

(a) In Proposition 2.7a $\varphi \in L^2(\mathbb{R}^N)$ implies $T(t)\varphi \in L^p$ for almost all $t \in \mathbb{R}$ (for $p > 2$). It cannot be improved to "for all $t \ne 0$".

(b) Since the Schrödinger equation is time reversible, the presented smoothing effects are much more subtle than for the heat equation. The evolution also improves the local integrability of the solution ψ for almost all t. The smoothing effects in Propositions 2.3 and 2.7 are due to the dispersion in the Schrödinger equation. This means that waves of different frequencies (or wavelengths) travel at different velocities, when decomposing the solution ψ into plane waves.

(c) A remarkable aspect of Proposition 2.7b is that the index pairs (q,p) and (a',b') are uncorrelated.

3 Schrödinger–Poisson Analysis in $\mathbb{R}^3$

The goal of this section is to prove that the repulsive Schrödinger–Poisson (SP) equation (or *Hartree equation*) in $\mathbb{R}^3$ has a unique, global-in-time solution, first for initial data in H^1 and then in L^2. We shall mostly follow Sect. 6.3 of [Caz96] and [Cas97]; but see also [GV94, HO89]. We remark that extensions of this analysis to the Sobolev spaces H^k, $k \ge 2$ is straightforward [Caz96]. Extensions to space dimensions $N \ne 3$ require some modifications, since the used Sobolev embeddings depend on N [Caz96, AN91].

3.1 H^1-Analysis

A wave function $\psi \in H^1$ corresponds to a system with *finite mass* $\|\psi\|_{L^2}^2$ and *finite kinetic energy* $\frac{1}{2}\|\nabla\psi\|_{L^2}^2$. As we shall see, this property is propagated in time.

In the sequel we shall frequently need the following result on solutions to nonlinear Banach space-ODEs (i.e. an ordinary differential equation for a Banach space-valued function):

Proposition 3.1 (Local Lipschitz perturbations of generators [Paz83]). *Let A be the infinitesimal generator of the C_0-semigroup $T(t)$, $t \geq 0$ on the Banach space X, and let $f = f(t, u) : [0, \infty) \times X \to X$ be continuous in t and locally Lipschitz in u (uniformly in t on bounded intervals).*

(a) Then, $\forall \varphi \in X$, $\exists t_{max} = t_{max}(\varphi) \leq \infty$:

$$\begin{cases} \frac{du}{dt} = Au + f(t, u(t)), & t \geq 0 \\ u(0) = \varphi \end{cases}$$

has a unique mild solution $u \in C([0, t_{max}); X)$.
(b) If $t_{max} < \infty$ then $\lim_{t \nearrow t_{max}} \|u(t)\|_X = \infty$, i.e. blow-up in finite time.
(b') If $\|u(t)\|_X < \infty$ $\forall t \in [0, \infty)$ $\Rightarrow$ The solution exists global-in-time.

This theorem will now be applied to the repulsive *Schrödinger–Poisson* equation (or *Hartree equation*):

$$\begin{cases} i\psi_t = -\frac{1}{2}\Delta\psi + V\psi, & x \in \mathbb{R}^3, \ t \in \mathbb{R} \\ -\Delta_x V(x, t) = n(x, t) := |\psi(x, t)|^2 \\ \psi(0) = \varphi \end{cases} \tag{16}$$

We take the *Newton potential solution* of the Poisson equation:

$$\begin{aligned} V &= \frac{1}{4\pi|x|} * |\psi|^2 \\ \nabla V &= -\frac{x}{4\pi|x|^3} * |\psi|^2 \end{aligned} \tag{17}$$

Theorem 3.1. *Let $\varphi \in H^1(\mathbb{R}^3)$. Then (16) has a unique solution $\psi \in C(\mathbb{R}; H^1(\mathbb{R}^3))$.*

Proof.

1. $T(t) = e^{\frac{i}{2}\Delta t}$ is a C_0-group of isometries both on $L^2(\mathbb{R}^3)$ and $H^1(\mathbb{R}^3)$.

2. $f(\psi) := -iV[\psi]\psi = -i\left(\frac{1}{4\pi|x|} * |\psi|^2\right)\psi$ is locally Lipschitz in H^1 (but *not* in L^2; hence we analyze (16) in H^1) since:
   The weak Young inequality (cf. [RS75]) for (17) yields:

$$\begin{aligned} \|V\|_{L^p} &\leq C\|\psi\|_{L^q}^2, & 3 < p \leq \infty, & \quad \frac{1}{p} = \frac{2}{q} - \frac{2}{3}, \\ \|\nabla V\|_{L^p} &\leq C\|\psi\|_{L^q}^2, & \frac{3}{2} < p < \infty, & \quad \frac{1}{p} = \frac{2}{q} - \frac{1}{3}, \end{aligned} \tag{18}$$

and Hölder's inequality and the Sobolev embedding $H^1(\mathbb{R}^3) \hookrightarrow L^6(\mathbb{R}^3)$ yield:

$$\|f(\psi)\|_{L^2} \le \|V\|_{L^\infty} \|\psi\|_{L^2} \le C\|\psi\|_{H^1}^3 \,,$$
$$\|\nabla f(\psi)\|_{L^2} \le \|V\|_{L^\infty} \|\nabla\psi\|_{L^2} + \|\nabla V\|_{L^3} \|\psi\|_{L^6} \le C\|\psi\|_{H^1}^3 \,.$$

3. By Proposition 3.1 it holds: The Schrödinger–Poisson equation (16) has a *unique local solution* $\psi \in C([0, t_{max}); H^1(\mathbb{R}^3))$.

4. $\|\psi\|_{H^1}$ cannot blow up in finite time because of the following two estimates:

(a) L^2 – *a priori estimate (mass conservation):*
The Schrödinger equation (16) holds in $C([0, t_{max}); H^{-1})$. We test it against $\psi(t) \in H^1$:

$$\mathrm{i}\langle \psi_t, \psi \rangle = \frac{1}{2}\|\nabla\psi\|_{L^2}^2 + \int_{\mathbb{R}^3} V|\psi|^2 \,\mathrm{d}x \,.$$

Taking the imaginary part yields: $\frac{\mathrm{d}}{\mathrm{d}t}\|\psi\|_{L^2}^2 = 0$.

(b) $\dot{H}^1$ – *a priori estimate (energy conservation):*
We test the Schrödinger equation (16) against ψ_t and integrate by parts. A formal calculation yields:

$$\mathrm{i}\|\psi_t\|_{L^2}^2 = \frac{1}{2}\int_{\mathbb{R}^3} \nabla\psi \cdot \nabla\bar{\psi}_t \,\mathrm{d}x + \int_{\mathbb{R}^3} V\psi\bar{\psi}_t \,\mathrm{d}x \,.$$

Taking the real part and using the Poisson equation yield:

$$0 = \tfrac{1}{2}\tfrac{\mathrm{d}}{\mathrm{d}t}\|\nabla\psi\|_{L^2}^2 + \int_{\mathbb{R}^3} Vn_t \,\mathrm{d}x = \tfrac{1}{2}\tfrac{\mathrm{d}}{\mathrm{d}t}\|\nabla\psi\|_{L^2}^2 + \int_{\mathbb{R}^3} \nabla V \cdot \nabla V_t \,\mathrm{d}x$$

$$= \frac{\mathrm{d}}{\mathrm{d}t}\Big[\ \underbrace{\frac{1}{2}\|\nabla\psi(t)\|_{L^2}^2}_{\text{kinetic energy}} + \underbrace{\frac{1}{2}\|\nabla V(t)\|_{L^2}^2}_{\text{self-consist. potential energy}}\ \Big]$$

Hence, $\|\nabla\psi(t)\|_{L^2}$ is uniformly bounded in t.

Remark: Here, the (self-consistent) potential energy is $\frac{1}{2}\|\nabla V\|_{L^2}^2 = \frac{1}{2}\int Vn \,\mathrm{d}x$, while it is $\int Vn \,\mathrm{d}x$ in the linear case (cf. (13)).

5. $\Rightarrow$ The solution exists $\forall t \in \mathbb{R}$. $\qquad\qquad\qquad\qquad\qquad\qquad\qquad\quad \square$

3.2 L^2-Analysis

If a wave function $\psi \in L^2$ but *not* in H^1, the corresponding quantum system has *finite mass* but *infinite kinetic energy*.

Since the nonlinearity $f(\psi) := -\mathrm{i}V[\psi]\psi = -\mathrm{i}\left(\frac{1}{4\pi|x|} * |\psi|^2\right)\psi$ is *not* locally Lipschitz in L^2, our analysis is much more difficult than the H^1-analysis above. Here we shall use that $f(\psi)$ is still (somehow) locally Lipschitz in the following space of t-dependent functions: $L_T^{q,p} := L^q((-T,T); L^p(\mathbb{R}^3))$, for some fixed $T > 0$.

We split the self-consistent potential $V[\psi]$ into a short and a long range potential: $V[\psi] = V_1[\psi] + V_2[\psi]$, with $V_2[\psi] := \frac{1}{4\pi}\min(1, \frac{1}{|x|}) * |\psi|^2$. And we split the nonlinearity $f(\psi)$ analogously: $f(\psi) = f_1(\psi) + f_2(\psi)$, with $f_j(\psi) := -\mathrm{i}V_j[\psi]\psi$.

The following result shows that $f_{1,2}(\psi)$ are locally Lipschitz, however, in different spaces $L_T^{q,p}$.

Lemma 3.1 ([Cas97]). *Let* $0 < T < 1$; $3 < p < 6$; $q = q(p)$ *with* $\frac{2}{q} = 3(\frac{1}{2} - \frac{1}{p})$; $\psi, \phi \in C([-T,T]; L^2(\mathbb{R}^3)) \cap L_T^{q,p}$; $M := \max_{[-T,T]}(\|\psi(t)\|_{L^2}, \|\phi(t)\|_{L^2})$. *Then*

(a)

$$\|f_1(\psi(t)) - f_1(\phi(t))\|_{L_T^{q',p'}} \le C(p)M^2 T^{1-\frac{2}{q}}\|\psi(t) - \phi(t)\|_{L_T^{q,p}},$$

(b)

$$\|f_2(\psi(t)) - f_2(\phi(t))\|_{L_T^{1,2}} \le CM^2 T \|\psi(t) - \phi(t)\|_{L_T^{\infty,2}}$$

$$\le CM^2 T^{1-\frac{2}{q}}\|\psi(t) - \phi(t)\|_{L_T^{\infty,2}}.$$

Theorem 3.2. *Let* $\varphi \in L^2(\mathbb{R}^3)$. *Then the Schrödinger–Poisson equation* (16) *has a unique mild solution* $\psi \in C(\mathbb{R}; L^2(\mathbb{R}^3)) \cap L_{loc}^{q,p}$ *with* $3 < p < 6$, $\frac{2}{q} = 3(\frac{1}{2} - \frac{1}{p})$.

Proof.

1. *Approximating H^1-sequence to construct a solution:*
   Let $\{\varphi_m\}_{m\in\mathbb{N}} \subset H^1(\mathbb{R}^3)$ with $\varphi_m \xrightarrow{m\to\infty} \varphi$ in L^2, $\|\varphi_m\|_{L^2} = \|\varphi\|_{L^2}$.
   By Theorem 3.1, for each φ_m, $m \in \mathbb{N}$, the SP-problem then has a unique solution $\psi_m \in C(\mathbb{R}; H^1(R^3))$, satisfying $\|\psi_m(t)\|_{L^2} = \|\varphi\|_{L^2} =: M$ $\forall t \in \mathbb{R}$.

2. $\{\psi_m(t)\}$ *is a Cauchy sequence in* $L_T^{a,b}$ *for T small:*

$$\begin{aligned}
\psi_m(t) - \psi_k(t) = &\, T(t)(\varphi_m - \varphi_k) \\
&+ \int_0^t T(t-s)[f_1(\psi_m(s)) - f_1(\psi_k(s))]\,\mathrm{d}s \quad (19)\\
&+ \int_0^t T(t-s)[f_2(\psi_m(s)) - f_2(\psi_k(s))]\,\mathrm{d}s
\end{aligned}$$

The homogeneous Strichartz inequality (Proposition 2.7a) yields:

$$\|T(t)(\varphi_m - \varphi_k)\|_{L_T^{a,b}} \le C(a)\|\varphi_m - \varphi_k\|_{L^2}.$$

The inhomogeneous Strichartz inequality (Proposition 2.7b) and Lemma 3.1a yield for the first nonlinearity:

$$\left\| \int_0^t T(t-s)[f_1(\psi_m(s)) - f_1(\psi_k(s))]\,ds \right\|_{L_T^{a,b}}$$
$$\leq C(a,q)\|f_1(\psi_m(t)) - f_1(\psi_k(t))\|_{L_T^{q',p'}}$$
$$\leq C(a,q)M^2 T^{1-\frac{2}{q}}\|\psi_m(t) - \psi_k(t)\|_{L_T^{q,p}}.$$

Here, (a,b) is any admissible pair.
Similarly, the inhomogeneous Strichartz inequality (Proposition 2.7b) and Lemma 3.1b yield for the second nonlinearity:

$$\left\| \int_0^t T(t-s)[f_2(\psi_m(s)) - f_2(\psi_k(s))]\,ds \right\|_{L_T^{a,b}}$$
$$\leq C(a)M^2 T^{1-\frac{2}{q}}\|\psi_m(t) - \psi_k(t)\|_{L_T^{\infty,2}}$$

We collect the last three inequalities and add the resulting estimates for the two index-choices $(a,b) = (q,p)$, $(a,b) = (\infty,2)$. This yields the following estimate for (19):

$$\|\psi_m(t) - \psi_k(t)\|_{L_T^{q,p}} + \|\psi_m(t) - \psi_k(t)\|_{L_T^{\infty,2}}$$
$$\leq C(q)\|\varphi_m - \varphi_k\|_{L^2}$$
$$+ C(q)M^2 T^{1-\frac{2}{q}}\left[\|\psi_m(t) - \psi_k(t)\|_{L_T^{q,p}} + \|\psi_m(t) - \psi_k(t)\|_{L_T^{\infty,2}}\right]$$

$$\Rightarrow \exists T_0 = T_0(q,M) > 0, \text{ small enough such that:}$$

$$\|\psi_m(t) - \psi_k(t)\|_{L_{T_0}^{q,p}} + \|\psi_m(t) - \psi_k(t)\|_{L_{T_0}^{\infty,2}} \leq C(q,M)\|\varphi_m - \varphi_k\|_{L^2}. \quad (20)$$

This implies the following properties of the approximating sequence $\{\psi_m\}$:

- $\{\psi_m\}$ is a Cauchy sequence in $L_{T_0}^{q,p} \cap L_{T_0}^{\infty,2}$.
- $\{\psi_m\} \subset C([-T_0,T_0]; L^2(\mathbb{R}^3))$.
- $\psi_m \to \psi$ in $L_{T_0}^{q,p} \cap C([-T_0,T_0]; L^2(\mathbb{R}^3))$.
- $\|\psi_m(t)\|_{L^2} = \|\psi(t)\|_{L^2} = \|\varphi\|_{L^2} = M, \quad \forall m \in \mathbb{N}, \forall t \in \mathbb{R}$.

Since $T_0 = T_0(q,M)$ only depends on the index q and $M = \|\varphi\|_{L^2}$, the solution ψ can be extended up to $2T_0$, $3T_0$, ..., $-T_0$, $-2T_0$, ... Hence, $\psi \in C(\mathbb{R}; L^2(\mathbb{R}^3)) \cap L_{loc}^{q,p}$.
The estimate (20) also implies uniqueness of the limit ψ and its continuous dependence on the data φ. The constructed limit ψ is the mild solution of (16). To verify this, choose $\psi_k := 0$ and pass to the limit $(m \to \infty)$ in the integral equation (19). $\qquad\Box$

3.3 Schrödinger–Poisson Systems

Up to now we considered just one Schrödinger equation that is coupled to the Poisson equation. This would describe a *pure quantum state*. In most realistic application, however, one has to deal with a *mixed quantum state*, which can be described by a sequence of wave functions:

$$\psi_j(x,t) \in \mathbb{C},\ j \in \mathbb{N};\ x \in \mathbb{R}^3,\ t \in \mathbb{R}.$$

For a system of many particles, this mixed quantum state describes a statistical mixture, and each ψ_j has an *occupation probability* $\lambda_j \geq 0$, $j \in \mathbb{N}$; $\sum_j \lambda_j = 1$. Here, λ_j are given data and it depends on the initial state of the system.

In this section we only consider *closed quantum systems*, i.e. a system without interaction to an (infinitely large) "environment" or "heat bath". Its dynamics is time-reversible and fully described by a Hamiltonian. In this case the above occupation probabilities λ_j are constant in time.

Open quantum systems, being the opposite of closed quantum systems will be discussed in Sect. 6.

We consider now the time evolution of this mixed quantum state, given by the *repulsive Schrödinger–Poisson system* (SPS):

$$\begin{cases} i\frac{\partial}{\partial t}\psi_j = -\frac{1}{2}\Delta\psi_j + V\psi_j, & x \in \mathbb{R}^3,\ t \in \mathbb{R},\ j \in \mathbb{N} \\ V(x,t) = \frac{1}{4\pi|x|} * n(x,t), & n(x,t) := \sum_{j=1}^{\infty} \lambda_j|\psi_j(x,t)|^2 \qquad (21) \\ \psi_j(0) = \varphi_j, & j \in \mathbb{N} \end{cases}$$

In the special case $\lambda_j := \delta_j^1$ (δ_j^i is the Kronecker-Delta) the SPS reduces to the scalar Hartree equation of Sect. 3.1. In the subsequent analysis we follow mostly [Cas97].

Notation:
For any fixed sequence $\lambda := \{\lambda_j\}_{j\in\mathbb{N}} \in \ell^1$ with $\lambda_j \geq 0$ we define:

$$H^1(\lambda) := \{\Phi(x) = (\varphi_j(x))_{j\in\mathbb{N}},\ \|\Phi\|_{H^1(\lambda)}^2 = \sum_j \lambda_j\|\varphi_j(x)\|_{H^1(\mathbb{R}^3)}^2 < \infty\}$$

$$L^p(\lambda) := \{\Phi(x) = (\varphi_j(x))_{j\in\mathbb{N}},\ \|\Phi\|_{L^p(\lambda)}^2 = \sum_j \lambda_j\|\varphi_j(x)\|_{L^p(\mathbb{R}^3)}^2 < \infty\}$$

$$L_{loc}^{q,p}(\lambda) := L_{loc}^q(\mathbb{R}; L^p(\lambda))$$

Theorem 3.3.

(a) Let $\Phi \in H^1(\lambda)$. Then (21) has a unique solution $\Psi \in C(\mathbb{R}; H^1(\lambda))$.

(b) [Cas97]: Let $\Phi \in L^2(\lambda)$. Then (21) has a unique mild solution $\Psi \in C(\mathbb{R}; L^2(\lambda) \cap L_{loc}^{q,p}(\lambda))$ for all admissible pairs (q,p) with $3 < p < 6$, $\frac{2}{q} = 3(\frac{1}{2} - \frac{1}{p})$.

Proof.

(a) This is a straightforward generalization of Theorem 3.1 for the Hartree
    equation (cf. [ILZ94], e.g.):

    $f(\Psi) := -iV[\Psi]\Psi$ is locally Lipschitz in $H^1(\lambda)$. The required a-priori
    estimates are provided by $\|\Psi(t)\|_{L^2(\lambda)}^2 = \|\Phi\|_{L^2(\lambda)}^2$ (mass conservation)
    and $\frac{1}{2}\|\nabla\Psi(t)\|_{L^2(\lambda)}^2 + \frac{1}{2}\|\nabla V(t)\|_{L^2}^2 = \text{const.}$ (energy conservation).

(b) This part is based on *vector valued Strichartz inequalities* for mixed quan-
    tum states which are non-trivial extensions of Proposition 2.7. E.g., the
    extension of the homogeneous estimate (15) reads ([Cas97]):

$$\|T(t)\Phi\|_{L_T^{q,p}(\lambda)} \leq C(q,T)\|\Phi\|_{L^2(\lambda)} \quad \forall \text{ admissible pairs } (q,p)$$

with

$$\|T(t)\Phi\|_{L_T^{q,p}(\lambda)}^q = \int_{-T}^{T} \left(\sum_j \lambda_j \|T(t)\varphi_j\|_{L^p}^2 \right)^{\frac{q}{2}} \mathrm{d}t.$$

In contrast, a trivial extension of Proposition 2.7 would be

$$\sum_j \lambda_j \|T(t)\varphi_j\|_{L^{q,p}}^2 \leq C(q) \sum_j \lambda_j \|\varphi_j\|_{L^2}^2,$$

but it is not useful here.

$\square$

4 Density Matrices

In this section we present an alternative description of mixed quantum states
which is (formally) equivalent to the Schrödinger system of Sect. 3.3 (see
[AF01, DL88, DL88a] for more details).

4.1 Framework, Trace Class Operators

Let $\mathcal{J}_1(L^2(\mathbb{R}^N))$ denote the Banach space of trace class operators on $L^2(\mathbb{R}^N)$,
and $\tilde{\mathcal{J}}_1(L^2(\mathbb{R}^N))$ its closed subspace of self-adjoint operators.

Definition 4.1. A *density matrix (operator)* is a positive, self-adjoint trace
class operator on $L^2(\mathbb{R}^N)$, i.e. $\hat{\varrho} \in \tilde{\mathcal{J}}_1(L^2(\mathbb{R}^N))$ with $\hat{\varrho} \geq 0$.

Since $\hat{\varrho}$ is self-adjoint and compact, there exists a complete ONS $\{\psi_j\}_{j\in\mathbb{N}} \subset$
$L^2(\mathbb{R}^N)$ of eigenvectors. Since $\hat{\varrho}$ is positive and trace class, its eigenvalues
satisfy $\lambda_j \geq 0$, $\{\lambda_j\}_{j\in\mathbb{N}} \subset \ell^1$.

We remark that the eigenvectors ψ_j are exactly the pure state wave functions from Sect. 3.3, and the eigenvalues λ_j are the their occupation probabilities.

A typical normalization (on the total mass) is: $\operatorname{Tr} \hat{\varrho} = \sum_j \lambda_j = 1$. The norm of a self-adjoint (but non necessarily positive) trace class operators is given by:

$$\|\hat{\varrho}\|_1 := \operatorname{Tr} |\hat{\varrho}| \overset{\hat{\varrho} \text{ s.a.}}{=} \sum_j |\lambda_j| \, .$$

Each density matrix operator has a unique integral representation:

$$(\hat{\varrho}f)(x) = \int_{\mathbb{R}^N} \varrho(x, y) f(y) \, \mathrm{d}y \qquad \forall f \in L^2(\mathbb{R}^N)$$

with the *density matrix function*

$$\varrho(x, y) = \sum_j \lambda_j \psi_j(x) \bar{\psi}_j(y) \in L^2(\mathbb{R}^{2N}) \, . \tag{22}$$

Here, $x, y \in \mathbb{R}^N$ are position variables. The self-adjointness of $\hat{\varrho}$ implies $\bar{\varrho}(x, y) = \varrho(y, x)$.

The L^2-norm of ϱ and the Hilbert–Schmidt norm $\|\hat{\varrho}\|_2$ of $\hat{\varrho}$ are related by

$$\|\varrho\|_2 = \|\hat{\varrho}\|_2 := (\operatorname{Tr} |\hat{\varrho}|^2)^{\frac{1}{2}} \le \|\hat{\varrho}\|_1 \, .$$

4.2 Macroscopic Quantities

We shall now define the most important macroscopic quantities of a quantum state that is modeled by a density matrix. We give two parallel definitions, both when the system is described by a density matrix *function* and a density matrix *operator*.

(a) Definition from the integral kernel $\varrho(x, y)$: The following formulae can be obtained from the analogous expressions for a wave function (see Sect. 1.1) and the eigenfunction expansion (22).

Particle density:
For $0 \le \hat{\varrho} \in \tilde{\mathcal{J}}_1$ it holds:

$$n(x) := \varrho(x, x) = \sum_j \lambda_j |\psi_j(x)|^2 \in L^1(\mathbb{R}^N), \quad n(x) \ge 0 \text{ since } \lambda_j \ge 0 \, ,$$

$$\|n\|_{L^1(\mathbb{R}^N)} \le \sum_j |\lambda_j| \, \|\psi_j\|_{L^2}^2 = \|\hat{\varrho}\|_1 \overset{\hat{\varrho} \ge 0}{=} \operatorname{Tr} \hat{\varrho} \, .$$

$$\tag{23}$$

Remark 4.1. *While $n(x)$ is defined a.e. for $\hat{\varrho} \in \mathcal{J}_1$, the definition $n(x) := \varrho(x, x)$ is meaningless for Hilbert–Schmidt operators $\hat{\varrho} \in \mathcal{J}_2$. Moreover, then there is no natural estimate of n in terms of the density matrix function ϱ.*

This leads to the following problem for an evolution equation of $\hat{\varrho}$: For an operator $\hat{\varrho} \in \mathcal{J}_2$ there is a simple functional representation of the corresponding integral kernel: $\hat{\varrho} \in \mathcal{J}_2(L^2) \Leftrightarrow \varrho \in L^2(\mathbb{R}^{2N})$, but for $\hat{\varrho} \in \mathcal{J}_1$ no 'nice' equivalent space exists for the kernel $\varrho(x, y)$. Now, if one wants to describe the time evolution of a density matrix $\hat{\varrho} \in \mathcal{J}_2(L^2(\mathbb{R}^N))$, a PDE for $\varrho \in L^2(\mathbb{R}^{2N})$ is the natural choice (see (27), below). However, in this framework the particle density $n(x)$ cannot be defined. For a self-consistent model we therefore need to consider the time evolution of a density matrix $\hat{\varrho} \in \mathcal{J}_1(L^2(\mathbb{R}^N))$. Due to the lack of a corresponding function space for its kernel, this must be considered as abstract evolution problem (Banach space-ODE) for $\hat{\varrho}(t) \in \mathcal{J}_1$ (see (28), below) instead of a PDE for its kernel $\varrho(t)$!

Higher order macroscopic quantities are formally defined as:
Current density:

$$j(x) := \Im \nabla_x \varrho \Big|_{x=y} = \sum_j \lambda_j \Im [\nabla \psi_j(x) \bar{\psi}_j(x)] \, .$$

Kinetic energy density:

$$e_{kin}(x) := \frac{1}{2}(\nabla_x \cdot \nabla_y)\varrho \Big|_{x=y} = \frac{1}{2}\sum_j \lambda_j |\nabla \psi_j(x)|^2 \geq 0 \quad \text{since } \lambda_j \geq 0 \, .$$

(b) Definition from the trace class operator $\hat{\varrho} \in \mathcal{J}_1$:

Particle density:
$n[\hat{\varrho}]$ can be defined by duality as

$$\int \varphi(x)n(x)\,\mathrm{d}x = \mathrm{Tr}\,(\varphi\,\hat{\varrho}) = \mathrm{Tr}\,(\hat{\varrho}\,\varphi) \qquad \forall \varphi \in L^\infty(\mathbb{R}^N)\,, \qquad (24)$$

where φ inside the operator trace Tr means the bounded multiplication operator by the function $\varphi \in L^\infty$. If (24) holds $\forall \varphi \in C_0(\mathbb{R}^N)$, n is defined as a *Radon measure* on $\mathbb{R}^N$ (cf. [Bre87]).
Kinetic energy:
A formal calculation shows

$$E_{kin}(\hat{\varrho}) := -\frac{1}{2}\mathrm{Tr}\,(\Delta_x \hat{\varrho}) = \frac{1}{2}\mathrm{Tr}\,(|\nabla|\hat{\varrho}|\nabla|) \geq 0 \qquad \text{since } \hat{\varrho} \geq 0 \, .$$

Table 2 Macroscopic quantities of a density matrix $\hat{\varrho}$

Observables: operator A	Expectation value $\mathrm{Tr}\,(A\hat{\varrho})$	Expectation value $\int_{\mathbb{R}^N} A_x \varrho\big\|_{x=y} \, \mathrm{d}x$
x ... position	$\mathrm{Tr}\,(x\,\hat{\varrho})$	$\int x\varrho(x,x)\,\mathrm{d}x$
$-i\nabla_x$ ... momentum	$-i\mathrm{Tr}\,(\nabla_x\hat{\varrho})$	$-i\int(\nabla_x\varrho)(x,x)\,\mathrm{d}x$
$-\frac{1}{2}\Delta_x$ ... kinetic energy	$-\frac{1}{2}\mathrm{Tr}\,(\Delta_x\hat{\varrho})$	$\frac{1}{2}\int(\nabla_x\cdot\nabla_y\varrho)(x,x)\,\mathrm{d}x$
$V(x)$ ... potential energy	$\mathrm{Tr}\,(V(x)\,\hat{\varrho}) \geq 0$ if $V \geq 0$	$\int V(x)\varrho(x,x)\,\mathrm{d}x$

Here, $|\nabla| = \sqrt{-\Delta}$ is a pseudo-differential operator (ΨDO) with the symbol $|\xi|$, i.e. $(|\nabla|f)(x) = \mathcal{F}^{-1}(|\xi|(\mathcal{F}f)(\xi))\ \forall f \in H^1(\mathbb{R}^n)$.

More generally, we now illustrate how to compute from a density matrix its macroscopic observable that corresponds to a self-adjoint operator A (cf. also Sect. 1.1 and Table 1). If a quantum system is in state $\hat{\varrho}$ (or analogously described by $\varrho(x,y)$), the expectation value of the observable A is given by $\mathrm{Tr}\,(A\hat{\varrho})$ or $\int_{\mathbb{R}^N} A_x \varrho\big\|_{x=y}\,\mathrm{d}x$, resp. In the latter case A_x denotes the realization of the operator A, acting on the x-variable. The most important examples are summarized in Table 2.

We now collect some analytic tools needed for the self-consistent problem. For mixed quantum states, $\hat{\varrho} \in \tilde{\mathcal{J}}_1$ is the analogue of $\psi \in L^2$ in the pure state case. For the related Schrödinger–Poisson equation, a simple analysis was possible in the *energy space* $\psi \in H^1$ (cf. Sect. 3.1). We now give a corresponding density matrix framework that again allows to control the kinetic energy (as needed for the self-consistent potential).

With the ΨDO $\sqrt{1 - \Delta}$ we define the *energy space*

$$\mathcal{E} := \left\{ \hat{\varrho} \in \tilde{\mathcal{J}}_1 \mid \sqrt{1 - \Delta}\,\hat{\varrho}\,\sqrt{1 - \Delta} \in \mathcal{J}_1 \right\},$$

which is a Banach space with the norm

$$\|\hat{\varrho}\|_{\mathcal{E}} := \|\sqrt{1 - \Delta}\hat{\varrho}\sqrt{1 - \Delta}\|_1\,.$$

For $\hat{\varrho} \geq 0$ it holds

$$\|\hat{\varrho}\|_{\mathcal{E}} = \mathrm{Tr}\,((1 - \Delta)\hat{\varrho}) = \mathrm{Tr}\,\hat{\varrho} + 2E_{kin}(\hat{\varrho})\,.$$

In order to estimate the particle density n, and hence the self-consistent potential V in terms of $\hat{\varrho}$ (analogously to (18)) we shall need the following *Lieb–Thirring-type inequality* [LT76, LP93, Arn96]. This is a collective Sobolev or Gagliardo–Nierenberg inequality.

Lemma 4.1 ([Arn96]). *Let* $1 \leq p \leq \frac{N}{N-2}$ *(or* $1 \leq p \leq \infty$ *if* $N = 1$; $1 \leq p < \infty$ *if* $N = 2$*) and* $\theta := \frac{N-p(N-2)}{2p} \in [0,1]$. *Then*

$$\|n[\hat{\varrho}]\|_{L^p(\mathbb{R}^N)} \leq C_p \|\hat{\varrho}\|_1^\theta \, E_{kin}(|\hat{\varrho}|)^{1-\theta}, \qquad \forall \hat{\varrho} \in \mathcal{E}$$

Proof (for $N \geq 3$*).*
Consider $p = 1$, $\theta = 1$:

$$\|n\|_{L^1} \leq \|\hat{\varrho}\|_1 \quad \text{follows from (23)}.$$

Consider $p = p_* := \frac{N}{N-2}$, $\theta = 0$: Then,

$$\|n\|_{L^{p_*}} \leq \sum_j |\lambda_j| \, \|\psi_j\|_{L^{2p_*}}^2 \leq C \sum_j |\lambda_j| \, \|\nabla\psi_j\|_{L^2}^2 = 2C E_{kin}(|\hat{\varrho}|)$$

follows by the Sobolev inequality. And the general case follows by interpolation. □

Remark 4.2. *A similar result with* $\|\hat{\varrho}\|_q^\theta$, $q > 1$ *on r.h.s. was obtained in [LP93]. Its proof is much harder.*

4.3 Time Evolution of Closed/Hamiltonian Systems

Assume that the time evolution of a wave function is determined by the *Hamiltonian* H, e.g. $H(t) = -\frac{1}{2}\Delta + V(t)$. The Schrödinger equation for a general pure state then reads

$$i\psi_t = H\psi, \quad t \in \mathbb{R}. \tag{25}$$

Next we consider the eigenfunction decomposition for ϱ:

$$\varrho(x, y, t) = \sum_j \lambda_j \psi_j(x, t) \bar{\psi}_j(y, t). \tag{26}$$

Using (25) in (26) yields the evolution equation for the density matrix. If $\hat{\varrho} \in \tilde{\mathcal{J}}_2$ or equivalently $\varrho(.,.,t) \in L^2(\mathbb{R}^{2N})$, this evolution can be written as a PDE for the kernel $\varrho(t)$ (cf. Remark 4.1):

$$i\varrho_t = (H_x - H_y)\varrho, \quad t \in \mathbb{R}. \tag{27}$$

Here, H_x and H_y denote copies of the Hamiltonian H acting, resp., on the x- and y-variable. This is called the *quantum Liouville* or *von Neumann equation* (in coordinate representation).

However, if $\hat{\varrho} \in \tilde{\mathcal{J}}_1$, its evolution cannot be written as a PDE (cf. Remark 4.1). Instead, one has to write it as an abstract evolution problem (Banach space-ODE) for $\hat{\varrho}(t)$:

$$i\frac{d}{dt}\hat{\varrho} = [H, \hat{\varrho}] := H\hat{\varrho} - \hat{\varrho}H, \quad t \in \mathbb{R}. \tag{28}$$

Also this variant of (27) is called *quantum Liouville* or *von Neumann equation*.

In order to solve it, we first consider the free Hamiltonian $H_0 := -\frac{1}{2}\Delta$. According to (28), the corresponding infinitesimal generator of the C_0-group $G_0(t)$, $t \in \mathbb{R}$, for the $\hat{\varrho}$-evolution formally reads $h_0 = -i[H_0, \hat{\varrho}]$. $G_0(t)$ has the following explicit representation:

$$G_0(t)\hat{\varrho} = e^{-iH_0t}\hat{\varrho}e^{iH_0t}, \quad t \in \mathbb{R}, \tag{29}$$

with the kernel

$$\sum_j \lambda_j (e^{-iH_0t}\psi_j)(x)\,(e^{iH_0t}\overline{\psi_j})(y).$$

One can see that the density matrix $\hat{\varrho}(t)$ solving (28) has eigenvalues that are constant in time. The corresponding eigenvectors obey the Schrödinger equation (25) and stay orthonormal during the evolution (which implies that they can only "rotate" in $L^2(\mathbb{R}^N)$).

The following lemma gives additional properties of the evolution group $G_0(t)$ and its generator h_0:

Lemma 4.2 ([DL88a]).

(a) $G_0(t)$ is a C_0-group of isometries on $\mathcal{J}_1$ It preserves self-adjointness and positivity.

(b) Its generator is characterized by

$$\mathcal{D}(h_0) = \{\hat{\varrho} \in \mathcal{J}_1 \mid \hat{\varrho}\mathcal{D}(H_0) \subset \mathcal{D}(H_0), (H_0\hat{\varrho} - \hat{\varrho}H_0) \text{ is an operator with}$$
$$\text{domain } H^2(\mathbb{R}^N) \text{ and it can be extended to } L^2(\mathbb{R}^N),$$
$$\text{such that } \overline{H_0\hat{\varrho} - \hat{\varrho}H_0} \in \mathcal{J}_1\},$$

$$h_0(\hat{\varrho}) := -i(\overline{H_0\hat{\varrho} - \hat{\varrho}H_0}).$$

Proof (of part (a)). The strong $\mathcal{J}_1$-continuity of $G_0(t)$ at $t = 0$ follows from the following two ingredients:

$$\|G_0(t)\hat{\varrho}\|_1 = \|\hat{\varrho}\|_1 \quad \text{since } \lambda_j = \text{const. in } t \quad (\text{convergence of the } \mathcal{J}_1-\text{norm}),$$

$$\langle f, (G_0(t)\hat{\varrho})\,g \rangle = \langle \underbrace{e^{iH_0t}f}_{\in C(\mathbb{R};L^2)}, \underbrace{\hat{\varrho}}_{\in \mathcal{B}(L^2)}(e^{iH_0t}g)\rangle \xrightarrow{t\to 0} \langle f, \hat{\varrho}g \rangle \quad \forall f, g \in L^2$$

(weak operator convergence).

These two properties imply the desired $\mathcal{J}_1$-convergence. This is a corollary to Grümm's Theorem (cf. [Sim79]).

The preservation of self-adjointness and positivity follows directly from (29). □

The above result for the evolution in $\mathcal{J}_1$ can easily be modified to an analogous result for the evolution in the energy space $\mathcal{E}$:

Corollary 4.1. h_0 *generates a C_0-group of isometries on $\mathcal{E}$.*

Proof. The operators $\sqrt{1 - \Delta}$ and $e^{-iH_0 t}$ commute. Hence:

$$\sqrt{1 - \Delta}\,(G_0(t)\hat{\varrho})\,\sqrt{1 - \Delta} = e^{-iH_0 t}\,\underbrace{\left(\sqrt{1 - \Delta}\hat{\varrho}\sqrt{1 - \Delta}\right)}_{\in \mathcal{J}_1}\,e^{iH_0 t} \in C(\mathbb{R};\mathcal{J}_1)\,.$$

□

4.4 Von Neumann–Poisson Equation in $\mathbb{R}^3$

Here, we present the density matrix analogue of the H^1-analysis for a Schrödinger–Poisson system (SPS) (cf. Sect. 3.3). The von Neumann–Poisson equation for $\hat{\varrho}(t)$ discussed here is almost equivalent to the SPS-analysis in H^1: Via the SPS-analysis one constructs a corresponding solution $\hat{\varrho} \in C(\mathbb{R};\mathcal{E})$, hence the existence of a solution is guaranteed. However, its uniqueness in $\mathcal{J}_1$ or $\mathcal{E}$ would stay open.

Since the SPS-analysis is technically much simpler than the density matrix analysis, a $\mathcal{J}_1$-analysis would hence (almost) not be worth the effort for closed, i.e. Hamiltonian systems. However, the time evolution of open quantum systems (cf. Sect. 6) cannot be rewritten as a SPS. For such models, the $\hat{\varrho}$-analysis seems therefore unavoidable.

We start with the analysis of the *repulsive von Neumann–Poisson equation*:

$$\begin{cases} i\frac{d}{dt}\hat{\varrho}(t) = \overline{[-\frac{1}{2}\Delta + V(t), \hat{\varrho}(t)]}, & t \in \mathbb{R} \\ V(t) = \frac{1}{4\pi|x|} * n[\hat{\varrho}(t)] \\ \hat{\varrho}(0) = \hat{\sigma} \end{cases} \tag{30}$$

Theorem 4.3. *Let $\hat{\sigma} \in \mathcal{E}$. Then (30) has a unique solution $\hat{\varrho} \in C(\mathbb{R};\mathcal{E})$. It satisfies $\|\hat{\varrho}(t)\|_1 = \|\hat{\sigma}\|_1$.*

Proof. Morally, we follow the proof of Theorem 3.1. But since we are dealing with the evolution of operators instead of functions, we have to cope with many technical difficulties (Lieb–Thirring-type inequality instead of Sobolev inequality, e.g.):

1. $G_0(t)$ is a C_0-group of isometries on $\mathcal{J}_1$, $\tilde{\mathcal{J}}_1$, $\mathcal{E}$ (see Sect. 4.3).
2. $f(\hat{\varrho}) := -\mathrm{i}[V[\hat{\varrho}], \hat{\varrho}]$ is locally Lipschitz in $\mathcal{E}$ (but *not* in $\tilde{\mathcal{J}}_1$):

   - $\hat{\varrho} \in \mathcal{E} \Rightarrow V[\hat{\varrho}] \in L^\infty(\mathbb{R}^3)$ by (18) and the Lieb–Thirring-type inequality from Lemma 4.1.
   - We need to show that $\sqrt{1 - \Delta}\,(V\hat{\varrho})\,\sqrt{1 - \Delta} \in \mathcal{J}_1$:

     (a) First we decompose the ΨDO $\sqrt{1 - \Delta}$ as follows:

     $$\sqrt{1 - \Delta} = 1 + \sum_j \underbrace{K_j}_{\in \mathcal{B}(L^2)} \partial_{x_j} \,.$$

     This allows to use the product rule for $(V\hat{\varrho})$. Here, K_j is a ΨDO with the symbol

     $$\mathrm{i}\frac{\xi_j}{|\xi|^2}\left(\sqrt{1 + |\xi|^2} - 1\right)$$

     (b) Next we need to show that $V\hat{\varrho}\,\sqrt{1 - \Delta}$, $\nabla(V\hat{\varrho}\,\sqrt{1 - \Delta}) \in \mathcal{J}_1$. For the first term we use that

     $$\sqrt{1 - \Delta}\,\hat{\varrho}\,\sqrt{1 - \Delta} \in \mathcal{J}_1 \overset{\hat{\varrho} \geq 0}{\Longrightarrow} \hat{\varrho}^{\frac{1}{2}}\,\sqrt{1 - \Delta}, \ \hat{\varrho}^{\frac{1}{2}} \in \mathcal{J}_2 \,.$$

     For simplicity we assumed here first that $\hat{\varrho} \geq 0$ (but it can be generalized). Next, the "Hölder inequality" for the operator spaces $\mathcal{J}_p$ (cf. [RS75]) yields

     $$\underbrace{V}_{\in \mathcal{B}}\,\underbrace{\hat{\varrho}^{\frac{1}{2}}}_{\in \mathcal{J}_2}\,(\underbrace{\hat{\varrho}^{\frac{1}{2}}\,\sqrt{1 - \Delta}}_{\in \mathcal{J}_2}) \in \mathcal{J}_1 \,.$$

3. Proposition 3.1 yields: The von Neumann–Poisson equation (30) has a *unique local solution* $\hat{\varrho} \in C([0, t_{max}); \mathcal{E})$.
4. We have the a-priori estimates $\|\hat{\varrho}(t)\|_1 = $ const in t, $E_{kin}(\hat{\varrho}(t))$ is uniformly bounded (in t). Hence, the solution is global on $\mathbb{R}$.

$$\square$$

Remark 4.4. *The quantum attractive case (with $V(t) = -\frac{1}{4\pi|x|} * n[\hat{\varrho}(t)])$) can be included with the following estimate (using the Lieb–Thirring-type inequality):*

$$-E_{pot}(\hat{\varrho}) := \|\nabla V\|_{L^2}^2 \leq C\|n\|_{\frac{6}{5}}^2 \leq C\|\hat{\varrho}\|_1^{\frac{3}{2}}\,E_{kin}(\hat{\varrho})^{\frac{1}{2}}$$

(cf. [Arn96]). A similar strategy also works for SP-analysis in Sect. 3.

5 Wigner Function Models

5.1 Wigner Functions

A Wigner function is obtained from the corresponding density matrix function by the *Wigner–Weyl transformation* (cf. [W32, SS87]):

$$w(x, v, t) = (2\pi)^{-N/2} \mathcal{F}_{\eta \to v}\, \varrho\left(x + \frac{\hbar\eta}{2m}, x - \frac{\hbar\eta}{2m}, t\right)$$

$$= (2\pi)^{-N} \sum_j \lambda_j \int_{\mathbb{R}^N} \psi_j\left(x + \frac{\hbar\eta}{2m}, t\right)\bar{\psi}_j\left(x - \frac{\hbar\eta}{2m}, t\right) e^{-iv \cdot \eta}\, \mathrm{d}\eta\,.$$

Since $\varrho(x, y) = \overline{\varrho(y, x)}$, we have $w(x, v) \in \mathbb{R}$. Moreover,

$$w \in L^2(\mathbb{R}^{2N}) \Leftrightarrow \varrho \in L^2(\mathbb{R}^{2N}) \Leftrightarrow \hat{\varrho} \in \mathcal{J}_2(L^2(\mathbb{R}^N))\,.$$

We call w a *physical Wigner function*, if it corresponds to a density matrix $0 \leq \hat{\varrho} \in \tilde{\mathcal{J}}_1$.

Following [SS87] we now also give the direct transformation from the density matrix operator $\hat{\varrho}$ to the Wigner function w: With the *Weyl operators*

$$W(\xi, \eta) := e^{-i(\xi \cdot x - i\frac{\hbar}{m}\eta \cdot \nabla_x)}, \quad \text{for each } \xi, \eta \in \mathbb{R}^N$$

we have

$$w(x, v, t) = (2\pi)^{-2N} \iint_{\mathbb{R}^{2N}} \mathrm{Tr}\left(\hat{\varrho}(t) W(\xi, \eta)\right) e^{i(\xi \cdot x + \eta \cdot v)}\, \mathrm{d}\xi \mathrm{d}\eta\,.$$

Since the operators $\xi \cdot x - i\frac{\hbar}{m}\eta \cdot \nabla_x$ are essentially self-adjoint on $C_0^\infty(\mathbb{R}^N)$ $\forall \xi, \eta \in \mathbb{R}^N$ (cf. Lemma 5.1, Example 5.1, below), $W(\xi, \eta)$ is a unitary operator on $L^2(\mathbb{R}^N)$ by Stone's Theorem (cf. [Paz83]). Hence, $\mathrm{Tr}\left(\hat{\varrho}(t) W(\xi, \eta)\right)$ is well-defined for $\hat{\varrho} \in \mathcal{J}_1$ and each $\xi, \eta \in \mathbb{R}^N$.

The time evolution of w follows from the von Neumann equation: The *Wigner equation* reads

$$w_t + v \cdot \nabla_x w - e\,\Theta[V]w = 0, \quad x, v \in \mathbb{R}^N, t \in \mathbb{R}, \tag{31}$$

with the pseudo-differential operator (ΨDO)

$$\Theta[V]w(x, v)$$
$$= \frac{i}{\hbar}(2\pi)^{-N} \iint_{\mathbb{R}^{2N}} \left[V(x + \frac{\hbar\eta}{2m}) - V(x - \frac{\hbar\eta}{2m})\right] w(x, \tilde{v})e^{i(v - \tilde{v}) \cdot \eta}\, \mathrm{d}\tilde{v}\mathrm{d}\eta\,.$$

In the *classical limit*, $\Theta[V]$ converges formally to its classical counterpart (see [LP93] for rigorous results). For a fixed function $w = w(x, v)$ and a fixed potential $V = V(x)$, we have:

$$\Theta[V]w \xrightarrow{\hbar \to 0} \frac{1}{m} \nabla_x V \cdot \nabla_v w.$$

For a quadratic potential V, the operator $\Theta[V]$ takes exactly the form of its classical counterpart:

$$\Theta[V]w = \frac{1}{m} \nabla_x V \cdot \nabla_v w. \tag{32}$$

In this special case the Wigner equation (formally) looks like the *classical Liouville equation*

$$f_t + v \cdot \nabla_x f - \frac{e}{m} \nabla_x V \cdot \nabla_v f = 0, \quad x, v \in \mathbb{R}^N. \tag{33}$$

Note that the Liouville equation and also the nonlinear Liouville–Poisson equation (also called *Vlasov–Poisson* equation) preserve all L^p-norms in time, i.e.

$$\|f(.,.,t)\|_{L^p(\mathbb{R}^{2N})} = \text{const. in } t \quad \forall 1 \leq p \leq \infty, \ t \in \mathbb{R}.$$

This is implied by the fact that the solution of (33) is constant along its characteristics. In contrast, the Wigner equation (in general) only preserves the $L^2(\mathbb{R}^{2N})$-norm, since $v \cdot \nabla_x - e\,\Theta[V]$ is skew-adjoint.

If $V \in L^\infty(\mathbb{R}^N)$, then $\|\Theta[V]\|_{\mathcal{B}(L^2)} \leq \frac{2}{\hbar}\|V\|_{L^\infty}$. Hence, in this case there exists a C_0-evolution group of isometries for the Wigner equation on $L^2(\mathbb{R}^{2N})$. This follows from Stone's theorem. Moreover, $\Theta[V]$ is a bounded perturbation of $v \cdot \nabla_x$.

From now on we set the parameters $e = m = \hbar = 1$ and we recall the definition of the *macroscopic quantities* for a Wigner function w:

- Typical normalization of total mass: $\iint_{\mathbb{R}^{2N}} w \, dx dv = 1$
- Particle density: $n(x, t) := \int_{\mathbb{R}^N} w(x, v, t) \, dv$ (≥ 0 for a physical Wigner function)
- Current density: $j(x, t) := \int_{\mathbb{R}^N} vw(x, v, t) \, dv$
- Kinetic energy density: $e_{kin}(x, t) := \int_{\mathbb{R}^N} \frac{|v|^2}{2} w(x, v, t) \, dv$ (≥ 0 for a physical Wigner function)

Note that these definitions are *purely formal* since a Wigner function satisfies $w(.,.,t) \in L^2(\mathbb{R}^{2N})$ but typically $w \notin L^1(\mathbb{R}^{2N})$. Hence, the "definition" $n := \text{"} \int w \, dv \text{"}$ is *meaningless*!

This is a key problem for analyzing the self-consistent Wigner–Poisson equation, i.e. (31) with the Coulomb potential obtained from $-\Delta V = n = \int w \, dv$. In other words, the quadratically nonlinear term $\Theta[V[w]]w$ is *not defined pointwise in t* on the state space of the Wigner function ($w(t) \in L^2$, e.g.). This is the same problem like in the L^2-analysis of SP in Sect. 3.2. There are two simple solutions to this problem:

- Change the state space for w (even if it is not very physical): A weighted L^2-space with sufficient weight in the v-variable implies $w \in L^2_x(L^1_v)$ and hence $n \in L^2(\mathbb{R}^N_x)$. Possible options are:
  in 1D: $w \in L^2(\mathbb{R}^2; (1 + v^2)\, dx\, dv)$, cf. [ACD02]
  in 3D: $w \in L^2(\mathbb{R}^6; (1 + |v|^4)\, dx\, dv)$, cf. [ADM07]

- For Hamiltonian or closed systems (i.e. without collision operators in the Wigner equation) the Wigner–Poisson equation is (almost) equivalent to the SPS (Sect. 3.3). This allows for a much simpler analysis (cf. Theorem 3.3a, and [BM91, AN91, ILZ94, MRS90]).

5.2 Linear Wigner–Fokker–Planck: Well-Posedness

We shall now consider an *open quantum system* that includes a collision operator on the r.h.s. of (31). Such a model is *not* any more equivalent to a system of Schrödinger equations. Any mathematical analysis must hence be done on the level of Wigner functions or density matrices. In this and the next section we shall illustrate both approaches.

In this subsection and in Sect. 5.3 we analyze the linear *Wigner–Fokker–Planck equation* (WFP) with an external potential of the form $\frac{\mu}{2}|x|^2 + V(x)$, $\mu \in \mathbb{R}$, $V \in L^\infty(\mathbb{R}^N)$. Because of (32), the quadratic potential yields the classical potential term. Hence, the WFP equation reads:

$$
\begin{cases}
w_t + v \cdot \nabla_x w - \mu x \cdot \nabla_v w - \Theta[V]w = Qw, \qquad x, v \in \mathbb{R}^N, t > 0 \\
Qw = \underbrace{D_{pp}\Delta_v w}_{\text{class. diffusion}} + \underbrace{2\gamma \operatorname{div}_v(vw)}_{\text{friction}} + \underbrace{D_{qq}\Delta_x w + 2D_{pq} \operatorname{div}_x(\nabla_v w)}_{\text{quantum diffusion}} \\
w(x, v, t = 0) = w_0(x, v)
\end{cases}
\tag{34}
$$

- This model is quantum mechanically "correct" if the following *Lindblad condition* holds:

$$
D_{pp} D_{qq} - D_{pq}^2 \geq \frac{\gamma^2}{4} .
\tag{35}
$$

Exactly in this case, (34) can be rewritten as a Lindblad equation (see (50) below) for the corresponding density matrix operator [ALMS04, SS87]. As a consequence, the positivity of the particle density is preserved under the time evolution: $n(x, t) \geq 0$, $\forall x, t$. Note that the "classical Fokker–Planck-term", i.e. the so-called *Caldeira–Leggett model* [CL83], with $D_{qq} = D_{pq} = 0$ does *not* satisfy the Lindblad condition (35). Nevertheless it is frequently used in applications, yielding reasonable results.
- The collision operator Qw models *diffusive effects* (e.g. the electron–phonon-interaction). Hence, (34) has applications for the electron transport in quantum semiconductors and for quantum Brownian motion.

- A derivation of (34) from the coupling of electrons to a bath of harmonic oscillators was given in [CEFM00].

Next we give an existence result for the linear WFP equation (34). The following theorem crucially depends on the *Lumer–Phillips Theorem* (cf. [Paz83]):

Proposition 5.1. *Let the operator A be densely defined and closed on the Banach space X. Let A and A^* be dissipative. Then, A generates a C_0-semigroup of contractions on X.*

We rewrite (34) as an evolution problem on $L^2(\mathbb{R}^{2N})$:

$$\begin{cases} w_t = Aw + \Theta[V]w, & t > 0, \\ w(0) = w_0 \in L^2(\mathbb{R}^{2N}), \end{cases} \tag{36}$$

with the abbreviation

$$\begin{aligned} Aw := & -v \cdot \nabla_x w + \mu x \cdot \nabla_v w + Qw \\ = & -v \cdot \nabla_x w + \mu x \cdot \nabla_v w + 2\gamma \operatorname{div}_v(vw) \\ & + D_{pp}\Delta_v w + 2D_{pq}\operatorname{div}_v(\nabla_x w) + D_{qq}\Delta_x w . \end{aligned}$$

Theorem 5.1. *Let $V \in L^\infty(\mathbb{R}^N)$ or $V \in L^1_{loc}(\mathbb{R}^+; L^\infty(\mathbb{R}^N))$. Then (36) has a unique mild solution $w \in C([0,\infty); L^2(\mathbb{R}^{2N}))$.*

Proof.

- We define the operator $\tilde{A} := A - N\gamma$ on the domain $\mathcal{D}(\tilde{A}) := C_0^\infty(\mathbb{R}^{2N})$.
- Then, $\tilde{A}$ is dissipative on $\mathcal{D}(\tilde{A})$ and $\tilde{A}^*$ is (formally) *dissipative* in $L^2(\mathbb{R}^{2N})$, i.e. $\langle \tilde{A}w, w \rangle_{L^2} \leq 0 \quad \forall w \in \mathcal{D}(\tilde{A})$. The rigorous proof of dissipativity for $\tilde{A}^*$ will follow from Lemma 5.1 below.

- From the Lumer–Phillips Theorem then follows: $\bar{\tilde{A}}$ generates a C_0-semigroup:

$$\left\| e^{t\bar{\tilde{A}}} w \right\|_{L^2} \leq e^{N\gamma t} \|w\|_{L^2}, \quad t \geq 0. \tag{37}$$

- The final result follows, since $\Theta[V]$ is a bounded perturbation on $L^2(\mathbb{R}^{2N})$ (see Proposition 2.4 or 3.1, resp.).

$\square$

Now we still have to prove the dissipativity of $\tilde{A}^*$ on $\mathcal{D}(\tilde{A}^*)$. Just like for $\tilde{A}$, one immediately finds that $\tilde{A}^*|_{\mathcal{D}(\tilde{A})}$ is dissipative, with

$$\tilde{A}^* = v \cdot \nabla_x w - \mu x \cdot \nabla_v w - 2\gamma\, v \cdot \nabla_v w + D_{pp}\Delta_v w + 2D_{pq}\operatorname{div}_v(\nabla_x w) + D_{qq}\Delta_x w - N\gamma .$$

However, $\mathcal{D}(\tilde{A}^*)$ is not known explicitly. To verify that $\tilde{A}^*$ is the closure of $\tilde{A}^*|_{\mathcal{D}(\tilde{A})}$ we shall use the following lemma, since $\tilde{A}^*$ is a quadratic polynomial in x, v, ∇_x, ∇_v:

Lemma 5.1 ([ACD02], [AS04]). *Let the operator $P = p_2(x, -i\nabla)$ be a quadratic polynomial in x, $-i\nabla$. Define the* minimum realization *of P on*

$$\mathcal{D}(P_{\min}) := C_0^\infty(\mathbb{R}^N) \subseteq L^2(\mathbb{R}^N).$$

Then, $\overline{P_{\min}} = P_{\max}$, *the* maximum extension *of P, i.e.*

$$\mathcal{D}(P_{\max}) = \{f \in L^2 | Pf \in L^2\}.$$

Proof. For $f \in \mathcal{D}(P_{\max})$ we need to show, that it can be approximated in the graph norm $\|.\|_P$ by a sequence $\{f_n\} \subset C_0^\infty(\mathbb{R}^N)$. This is accomplished by the following (standard) approximation sequence (but the proof is lengthy):

$$f_n(x) := \underbrace{\chi_n(x)}_{C_0^\infty\text{-cutoff}} \cdot (f * \underbrace{\varphi_n}_{C_0^\infty\text{-mollifier}})(x) \overset{n\to\infty}{\longrightarrow} f \quad \text{in } \|.\|_P.$$

$\square$

The following examples illustrate applications and limitations of this lemma:

Example 5.1. We consider several Schrödinger operators. Their essential self-adjointness on $C_0^\infty(\mathbb{R}^N)$ is crucial for the existence of a corresponding evolution group (cf. [RS75]).

(a) $P = -\Delta - |x|^2$, $\mathcal{D}(P) = C_0^\infty(\mathbb{R}^N)$ is essentially self-adjoint in $L^2(\mathbb{R}^N)$.
(b) $P = -\partial_x^2 + x^4$, $\mathcal{D}(P) = C_0^\infty(\mathbb{R})$ is also essentially self-adjoint in $L^2(\mathbb{R})$ [RS75]. Hence, Lemma 5.1 can be extended to this case of a positive quartic potential.
(c) $P = -\partial_x^2 - x^4$, $\mathcal{D}(P) = C_0^\infty(\mathbb{R})$ is *not* essentially self-adjoint in $L^2(\mathbb{R})$ [RS75]. Hence, Lemma 5.1 cannot be extended to this negative quartic potential. Note, that for this potential (classical) particles would run off to $x = \infty$ in finite time. Hence, no reversible, mass conserving dynamics can exist.

So we conclude, that Lemma 5.1 cannot be extended to all operators of the form $P = p_4(x, -i\nabla)$ (i.e. quartic polynomials) – neither to all cubic polynomials, by the way.

5.3 Linear Wigner–Fokker–Planck: Large Time Behavior

First we consider the case of a quadratic external confinement potential $V(x) = \frac{\mu}{2}|x|^2$, $\mu \geq 0$. Because of (32), the linear *Wigner–Fokker–Planck equation* (WFP) then takes the form of a classical kinetic equation:

$$\begin{cases} w_t + v \cdot \nabla_x w - \mu x \cdot \nabla_v w = Qw \,, \qquad x,\, v \in \mathbb{R}^N,\, t > 0 \\ Qw := D_{pp}\Delta_v w + 2\gamma \operatorname{div}_v(vw) + D_{qq}\Delta_x w + 2D_{pq} \operatorname{div}_x(\nabla_v w) \qquad (38) \\ w(x,v,t=0) = w_0(x,v)w(x,v,t=0) = w_0(x,v) \in L^2(\mathbb{R}^{2N}) \end{cases}$$

Theorem 5.2 ([SCDM04]).

(a) (38) *has a Green's function* $G(x,v,x_0,v_0,t) \geq 0$ *(it is a non-isotropic Gaussian).*

(b) $\exists$*! mild (and actually classical) solution of* (38)*:*

$$w(x,v,t) = \iint G(x,v,x_0,v_0,t)w_0(x_0,v_0)\,\mathrm{d}x_0\,\mathrm{d}v_0 \in C([0,\infty); L^2(\mathbb{R}^{2N})) \,.$$
$$(39)$$

When transforming (x,v) *to the characteristic coordinates of the Liouville equation* $w_t + v \cdot \nabla_x w - \mu x \cdot \nabla_v w = 0$, *the integral in* (39) *becomes a convolution in* $x_0,\, v_0$.

(c) $w_0(x,v) \geq 0 \Rightarrow w(x,v,t) \geq 0 \quad \forall t \geq 0$.

The dissipativity introduced by the collision operator Q and the confinement of the potential V makes the system converge to the equilibrium. This steady state w_∞ is unique up to the normalization of mass:

Theorem 5.3 ([SCDM04]). *Let* $\gamma > 0$ *and* $\mu > 0$. *Then*

(a) The WFP equation (38) *has a unique steady state (up to normalization of mass):*
$$w_\infty(x,v) = \mathrm{e}^{-[\alpha|x|^2 + 2\beta x \cdot v + \gamma|v|^2]} \,.$$

It is a non-isotropic $2N$*-dimensional Gaussian.*

(b) $w(t) \overset{t\to\infty}{\longrightarrow} w_\infty(x,v)$ *in relative entropy (cf.* (52)*),* $L^1(\mathbb{R}^{2N})$, *and in* $L^2(\mathbb{R}^{2N})$ *with an exponential rate. Here,* w_∞ *is normalized as*

$$\iint w_\infty \mathrm{d}x\mathrm{d}v = \iint w_0 \mathrm{d}x\mathrm{d}v \,.$$

Proof (of (b)).
This is an application of the *entropy method* (cf. [AMTU01]) for uniformly parabolic drift-diffusion equations with a uniformly convex "potential". The method is applied separately for the positive and negative part of the Wigner function: $w^\pm(x,v,t)$. $\qquad\qquad\qquad\qquad\qquad\qquad\qquad\qquad\qquad\qquad\qquad\quad \square$

Remark 5.4. *In contrast to classical kinetic models,* w_∞ *does not separately annihilate the collision term* Qw *and the transport term* $v \cdot \nabla_x w - \mu x \cdot \nabla_v w$ *in* (38). *This reflects the non-local flavor of quantum mechanics.*

Next we present a recent extension of Theorem 5.3 for (small) perturbations λV_0 of the harmonic oscillator potential (cf. [AGGS07]). We consider the linear WFP equation (36) with the identity as diffusion matrix (just for notational simplicity):

$$\begin{cases} w_t = Aw + \lambda\Theta[V_0]w, & t > 0, \\ w(0) = w_0, \end{cases} \tag{40}$$

with the abbreviation

$$Aw := -v \cdot \nabla_x w + x \cdot \nabla_v w + Qw$$
$$= -v \cdot \nabla_x w + x \cdot \nabla_v w + 2\operatorname{div}_v(vw) + \Delta_v w + \Delta_x w.$$

Let $w_\infty(x,v) > 0$ denote the unperturbed steady state (i.e. for $\lambda = 0$) from Theorem 5.3. It is unique when imposing the normalization $\iint_{\mathbb{R}^{2N}} w_\infty \mathrm{d}x\mathrm{d}v = 1$. We introduce the weighted Hilbert space $\mathcal{H} := L^2(\mathbb{R}^{2N}, w_\infty^{-1}\mathrm{d}x\mathrm{d}v)$.

Then we have

Theorem 5.5 ([AGGS07]). *Let $V_0 \in C^\infty(\mathbb{R}^N)$, such that $\hat{V}_0$ decays sufficiently fast (see [AGGS07] for the details), and let $|\lambda| > 0$ be sufficiently small. Then*

(a) *The WFP equation (40) has a unique steady state $\tilde{w}_\infty \in \mathcal{H}$, satisfying the normalization condition $\iint_{\mathbb{R}^{2N}} \tilde{w}_\infty \, \mathrm{d}x\mathrm{d}v = 1$.*

(b) *For any initial function $w_0 \in \mathcal{H}$ with $\iint_{\mathbb{R}^{2N}} w_0 \, \mathrm{d}x\mathrm{d}v = 1$ we have exponential convergence towards the steady state:*

$$\|w(t) - \tilde{w}_\infty\|_{\mathcal{H}} \leq \mathrm{e}^{-\varepsilon t}\|w_0 - \tilde{w}_\infty\|_{\mathcal{H}}, \qquad t \geq 0,$$

with some $\varepsilon > 0$.

Proof.

(a) We rewrite the stationary version of (40) as the fixed point problem

$$Aw = -\lambda\Theta[V_0]w$$

in order to use Banach's fixed point theorem. Since A has a non-trivial kernel (in fact $Aw_\infty = 0$ by Theorem 5.3), we cannot invert A. Hence we define $\mathcal{H}^\perp := \{f \in \mathcal{H} : f \perp w_\infty\}$ and modify the fixed point problem to

$$Az = -\lambda\Theta[V_0](z + w_\infty) \tag{41}$$

for $z := w - w_\infty \in \mathcal{H}^\perp$. Now, $\lambda A^{-1}\Theta[V_0]$ is contractive on $\mathcal{H}^\perp$ and its unique fixed point z^* yields the unique normalized steady state $\tilde{w}_\infty = z^* + w_\infty \in \mathcal{H}$ of (40).

(b) Consider on $\mathcal{H}^\perp$ the evolution of $v(t) := w(t) - \tilde{w}_\infty$, with $w(t)$ satisfying (40). Then, $\|v(t)\|_{\mathcal{H}}^2$ is a Lyapunov functional with exponential decay: For computing $\frac{\mathrm{d}}{\mathrm{d}t}\|v(t)\|_{\mathcal{H}}^2$ we use that the operator A has a spectral gap of size $\sigma = 1 - 1/\sqrt{2}$ on $\mathcal{H}$. Hence, its symmetric part A^s satisfies on $\mathcal{H}^\perp$: $A^s \leq -\sigma$. The idea is now that the perturbation potential $\lambda\Theta[V_0]$ can be compensated by this spectral gap. $\qquad\square$

5.4 Wigner–Poisson–Fokker–Planck: Global Solutions in $\mathbb{R}^3$

The Wigner–Poisson–Fokker–Planck (WPFP) system reads:

$$\begin{cases} w_t + v \cdot \nabla_x w - \Theta[V]w = Qw, \quad x, v \in \mathbb{R}^3, \ t > 0 \\ Qw = \underbrace{D_{pp}\Delta_v w}_{\text{class. diffusion}} + \underbrace{2\gamma \operatorname{div}_v(vw)}_{\text{friction}} + \underbrace{D_{qq}\Delta_x w + 2D_{pq}\operatorname{div}_x(\nabla_v w)}_{\text{quantum diffusion}} \\ -\Delta V(x,t) = n(x,t) := \int w(x,v,t)\,\mathrm{d}v \\ w(x,v,t=0) = w_0(x,v) \end{cases} \qquad (42)$$

This is an open quantum systems with mean-field potential. For proving the existence of a global-in-time solution to (42) we face *two analytical problems:*

1. For $w \in L^2(\mathbb{R}^6)$ the nonlinear potential term is not locally Lipschitz. This makes the construction of a local-in-time solution difficult. Actually, the particle density $n(x,t)$ is not defined pointwise in time.
2. We lack enough (physical) a-priori estimates to establish a global-in-t solution.

First we discuss the *difficulties with the a-priori estimates:* For WPFP, the only simple a-priori estimate is (cf. (37))

$$\|w(t)\|_2 \le e^{3\gamma t}\|w_0\|_2, \ t \ge 0,$$

which follows from the skew-adjointness of $v \cdot \nabla_x - \Theta[V]$. But we have *no* other L^p-estimates.

The physical conservation laws (like mass conservation $\iint w(x,v,t)\,\mathrm{d}x\,\mathrm{d}v = const$, or an energy balance involving the total kinetic energy $E_{kin}(t) = \frac{1}{2}\int\int |v|^2 w(x,v,t)\,\mathrm{d}x\,\mathrm{d}v$, cf. [ALMS04]) are *not* usable here, since $w \in \mathbb{R}$ (unless $\hat{\varrho}(t) \ge 0$ is used, as in [CLN04]).

The main idea for tackling both of the above problems is to find a *new a-priori estimate for the electric field* $E = E(x,t)$ in the WPFP equation. This will be used both:

- For the construction of the local-in-time solution via a fixed point map
- To establish the global-in-time solution

Since the particle density $n[w] = \int w\,\mathrm{d}v$ cannot be rigorously defined *pointwise in t*, we shall somehow eliminate it from the WPFP system. Instead, we shall define the electric field $E = -\nabla V = \frac{1}{4\pi}\frac{x}{|x|^3} * \int w\,\mathrm{d}v$ a.e. in t using the dispersive regularization of the free transport equation $w_t + v \cdot \nabla_x w = 0$.

A-Priori Estimate for Electric Field in Wigner–Poisson

We proceed similarly to [Per96] for the Vlasov–Poisson (VP) system, or [Cas98] for the Vlasov–Poisson–Fokker–Planck system.

To keep the notation simple we first illustrate the strategy for the Wigner–Poisson system, which reads in Duhamel form:

$$w(x,v,t) = w_0(x - vt, v) - \int_0^t (\Theta[V]w)(x-vs, v, t-s)\, ds. \qquad (43)$$

Using the particle density $n = \int w\, dv$ from (43) we split the field as $E(x,t) = -\nabla V(x,t) = \frac{1}{4\pi}\frac{x}{|x|^3} * n(x,t) = E_0 + E_1$:

$$E_0(x,t) := \frac{1}{4\pi}\frac{x}{|x|^3} *_x \int w_0(x-vt, v)\, dv,$$

$$E_1(x,t) := -\frac{1}{4\pi}\frac{x}{|x|^3} *_x \int_0^t \int_{\mathbb{R}_v^3} (\Theta[V]w)(x-vs, v, t-s)\, dv\, ds. \qquad (44)$$

For VP (studied in [Per96]) the corresponding term has the form

$$E_1(x,t) = -\frac{1}{4\pi}\frac{x}{|x|^3} *_x \int_0^t \int (\nabla_x V \cdot \nabla_v w)(x-vs, v, t-s)\, dv\, ds$$

$$= -\frac{1}{4\pi}\frac{x}{|x|^3} *_x \operatorname{div}_x \int_0^t s \int (\nabla_x V\, w)(x-vs, v, t-s)\, dv\, ds. \qquad (45)$$

Here, the key issue is that the second factor of this last convolution is in divergence form, in order to pass that div_x to the first convolution factor later on. Obviously, this is not the case in (44). However, with a tricky reformulation, $\Theta[V]$ can indeed be written in divergence form:

$$\Theta[V]w = \mathcal{F}_{\eta \to v}^{-1}\left(\delta V(x,\eta)\hat{w}(x,\eta)\right) = \nabla_x V *_x \Phi(x,v) *_v \nabla_v w,$$

with some distribution $\Phi(x,v)$ and

$$\delta V(x,\eta) = V\left(x + \frac{\eta}{2}\right) - V\left(x - \frac{\eta}{2}\right) = \int_{-\frac{1}{2}}^{\frac{1}{2}} \eta \cdot \nabla_x V(x - r\eta)\, dr.$$

We illustrate this computation for the (simpler) 1D case:

$$\delta V(x,\eta)\hat{w}(x,\eta) = \eta \nabla_x V *_x \underbrace{\left(\frac{1}{|\eta|}\chi_{[-|\eta|/2, |\eta|/2]}\right)(x)}_{=:\hat{\Phi}(x,\eta)} \hat{w}(x,\eta).$$

Following the strategy from (45), we can rewrite the components $j = 1, 2, 3$ of the field E_1 as

$$(E_1)_j(x,t) = \frac{1}{4\pi} \sum_{k=1}^{3} \frac{3x_j x_k - \delta_{jk}|x|^2}{|x|^5} *_x \int_0^t s \int (\partial_{x_k} V *_x \Phi *_v w)(x{-}vs, v, t{-}s)\, \mathrm{d}v\, \mathrm{d}s\,.$$

This yields the following a-priori estimate for the WP case on any time interval $(0,T)$ [ADM07]:

Lemma 5.2. *Let $w_0 \in L^2(\mathbb{R}^6)$ and $\left\| \int w_0(x - vt, v)\, \mathrm{d}v \right\|_{L_x^q(\mathbb{R}^3)} \leq t^{-\omega_q}$, $t \leq T$. Then, it holds for $0 < t \leq T$:*

$$\|E_1(t)\|_{L^2(\mathbb{R}^3)} \leq C \int_0^t s\, \mathrm{d}s \left\| \int (\nabla V *_x \Phi *_v w)(x - vs, v, t-s)\, \mathrm{d}v \right\|_{L_x^2(\mathbb{R}^3)}$$

$$\leq C \int_0^t \frac{s\, \mathrm{d}s}{s^{3/2}} \|(E_0 + E_1)(t-s)\|_{L^2(\mathbb{R}^3)} \|w_0\|_{L^2(\mathbb{R}^6)}\,.$$

Hence,

$$\|E_0(t)\|_{L^2(\mathbb{R}^3)} + \|E(t)\|_{L^2(\mathbb{R}^3)} \leq C t^{-\frac{3}{2q} + \frac{5}{4} - \omega_q}\,.$$

This lemma only provides an L^2-estimate on $E(t)$ for the WP system. For the VP system, however, one obtains a whole interval of L^p-estimates. This is due to the conservation of all norms $\|f(t)\|_{L^p(\mathbb{R}^6)}$, $1 \leq p \leq \infty$ in the VP case, while WP only conserves the L^2-norm in time.

A-Priori Estimate for Electric Field in WPFP

The WPFP system reads

$$w_t = Aw + \Theta[V]w\,,$$

$$-\Delta V(x,t) = n(x,t) := \int w(x,v,t)\, \mathrm{d}v\,,$$

with the abbreviation

$$Aw := -v\nabla_x w + 2\gamma \operatorname{div}_v(vw) + D_{pp}\Delta_v w + D_{qq}\Delta_x w + 2D_{pq} \operatorname{div}_x(\nabla_v w)\,.$$

Using the Green's function $G(x, v, x_0, v_0, t)$ from Theorem 5.2, the WPFP solution can be written in Duhamel form:

$$w(x,v,t) = \iint_{\mathbb{R}^6} G(x, x, x_0, v_0, t) w_0(x_0, v_0)\, \mathrm{d}x_0\, \mathrm{d}v_0$$

$$- \int_0^t \iint_{\mathbb{R}^6} G(x, x, x_0, v_0, s)(\Theta[V]w)(x_0, v_0, t-s)\, \mathrm{d}x_0\, \mathrm{d}v_0\, \mathrm{d}s\,.$$

Proceeding like for the WP case, we obtain an a-priori estimate on the field for the WPFP system [ADM07]:

Lemma 5.3. *Let $w_0 \in L^2$ and $\left\| \int w_0(x-vt,v)\,dv \right\|_{L_x^q} \leq t^{-\omega_q}$, $t \leq T$. Then, it holds for $0 < t \leq T$:*

$$\|E_0(t)\|_p + \|E(t)\|_p \leq Ct^{\frac{3}{2p}-\frac{3}{2q}+\frac{1}{2}-\omega_q}, \quad 2 \leq p < 6.$$

For $p > 2$, this result is obtained by the parabolic regularization of $G(t)$. Hence, this result goes beyond the WP-result in Lemma 5.2. For the iterative construction of the local-in-time solution we shall need in fact: $E \in L^1((0,T), L^3(\mathbb{R}^3))$. Hence, our procedure works for WPFP but not (yet) for WP.

Next we illustrate that the assumptions on w_0 in Lemmas 5.2, 5.3 can be obtained quite naturally. With the

Notation:

$$L_x^1(L_v^q) := L^1(\mathbb{R}_x^3; L^q(\mathbb{R}_v^3)), \quad 1 \leq q \leq \infty$$

we use the *Strichartz estimate* for the free transport equation [CP96]:

$$\left\| \int w_0(x-vt,v)\,dv \right\|_{L_x^q(\mathbb{R}^3)} \leq C\,|t|^{-3(1-\frac{1}{q})}\,\|w_0\|_{L_x^1(L_v^q)}, \quad t \in \mathbb{R}. \quad (46)$$

Hence, $w_0 \in L_x^1(L_v^q)$ yields exactly the needed assumption. When choosing some $q \in [1, \frac{6}{5})$, (46) and Lemma 5.3 imply $\|E(t)\|_3 \in L^1(0,T)$, as needed.

Once we have an a-priori estimate for the field E on any $(0,T]$, we can define the potential V using the Poisson equation $\mathrm{div}E = n$:

$$V = \frac{1}{4\pi|x|} * n = \frac{1}{4\pi|x|} * \mathrm{div}E = -\frac{1}{4\pi}\sum_j \frac{x_j}{|x|^3} * E_j, \quad (47)$$

and analogously for the "split quantities" V_0, V_1, n_0, n_1 (cf. (44) for the splitting of E). This implies the following a-priori estimate on the self-consistent potential V:

Lemma 5.4. *Let $w_0 \in L^2$ and $\left\| \int w_0(x-vt,v)\,dv \right\|_{L_x^q} \leq t^{-\omega_q}$, $t \leq T$. Then, it holds for $0 < t \leq T$:*

$$\|V(t)\|_p \leq Ct^{\frac{3}{2p}-\frac{3}{2q}+1-\omega_q}, \quad 6 \leq p \leq \infty,$$

and hence

$$\Theta[V(t)] \in \mathcal{B}(L^2(\mathbb{R}^6)), \quad t > 0.$$

Definition of the Nonlinear Term $\Theta[V[w]]w$

The *standard definition* of the macroscopic quantities: density, self-consistent potential, and self-consistent field are made pointwise in time:

$$n(x,t) = \int w(x,v,t)\,dv, \quad V(x,t) = \frac{1}{4\pi|x|} *_x n(x,t), \quad E(x,t) = -\nabla_x V(x,t).$$

However, this is unfeasible for $w \in L^2$.

Therefore, we shall use the following *alternative definition*, which is nonlocal in t, via an integral equation for E_1. This is based on the above derivation of the a-priori estimate for E_1, and it yields a nonlinear map $w \mapsto E_1[w]$. We present it here for the WP case (to keep the notation simple), but the WPFP case is analogous:

$$E_1[w]_j(x,t) \tag{48}$$

$$= -\frac{1}{4\pi}\sum_{k=1}^{3}\frac{3x_jx_k - \delta_{jk}|x|^2}{|x|^5} *_x \int_0^t s\int (E[w]_k *_x \Phi *_v w)(x-vs, v, t-s)\,dv\,ds.$$

Using (47) we then obtain (the inhomogeneous parts of) the potential and density:

$$V_1[w] = -\frac{1}{4\pi}\sum_j \frac{x_j}{|x|^3} * E_1[w]_j, \qquad n_1[w] = -\Delta V_1[w]. \tag{49}$$

Note that the a-priori estimate on $E_1[w]$ only depends on w_0 and $\|w(t)\|_2$ (cf. Lemmas 5.2, 5.3). Hence, $E_1[w] \in L^1((0,T); L^2(\mathbb{R}^3))$ is well-defined from (48) for all *Wigner trajectories* $w \in C([0,T]; L^2(\mathbb{R}^6))$. For this definition $w(t)$ need *not* be the self-consistent Wigner function solving WPFP.

While the two above definitions of $E = E_0 + E_1$ will generally be different, they coincide for the self-consistent solution w.

Iterative Construction of the WPFP Solution

For any time interval $[0,T]$ we now construct the WPFP solution by an iteration map M, defined on the following ball in a Banach space:

$$B_R = \left\{ w \in C([0,T]; L^2(\mathbb{R}^6)) \,\middle|\, \|w(t)\|_{L^2} \leq R \right\},$$

with $R := e^{3\gamma T}\|w_0\|_2$ reflecting the exponential growth of the L^2-norm: $\|w(t)\|_2 \leq e^{3\gamma t}\|w_0\|_2$.

In the (nonlinear) map M we first associate to w a potential $V[w]$ by (48), (49). Then, $\tilde{w}$ is the solution of the linear WFP (by Theorem 5.1):

$$\tilde{w}_t = A\tilde{w} + \Theta[V[w]]\tilde{w}, \quad t \in (0, T],$$
$$\tilde{w}(t=0) = w_0.$$

Summarizing we have

$$w \longmapsto \underbrace{V[w] = V_0 + V_1[w]}_{\in L^1((0,T);L^\infty(\mathbb{R}^3))} \longmapsto Mw := \tilde{w}.$$

Lemma 5.5. *Let $w_0 \in L^2$ and $\left\| \int w_0(x-vt, v)\, \mathrm{d}v \right\|_{L_x^q} \leq t^{-\omega_q}$, $t \leq T$. Then, M^n is a contraction on B_R for $n = n(T)$ large enough.*

This lemma implies that the WPFP system has a unique global-in-time solution:

Theorem 5.6 ([ADM07a]). *Let $w_0 \in L_{x,v}^2 \cap L_x^1(L_v^q)$ for some $1 \leq q < \frac{6}{5}$. Then, WPFP has a unique mild solution:*

$$w \in C([0,\infty); L^2(\mathbb{R}^6)) \cap C((0,\infty); C_B^\infty(\mathbb{R}^6)),$$

$$n, V, E \in C((0,\infty); C_B^\infty(\mathbb{R}^3)).$$

Proof. For the existence of the mild solution combine Lemma 5.5 and (46). The regularity of w, n, V, E follows a-posteriori by using the regularization of the WPFP–Green's function.   □

For other analytical approaches for the WPFP system we refer to [ALMS04, CLN04].

6 Open Quantum Systems in Lindblad Form

In a closed quantum system only interactions *within* that system exist. For an *open quantum system*, however, there actually exist two coupled systems: Firstly, our (finite) quantum system of interest, labeled S. And secondly, there exists a large reservoir (or "environment"), labeled R. If these two systems only interact with each other (but with no other external system), the coupled system $S + R$ forms a closed quantum system.

The idea of an open quantum system is now to model only the evolution of the small quantum system S: One includes the *uni-directional interaction* of R onto S, but one disregards the influence of S onto R. This is a reasonable approximation, if R is much larger than S. Henceforth we shall assume that R is in thermodynamic equilibrium (a "heat bath") that does not undergo a time evolution while interacting with S. As an example motivated by the discussion of Sect. 1.2, S might model an ensemble of electrons in a semiconductor crystal, while R models a phonon bath. For a general introduction to open quantum systems we refer to [AF01, AJP06, Dav76].

For the description of an open quantum system and its dynamics we make the following postulates:

- The open system S at time t is described by a density matrix $\hat{\varrho}(t) \in \tilde{\mathcal{J}}_1(\mathcal{H})$ with some Hilbert space $\mathcal{H}$.
- The dynamics of S is Markovian (i.e. there are no memory effects) and it is described by a conservative quantum dynamical semigroup (QDS) on $\mathcal{J}_1(\mathcal{H})$ (see below).

Definition 6.1. Let $G(t)$, $t \geq 0$ be a C_0-semigroup on some Hilbert space $\mathcal{H}$. Its dual map $G(t)^*$ on $\mathcal{B}(\mathcal{H})$ is called *completely positive*, if the tensor product

$$G(t)^* \otimes I_n \quad \text{defined on } \mathcal{B}(\mathcal{H}) \otimes \mathcal{B}(\mathcal{H}_n)$$

is positivity preserving $\forall n \in \mathbb{N}$ (i.e. it maps positive operators to positive operators). Here, $\mathcal{H}_n$ denotes any n-dimensional Hilbert space, and I_n the n-dimensional unit matrix.

Definition 6.2. Let $G(t)$, $t \geq 0$ be a C_0-semigroup on $\mathcal{J}_1(\mathcal{H})$, with some Hilbert space $\mathcal{H}$. $G(t)$ is called a *conservative quantum dynamical semigroup*, if

- $G(t)$ is trace preserving, i.e. $Tr\hat{\varrho}(t) = Tr\hat{\varrho}(0)$, $t \geq 0$.
- $G(t)^*$ is *completely positive*.

Remark 6.1. *Complete positivity is clearly stronger than the positivity preservation. Moreover it is invariant under forming tensor products. This is important, if one wants to recover the dynamics of the coupled system $S + R$ from the dynamics of the open quantum system S.*

6.1 Lindblad Form

Theorem 6.2 (Structure of QDS-generators, [Lin76]). *Let $\mathcal{L}$ be the bounded generator of a QDS on $\mathcal{J}_1(\mathcal{H})$. Then, $\mathcal{L}$ is in Lindblad form:*

$$\mathcal{L}(\hat{\varrho}) = -i[H, \hat{\varrho}] + \sum_k \left(L_k \hat{\varrho} L_k^* - \frac{1}{2} \left(L_k^* L_k \hat{\varrho} + \hat{\varrho} L_k^* L_k \right) \right) ,$$

with some Hamiltonian $H \in \mathcal{B}(\mathcal{H})$ and (maybe countably many) Lindblad operators $L_k \in \mathcal{B}(\mathcal{H})$.

Remark 6.3. *The boundedness of $\mathcal{L}$ is important for the proof of Theorem 6.2. The extension of this result to arbitrary unbounded generators $\mathcal{L}$ is unknown. Nevertheless, this Lindblad structure of a QDS-generator is a useful and generally accepted assumption – also for the unbounded case!*

The most general type of a Markovian master equation describing the evolution of an open quantum system is the following *von Neumann equation* (also called *Lindblad equation* or *master equation in Lindblad form*):

$$\frac{\mathrm{d}}{\mathrm{d}t}\hat{\varrho} = \mathcal{L}(\hat{\varrho})$$

$$= -\mathrm{i}[H, \hat{\varrho}] + \underbrace{\sum_k \left(L_k \hat{\varrho} L_k^* - \frac{1}{2}\left(L_k^* L_k \hat{\varrho} + \hat{\varrho} L_k^* L_k \right) \right)}_{=:A(\hat{\varrho})}. \tag{50}$$

Here, the operators H, L_k may be unbounded. As we shall see, the evolution of the density matrix $\hat{\varrho}(t)$ is now in general *non-unitary*.

We recall that at each t fixed, $\hat{\varrho}(t)$ admits a spectral decomposition

$$\hat{\varrho}(t) = \sum_{j \in \mathbb{N}} \lambda_j(t)\, |\psi_j(t)\rangle \langle \psi_j(t)|,$$

where $\psi_j(t)$ are its eigenvectors and $\lambda_j(t)$ the eigenvalues. Here, $|\psi_j\rangle\langle\psi_j|$ denotes the projection operator onto the j-th eigenspace of $\hat{\varrho}(t)$. In contrast to a closed quantum system (cf. Sect. 4.3), the eigenvalues $\lambda_j(t)$ are *not* constant in time in the open quantum system (50). Hence, (50) cannot be reformulated as a Schrödinger system (cp. to Sects. 3.3, 4.3).

Moreover the dynamics is not time reversible, since the generator $\mathcal{L}$ is dissipative:

Definition 6.3. The operator A is called *dissipative*

(a)   On a Hilbert $\mathcal{H}$, if

$$\Re\langle Ax, x\rangle \le 0 \qquad \forall x \in \mathcal{D}(A) \subset \mathcal{H};$$

(b)   On the Banach space $\tilde{\mathcal{J}}_1(\mathcal{H})$, if

$$\Re\, Tr\left(A(\hat{\varrho}) \mathrm{sgn}(\hat{\varrho}) \|\hat{\varrho}\|_1 \right) \le 0 \qquad \forall \hat{\varrho} \in \mathcal{D}(A) \subset \tilde{\mathcal{J}}_1(\mathcal{H}).$$

Here, $\mathrm{sgn}(\hat{\varrho})$ is defined by the functional calculus.

Lemma 6.1. *A solution $\hat{\varrho}(t)$ of (50) formally satisfies:*

(a)   *Trace preservation;*
(b)   *Positivity preservation, i.e. $\hat{\varrho}(0) \ge 0 \Rightarrow \hat{\varrho}(t) \ge 0 \quad \forall t \ge 0$;*
(c)   *The generator $\mathcal{L}$ is dissipative in $\tilde{\mathcal{J}}_1(\mathcal{H})$.*

Proof (purely formal, since $\mathcal{L}$ might not generate an evolution semigroup).

(a)   From (50) it follows $\frac{d}{dt}\mathrm{Tr}\,\hat{\varrho}(t) = 0$, from the cyclic property of the trace.

(b)   Let $\hat{\varrho} \geq 0$, and decompose $\mathcal{L}$ as:

$$\mathcal{L}_1(\hat{\varrho}) := \sum_k L_k \hat{\varrho} L_k^* \geq 0\,,$$

$$\mathcal{L}_2(\hat{\varrho}) := -(\tilde{H}\hat{\varrho} + \hat{\varrho}\tilde{H}^*)\,, \quad \tilde{H} := iH + \frac{1}{2}\sum_k L_k^* L_k\,.$$

Then, $\hat{\varrho} \geq 0$ and $\frac{\mathrm{d}}{\mathrm{d}t}\hat{\varrho} = \mathcal{L}_1(\hat{\varrho}) \geq 0$ imply $e^{t\mathcal{L}_1}\hat{\varrho} \geq 0$, $\forall t \geq 0$. Moreover, $\mathcal{L}_2$ generates the evolution semigroup

$$e^{t\mathcal{L}_2}\hat{\varrho} = e^{-t\tilde{H}}\hat{\varrho}e^{-t\tilde{H}^*}\,,$$

which preserves positivity. Now we conclude the positivity preservation of $e^{t\mathcal{L}}$ by Trotter's product formula:

$$e^{t\mathcal{L}} = \lim_{n\to\infty}\left[e^{\frac{t}{n}\mathcal{L}_1}e^{\frac{t}{n}\mathcal{L}_2}\right]^n\,.$$

(c)   In the proof we split the Hamiltonian part from the Lindblad part $\dot{A}(\hat{\varrho})$:

$$\mathrm{Tr}\left([H,\hat{\varrho}]\,\mathrm{sgn}(\hat{\varrho})\right) = \mathrm{Tr}\left(H|\hat{\varrho}| - |\hat{\varrho}|H\right) = 0\,;$$

$$\begin{aligned}
&\Re\,\mathrm{Tr}\,\left(L_k\hat{\varrho}L_k^*\,\mathrm{sgn}(\hat{\varrho}) - L_k|\hat{\varrho}|L_k^*\right)\\
&= \mathrm{Tr}\,\left(L_k^*\mathrm{sgn}(\hat{\varrho})L_k\hat{\varrho} - L_k^* L_k|\hat{\varrho}|\right)\\
&= \sum_j \langle L_k\psi_j\,,\,(\mathrm{sgn}(\hat{\varrho})L_k\hat{\varrho} - L_k|\hat{\varrho}|)\,\psi_j\rangle\\
&= \sum_j \lambda_j \underbrace{\langle L_k\psi_j\,,\,\mathrm{sgn}(\hat{\varrho})L_k\,\psi_j\rangle}_{\leq\|L_k\psi_j\|_{L^2}^2} - \sum_j |\lambda_j|\,\|L_k\psi_j\|_{L^2}^2 \leq 0\,,
\end{aligned}$$

with (ψ_j, λ_j) denoting the eigenpairs of $\hat{\varrho}$.   □

Now we give several examples of Lindblad operators L_k and compute its corresponding Wigner collision operators Q (cf. the discussion in Sect. 1.2).

Example 6.1. $L_1 = x \qquad \Rightarrow Qw = \frac{1}{2}\Delta_v w$.

Example 6.2. $L_1 = \nabla_x \qquad \Rightarrow Qw = \frac{1}{2}\Delta_x w$.

Example 6.3. For the quantum Fokker–Planck term (in the Wigner formalism):

$$Qw = 2\gamma\,\mathrm{div}_v(vw) + D_{pp}\Delta_v w + 2D_{pq}\,\mathrm{div}_x(\nabla_v w) + D_{qq}\Delta_x w\,,$$

there exist $2N$ Lindblad operators L_k, such that the (linear) WFP equation (38) can be rewritten as a von Neumann equation in Lindblad form [ALMS04]. All of these L_k are in the form

$$L_k = \alpha_k \cdot x + \beta_k \cdot \nabla_x \, ; \quad k = 1, \ldots, 2N \, .$$

Note that, for a given Lindblad part $A(\hat{\varrho})$ in (50) (or for a given Wigner collision operator Q, as in this example), its "decomposition" into the Lindblad operators L_k are typically *never* uniquely defined.

Example 6.4. For the Wigner-relaxation term $Qw = \frac{w_{st} - w}{\tau}$ with a constant relaxation time τ, we assume the normalization $\iint_{\mathbb{R}^{2N}} w_{st} \, dx dv = 1$. The density matrix $\hat{\varrho}_{st}$ corresponding to w_{st} hence satisfies $\operatorname{Tr} \hat{\varrho}_{st} = 1$, and it has an eigenfunction expansion of the form

$$\hat{\varrho}_{st} = \sum_{j \in \mathbb{N}} \mu_j |\phi_j\rangle \langle \phi_j| \, .$$

Here, $|\phi_j\rangle \langle \phi_j|$ denote the projection operators onto the eigenfunctions ϕ_j of $\hat{\varrho}_{st}$. Since $\operatorname{Tr} \hat{\varrho}_{st} = 1$, we rewrite the relaxation term for the density matrix as:

$$\frac{\hat{\varrho}_{st} - \hat{\varrho}}{\tau} = \frac{\hat{\varrho}_{st} \operatorname{Tr} \hat{\varrho} - \hat{\varrho} \operatorname{Tr} \hat{\varrho}_{st}}{\tau} \, .$$

And now it is in homogeneous w.r.t. $\hat{\varrho}$. With the (countably many) Lindblad operators

$$L_{jk} = \sqrt{\frac{\mu_k}{\tau}} |\phi_k\rangle \langle \phi_j|, \quad j, k \in \mathbb{N}$$

this relaxation term can be represented in Lindblad form [Arn96a].

We now turn to some remarks on techniques applied to investigate the large-time behavior of quantum dynamical semigroups. Consider a system that admits an equilibrium state. An interesting issue then is, if the system would converge to that equilibrium as $t \to \infty$. An example of such a question was studied in [FR98] for a class of Lindblad equations. But it is yet unexplored if those results apply to the evolution problems discussed here and in Sect. 6.2. Anyway, we shall now briefly review the (relative) quantum entropy as a possible tool for a future analysis of the large-time behavior of (50).

Quantum Entropy (Von Neumann Entropy)

The *quantum entropy* of a density matrix $\hat{\varrho}$ is defined as

$$S(\hat{\varrho}) := \operatorname{Tr}(\hat{\varrho} \ln \hat{\varrho}) \, .$$

For $0 \leq \hat{\varrho} \leq 1$ it satisfies $-\infty \leq S(\hat{\varrho}) \leq 0$.

Proposition 6.1 ([BN88]). *Let the Lindblad operators in (50) satisfy* $\sum_k L_k L_k^* \leq \sum_k L_k^* L_k$. *Then*

$$\frac{\mathrm{d}}{\mathrm{d}t} S(\hat{\varrho}(t)) \leq 0 \, , \qquad t \geq 0 \, .$$

Proof.

$$\frac{\mathrm{d}}{\mathrm{d}t} S(\hat{\varrho}(t)) = \mathrm{Tr}\,(\hat{\varrho}_t \ln \hat{\varrho})$$

$$= \mathrm{Tr}\,\Big(\sum_k \big[L_k \hat{\varrho} L_k^* - L_k^* L_k \hat{\varrho} \big] \ln \hat{\varrho} \Big)$$

$$= \sum_{j,k} \langle \psi_j,\, L_k \ \hat{\varrho} L_k^* \psi_j \rangle \ln \lambda_j - \langle \psi_j,\, L_k^* \ L_k \psi_j \rangle \lambda_j \ln \lambda_j$$

(Insert here $\sum_l |\psi_l\rangle\langle\psi_l| = I$, the identity operator;

the sum is understood in the strong operator sense.)

$$= \sum_{j,k,l} \langle \psi_j,\, L_k \psi_l \rangle \langle \psi_l,\, L_k^* \psi_j \rangle \lambda_l \ln \lambda_j$$

$$- \langle \psi_l,\, L_k^* \psi_j \rangle \langle \psi_j,\, L_k \psi_l \rangle \lambda_l \ln \lambda_l \qquad (j,\, l \text{ interchanged})$$

$$\leq \sum_{j,k,l} |\langle \psi_j,\, L_k \psi_l \rangle|^2 \,(\lambda_j - \lambda_l) \qquad \Big(\text{use } \ln \frac{\lambda_j}{\lambda_l} \leq \frac{\lambda_j}{\lambda_l} - 1\Big)$$

$$= \mathrm{Tr}\,\Big(\sum_k \big[L_k L_k^* - L_k^* L_k \big] \hat{\varrho} \Big)$$

$$\leq 0\,,$$

with (ψ_j, λ_j) denoting the eigenpairs of $\hat{\varrho}$. $\qquad\qquad\qquad\qquad\square$

Remark 6.4. *The above result is sharp in the following sense. Assume that the condition $\sum_k L_k L_k^* \leq \sum_k L_k^* L_k$ does not hold. Then there exists an initial condition $\hat{\varrho}_0$, such that the corresponding trajectory $\hat{\varrho}(t)$ satisfies $\frac{\mathrm{d}}{\mathrm{d}t} S(\hat{\varrho}(t = 0)) > 0$.*

Applying this result to the Lindblad formulation of the WFP equation yields: $\sum_k L_k^* L_k - L_k^* L_k = -2N\gamma$. Hence, Proposition 6.1 can be applied only in the frictionless case ($\gamma = 0$) [ALMS04].

Relative Quantum Entropy

The *relative quantum entropy* is defined as

$$S(\hat{\varrho}|\hat{\sigma}) := \mathrm{Tr}\,(\hat{\varrho}(\ln \hat{\varrho} - \ln \hat{\sigma})) \qquad \text{for } \hat{\varrho},\, \hat{\sigma} \in \tilde{\mathcal{J}}_1\,;\quad \hat{\varrho},\, \hat{\sigma} \geq 0\,,$$

and it satisfies:

Proposition 6.2 ([AF01]). *Let dim $\mathcal{H} < \infty$, then:*

(a)

$$S(\hat{\varrho}|\hat{\sigma}) \geq 0\,,$$
$$S(\hat{\varrho}|\hat{\sigma}) = 0 \qquad \text{iff } \hat{\varrho} = \hat{\sigma}\,.$$

(b) Let $G(t)$, $t \geq 0$ be a conservative QDS. Then:

$$S(G(t)\hat{\varrho} \,|\, G(t)\hat{\sigma}) \leq S(\hat{\varrho}|\hat{\sigma}), \quad t \geq 0.$$

(c)

$$\frac{1}{2}\|\hat{\varrho} - \hat{\sigma}\|_1^2 \leq S(\hat{\varrho}|\hat{\sigma}). \tag{51}$$

Remark 6.5. *(b) + (c) imply the $\mathcal{J}_1$-stability of a steady state. For further results on the quantum entropy (production) in finite dimensional systems cf. [Spo76, Spo78].*

As a comparison we briefly recall the notion of *classical relative entropy*, which is frequently used for analyzing the large-time behavior of linear and non-linear parabolic equations and of (classical) kinetic equations via the so-called *entropy method* (cf. [AMTU01, Arn02]). For two probability densities $f, g \in L^1\left(\mathbb{R}^N\right)$ the *relative entropy* is defined as

$$e(f|g) := \int_{\mathbb{R}^N} \frac{f}{g} \ln\left(\frac{f}{g}\right) \, \mathrm{d}g \geq 0. \tag{52}$$

It satisfies the *Csiszár–Kullback* inequality:

$$\frac{1}{2}\|f - g\|_{L^1}^2 \leq e(f|g),$$

which is formally equivalent to inequality (51).

6.2 Quantum Fokker–Planck Equation

In this subsection we reconsider the Wigner–Poisson–Fokker–Planck system in $\mathbb{R}^6$:

$$\begin{cases} w_t + v \cdot \nabla_x w - \Theta[V]w = Qw, \\ Qw = \underbrace{D_{pp}\Delta_v w}_{\text{class. diffusion}} + \underbrace{2\gamma \, \mathrm{div}_v(vw)}_{\text{friction}} + \underbrace{D_{qq}\Delta_x w + 2D_{pq} \, \mathrm{div}_x(\nabla_v w)}_{\text{quantum diffusion}} \\ -\Delta V(x,t) = n(x,t) := \int w(x,v,t) \, \mathrm{d}v, \\ w(x,v,t=0) = w_0(x,v). \end{cases} \tag{53}$$

But now we shall analyze its well-posedness on the level of density matrices.

As already mentioned in Remark 4.1, the time evolution of a density matrix can either be consider in terms of a PDE for the density matrix function $\varrho(x,y)$ or as an abstract evolution problem (in the space of trace class operators) for the density matrix operator $\hat{\varrho}$:

We have $\hat{\varrho} \in \mathcal{J}_2(L^2(\mathbb{R}^N)) \Leftrightarrow \varrho(x,y) \in L^2(\mathbb{R}^{2N})$. Hence, we can write a PDE for the time evolution of the function $\varrho(x,y,t)$:

$$\varrho_t = -\mathrm{i}(H_x - H_y)\varrho - \gamma(x-y)\cdot(\nabla_x - \nabla_y)\varrho \qquad (54)$$
$$+ \left[D_{qq}|\nabla_x + \nabla_y|^2 - D_{pp}|x-y|^2 + 2\mathrm{i}D_{pq}(x-y)\cdot(\nabla_x + \nabla_y) \right]\varrho,$$

with the Hamiltonian $H_x = -\frac{1}{2}\Delta_x + V(x,t)$ (and analogously for H_y).

However, the L^2-setting is not enough since we also consider $\hat{\varrho} \in \mathcal{J}_1$. In this case, there exists no "nice" space to characterize the corresponding kernel $\varrho(x,y)$. And hence, we cannot use the PDE (54) for the time evolution of the function $\varrho(x,y)$. Instead we have to employ the following evolution equation of the time-dependent operator $\hat{\varrho}(t)$:

$$\frac{\mathrm{d}}{\mathrm{d}t}\hat{\varrho} = -\mathrm{i}[\tilde{H}, \hat{\varrho}]$$
$$+ \underbrace{\sum_k L_k \hat{\varrho} L_k^* - \frac{1}{2}\left(L_k^* L_k \hat{\varrho} + \hat{\varrho}L_k^* L_k\right)}_{=:A(\hat{\varrho})}, \qquad (55)$$

with the "adjusted" Hamiltonian

$$\tilde{H} = -\frac{1}{2}\Delta_x + V(x,t) - \mathrm{i}\frac{\gamma}{2}\{x, \nabla\},$$

and $2N$ Lindblad operators of the form

$$L_k = \alpha_k \cdot x + \beta_k \cdot \nabla_x, \quad k = 1, \ldots, 6 = 2N.$$

Equation (55) is equivalent to (53) with the appropriate L_k's (see [ALMS04] for details).

Since we shall eventually discuss here the self-consistent quantum Fokker–Planck–Poisson (QFPP) problem (54) coupled to the Poisson equation

$$-\Delta V(x,t) = n(x,t),$$

we need to use the operator framework, in order to properly define the particle density n and the potential V (cf. Remark 4.1).

Our subsequent analysis parallels the strategy of Theorem 3.1 (i.e. the H^1-analysis of SP in 3D), and we split it in three steps.

(a) Linear case:
   As before, we first consider the linear QFP equation (54) with V given.

For a general Lindblad equation

$$\begin{cases} \frac{\mathrm{d}}{\mathrm{d}t}\hat{\varrho} = \mathcal{L}(\hat{\varrho}) := -\mathrm{i}[\tilde{H}, \hat{\varrho}] + A(\hat{\varrho}), & t \geq 0 \\ \hat{\varrho}(t=0) = \hat{\varrho}_0 \end{cases} \qquad (56)$$

with the dissipative/Lindblad terms $A(\hat\varrho)$ introduced in (55), Davies [Dav77] constructed a linear C_0-evolution semigroup on $\mathcal{J}_1$. However, without special assumptions on the Hamiltonian $\tilde H$ and the Lindblad terms $A(\hat\varrho)$, this so-called "minimal solution" can exhibit the following problems, which are closely related to each other:

- The semigroup, and hence the solution to (56) may be *not* unique.
- The domain $\mathcal{D}(\mathcal{L})$ might be "too small".
- The semigroup may be *not* conservative, i.e. $\mathrm{Tr}\,(\hat\varrho(t)) < \mathrm{Tr}\,\hat\varrho_0$ is possible for some $t > 0$.

Indeed, there exist examples where these problems occur.

One possibility to check that they do not happen for a specific problem at hand, is to prove that $\mathcal{D}(\overline{\mathcal{L}})$ is "big enough". For our QFP equation (54) this can be accomplished with the following lemma. It is an extension of Lemma 5.1 to mappings on operator spaces.

Lemma 6.2 ([AS04]). *Let $\mathcal{L}$ be quadratic in x and ∇_x. Then $\overline{\mathcal{L}|_{\mathcal{D}_\infty}}$ is the "maximum extension" in $\mathcal{J}_1$, in the sense that*

$$\mathcal{D}\left(\overline{\mathcal{L}|_{\mathcal{D}_\infty}}\right) = \mathcal{D}(\mathcal{L}_{\max}) := \{\hat\varrho \in \mathcal{J}_1 \,|\, \mathcal{L}(\hat\varrho) \in \mathcal{J}_1\}.$$

Here, $\mathcal{D}_\infty$ is a dense subset of $\mathcal{J}_1$-operators with C_0^∞-kernels ϱ (for the precise definition cf. [AS04]). It plays the same role as the C_0^∞-functions in Lemma 5.1.

Proof. For $\hat\varrho \in \mathcal{D}(\mathcal{L}_{\max})$ we need to construct a sequence $\{\hat\sigma_n\}_{n\in\mathbb{N}} \subset \mathcal{D}_\infty$, such that

$$\hat\sigma_n \xrightarrow{n\to\infty} \hat\varrho \quad \text{in the graph norm } \|.\|_{\mathcal{L}}.$$

Their kernels can be obtained by

$$\sigma_n(x,y) := \underbrace{\chi_n(x)}_{C_0^\infty\text{-cutoff}} \left[\varphi_n(x) *_x \varrho(x,y) *_y \underbrace{\varphi_n(y)}_{C_0^\infty\text{-mollifier}}\right]\chi_n(y).$$

$\square$

This yields the following result for the linear QPF equation (54) on $\tilde{\mathcal{J}}_1(L^2(\mathbb{R}^N))$:

Theorem 6.6. *For the QFP equation, the C_0-semigroup $e^{\mathcal{L}t}$ constructed by Davies in [Dav77] is unique and trace preserving.*

As a next step we consider the linear QFP equation on the energy space $\mathcal{E}$, defined in Sect. 4.2. Indeed, $e^{\mathcal{L}t}$ is a C_0-semigroup also in $\mathcal{E}$. Since Davies' semigroup construction only holds in $\mathcal{J}_1$, one needs to prove explicitly the strong continuity of $e^{\mathcal{L}t}$ w.r.t. t in $\mathcal{E}$ (obtained by Grümm's Theorem, see [Sim79]).

(b) Nonlinear case – local solution:
Now we turn to the nonlinear QFPP equation in 3D. Like in Theorem 3.1 we prove that the nonlinear Hartree-term $[V[\hat{\varrho}], \hat{\varrho}]$ (in (55)) is a local Lipschitz map in $\mathcal{E}$ – but it is not in $\mathcal{J}_1$ (this was the reason for making the analysis in $\mathcal{E}$). To this end one uses the following two estimates on the particle density: (23) and

$$\|n[\hat{\varrho}]\|_{L^3(\mathbb{R}^3)} \leq C\|\hat{\varrho}\|_{\mathcal{E}}$$

from Lemma 4.1.
Proposition 3.1 then implies that the QFPP equation has a local-in-t solution $\hat{\varrho} \in C([0, t_{max}); \mathcal{E})$.

(c) Nonlinear case – global solution:
To show that the solution exists for all time, we use the following a-priori estimates on the total mass:

$$\text{Tr}\,\hat{\varrho}(t) = \text{const. in } t, \qquad \hat{\varrho}(t) \geq 0 \quad \text{(see Lemma 6.1)},$$

and on the total energy, defined as

$$E_{\text{tot}}(\hat{\varrho}) := E_{\text{kin}}(\hat{\varrho}) + \frac{1}{2}\|\nabla_x V[\hat{\varrho}]\|_{L^2}^2 .$$

The latter satisfies (cf. [ALMS04])

$$\frac{\mathrm{d}}{\mathrm{d}t} E_{\text{tot}} = 3\,D_{pp}\,\text{Tr}\,\hat{\varrho}_0 - 4\gamma\,E_{\text{kin}}(t) - D_{qq}\,\|n(t)\|_{L^2}^2 .$$

Finally we obtain:

Theorem 6.7 ([AS04]). *Let $\hat{\varrho}_0 \in \mathcal{E}$ and $\hat{\varrho}_0 \geq 0$. Then, there exists a global-in-time, positive, trace preserving, finite energy solution $\hat{\varrho} \in C([0, \infty); \mathcal{E})$ of the QFPP system (55).*

Let us briefly compare this theorem to the Wigner function approach of Sect. 5.4. The advantage of the density matrix approach for the QFPP system is that we work in the physical energy space $\mathcal{E} := \{\hat{\varrho} \in \mathcal{J}_1 | E_{\text{kin}}(\hat{\varrho}) < \infty\}$. Moreover, we only use physically important and meaningful a-priori estimates (for the total mass, kinetic energy, total energy). Hence, these estimates are much easier to derive than the rather technical estimate on $\|E(t)\|_{L^2}$ in Sect. 5.4.

However, if we would want to generalize the result to bounded domain problems, the density matrix formalism would not be appropriate, and we would need to pass to the Wigner formalism (53).

We finally remark, that a result similar to Theorem 6.7 was obtained in [Arn96] for the relaxation-time Wigner–Poisson system and its density matrix analogue.

7 Wigner Boundary Value Problems

In Sect. 1.2 we presented a relaxation-time Wigner–Poisson model for a resonant tunneling diode. This was an IBVP on the phase space slab $(x, v) \in (0, L) \times \mathbb{R}$ with inflow boundary conditions (BCs). In this section we shall discuss the linear Wigner boundary value problem (BVP), first in the stationary case in one spatial dimension and then the large-time behavior of the transient problem.

7.1 1D Stationary Boundary Value Problem

Here we shall discuss the existence and uniqueness of a solution $w(x, v)$ to the BVP for the linear, stationary Wigner equation in 1D with prescribed inflow BCs f^+, f^-:

$$\begin{cases} vw_x - \Theta[V]w &= 0, & 0 < x < L, \quad v \in \mathbb{R}, \\ w(0, v) &= f^+(v), & v > 0, \\ w(L, v) &= f^-(v), & v < 0. \end{cases} \tag{57}$$

The recall the definition of the operator $\Theta[V]$, which is non-local in v:

$$\Theta[V] = iV(x + \frac{\partial_v}{2i}) - iV(x - \frac{\partial_v}{2i}).$$

For a smooth and decaying potential $V(x)$ it can be rewritten as

$$\Theta[V]w(x, v) = \alpha(x, v) *_v w(x, v), \tag{58}$$

with

$$\alpha(x, v) = \sqrt{\frac{8}{\pi}} \Im[e^{2ixv} (\mathcal{F}V)(2v)].$$

In order for $\Theta[V]$ to be defined, the potential V has to be given on all of $\mathbb{R}$, although (57) is only posed on the spatial interval $(0, L)$. Again we assume that all constants are normalized: $e = m_* = \hbar = 1$.

As an introduction, we first consider the classical analogue of (57), i.e. the stationary BVP for the Liouville equation:

$$\begin{cases} vf_x - V_x f_x &= 0, & 0 < x < L, \quad v \in \mathbb{R}, \\ f(0, v) &= f^+(v), & v > 0, \\ f(L, v) &= f^-(v), & v < 0. \end{cases} \tag{59}$$

As soon as the potential V has a local minimum inside the interval $(0, L)$, (59) has closed characteristic curves (corresponding to particle trajectories). On such closed characteristics, the solution $f(x, v)$ cannot be "controlled" by

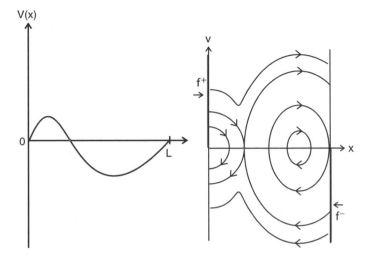

Fig. 10 The given potential $V(x)$ (*left*) gives rise to the phase space trajectories in *right* picture. Note the closed trajectories due to the potential well. There, the solution $f(x, v)$ cannot be "controlled" by the prescribed boundary values f^+, f^-

the boundary values f^+, f^- (cf. Fig. 10). Hence, the solution to (59) is in general non-unique!

For the Wigner equation the picture is very different: Due to the non-locality of $\Theta[V]$), the stationary Wigner equation (57) has *no characteristics* (except for a quadratic potential). Hence, compactly supported steady states cannot exist here (in contrast to the Liouville equation). As we shall see, the stationary solution $w(x, v)$ can be "controlled" by the boundary data.

"Standard" Transport Problems

First we compare (57) to the structure of conventional stationary transport problems, appearing in neutron transport, e.g. [GvMP87]. Those problems are typically of the form

$$v f_x - A f = 0, \qquad 0 < x < L,$$

with a positive Fredholm operator A. The positivity of A reflects the fact, that it models the particle interaction with a diffusive medium.

In the Wigner equation $A(x) := \Theta[V]$ is a skew-symmetric operator on $L^2(\mathbb{R}_v)$. Hence, it does not fall into the above class of standard transport problems.

Discrete Velocity Wigner Model

Since the continuous velocity problem (57) for $v w_x - \Theta[V]w = 0$ is not fully understood yet, we shall discuss here its semi-discretization in velocity. We make the approximation $w_j(x) \approx w(x, v_j)$ with the discrete velocities

$v_1 \geq \cdots \geq v_K > 0 > v_{K+1} \geq \cdots \geq v_N$. Hence, we consider the following BVP for the vector $w(x) \in \mathbb{R}^N$, $x \in [0, L]$:

$$
\begin{cases}
Tw_x - A(x)w = 0, \quad 0 < x < L, \\
w^+(0) = (w_1, \ldots, w_K)^T(x = 0) = f^+, \\
w^-(0) = (w_{K+1}, \ldots, w_N)^T(x = L) = f^-.
\end{cases}
\tag{60}
$$

Here, T is the diagonal matrix $T = diag(v_1, \ldots, v_N)$. In analogy to $\Theta[V]$, the space-dependent matrix $A(x) \in \mathbb{R}^{N \times N}$ is skew-symmetric $\forall x$. Moreover, $A(x)$ is typically a Toeplitz matrix (in analogy to (58)), but we shall not use this fact below.

Example 7.1. We briefly illustrate, why we exclude here 0 as a discrete velocity. Consider the simple case $N = 3$ with $v_1 = 1$, $v_2 = 0$, $v_3 = -1$. This yields the *algebro-differential equation* (ADE)

$$
\begin{cases}
\frac{d}{dx}w_1 - \alpha w_2 - \beta w_3 &=& 0, \\
\alpha w_1 - \gamma w_3 &=& 0, \\
-\frac{d}{dx}w_3 + \beta w_1 + \gamma w_2 &=& 0,
\end{cases}
\tag{61}
$$

with the (natural) BCs

$$
w_1(0) = f^+, \qquad w_3(L) = f^-.
\tag{62}
$$

Since w_2 does not appear in the algebraic constraint (second line of (61)), the index of this ADE is 2. Hence, the BCs (62) make the above system overdetermined and there is either no or infinitely many solutions. This problem of including 0 as a discrete velocity was already observed in numerical discretizations of the stationary Wigner equation [Fre90].

For the analysis of (60) we introduce the transformation: $\tilde{w} = \sqrt{|T|^{-1}}w \in \mathbb{R}^N$. Hence, $Tw_x - A(x)w = 0$ is transformed to

$$
\tilde{w}_x - \underbrace{\sqrt{|T|}^{-1} T^{-1} A(x) \sqrt{|T|}}_{:=B(x)} \tilde{w} = 0, \quad x \in (0, L).
$$

We make a block decomposition of the matrix $B(x)$ according to the positive and negative velocities v_j:

$$
B(x) = \underbrace{\begin{pmatrix} B^{++} & B^{+-} \\ B^{-+} & B^{--} \end{pmatrix}}_{K \quad N-K} \begin{matrix} \}K \\ \}N-K \end{matrix} \quad \begin{matrix} \ldots \text{ pos. velocities } v_j \\ \ldots \text{ neg. velocities } v_j \end{matrix}
$$

These blocks have the following symmetry which is the key structural property for the following theorem:

$$B^{++} = -(B^{++})^T, \quad B^{--} = -(B^{--})^T, \quad B^{+-} = (B^{-+})^T.$$

Theorem 7.1 ([ALZ00]). *Let 0 be none of the discrete velocities, and let $A \in L^1(0, L; \mathbb{R}^{N \times N})$. Then, the BVP (60) is well-posed.*

Remark 7.2. *In [Fre90] there is numerical evidence of this result.*

Proof. In the discrete velocity BVP $\tilde{w}_x - B(x)\tilde{w} = 0$, the *inflow boundary data* $\tilde{g}^+$, $\tilde{g}^-$ are prescribed. The *outflow boundary data* $\tilde{h}^+$, $\tilde{h}^-$ are obtained from the solution of the problem:

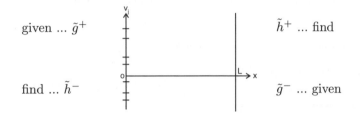

given ... $\tilde{g}^+$ $\tilde{h}^+$ ... find

find ... $\tilde{h}^-$ $\tilde{g}^-$ ... given

Now we convert the BVP into a forward a backward initial value problem (IVP) with propagator $U(x_1, x_2)$:

$$\binom{\tilde{h}^+}{0} = P^+ \underbrace{U(0, L)}_{\text{propagator}} \binom{\tilde{g}^+}{\tilde{h}^-}, \quad P^+ = \begin{pmatrix} I & 0 \\ 0 & 0 \end{pmatrix}, \tag{63}$$

$$\binom{0}{\tilde{h}^-} = P^- \underbrace{U(L, 0)}_{=U(0,L)^{-1}} \binom{\tilde{h}^+}{\tilde{g}^-}, \quad P^- = \begin{pmatrix} 0 & 0 \\ 0 & I \end{pmatrix}. \tag{64}$$

P^+ and P^- are projection matrices with the same block structure as $B(x)$. Eliminating $\tilde{h}^-$ from these systems yields an equation for the unknown $\tilde{h}^+$:

$$(\underbrace{I - P^+ U(0, L) P^- U(L, 0) P^+}_{=:K \geq 0}) \binom{\tilde{h}^+}{0}$$

$$= \underbrace{P^+ U(0, L) \left[\binom{\tilde{g}^+}{0} + P^- U(L, 0) \binom{0}{\tilde{g}^-} \right]}_{\text{given}}$$

The matrix K satisfies $K \geq 0$ since $A(x)$ is skew-symmetric. Hence, we can uniquely solve for $\tilde{h}^+$, and the solution $\tilde{w}(x)$ is obtained from a backward problem like (64). □

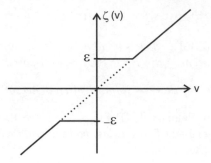

Fig. 11 Velocity cut-off function $\zeta(v)$ for the boundary value problem (65)

A similar strategy gives the following extensions:

- Countably many velocities $v_j \neq 0$.
- Continuous velocity problem ($v \in \mathbb{R}$) with a cut-off for small velocities in the Wigner equation (57) (cf. Fig. 11):

$$
\begin{cases}
\zeta(v)w_x - \Theta[V]w = 0, & |\zeta| \geq \varepsilon > 0 \\
w(0,v) = f^+; & w(L,v) = f^-
\end{cases}
\tag{65}
$$

Theorem 7.3. *Let $V \in L^\infty(\mathbb{R}_x)$, and let the inflow data $f^\pm(v) \in \mathcal{H} := L^2(\mathbb{R}^\pm, |v|dv)$. Then, the BVP (65) has a unique solution $w \in W^{1,1}((0,L); \mathcal{H})$.*

Proof (idea).

- $V \in L^\infty$ implies $A(x) := \Theta[V] \in L^1((0,L); \mathcal{B}(L^2(\mathbb{R}_v)))$.
- $\frac{1}{\zeta(v)}\Theta[V]$ is a bounded generator for the forward/backward IVP.

$\square$

Remark 7.4. *For the continuous Liouville equation the propagator $U(0,L)$ cannot exist. But Theorem 7.3 applies also to the v-discretized Liouville equation.*

7.2 Exponential Convergence to Steady State

We consider now the time dependent analogue of (60), i.e. the IBVP for the Wigner equation with discrete velocities. We make the approximation $w_j(x,t) \approx w(x, v_j, t)$, $j \in J \subset \mathbb{Z}$. The discrete velocity Wigner function $w(x,t) \in \mathbb{R}^N$ satisfies the following linear hyperbolic system:

$$\begin{cases} w_t + Tw_x - A(x)w = 0, & 0 < x < L, \quad t > 0 \\ w^+(0,t) = f^+; \ w^-(L,t) = f^-, \\ w(x,0) = w^I. \end{cases} \tag{66}$$

As before, we assume $T = diag(v_j)_{j \in J}$ and $0 \notin \{v_j\}_{j \in J}$. The matrix $A(x)$ is skew-symmetric for all x and (usually) some v-discretization of the operator $\Theta[V]$.

Theorem 7.5 ([ACT02]). *Let* $w^I \in L^2(0,L) \times \ell^2(J)$, $f^\pm \in \mathcal{H} := \ell^2(J; |v_j|)$, *and* $A \in L^\infty((0,L); \mathcal{B}(\ell^2(J)))$. *Then,*

$$w(.,t) \overset{t \to \infty}{\longrightarrow} w_\infty \quad \text{in } L^2(0,L) \times \ell^2(J) \quad \text{with some exponential rate } \varepsilon > 0.$$

w_∞ *is the unique steady state of* (66), *provided by the results of Sect. 7.1.*

Remark 7.6. *Since $A(x)$ is skew-symmetric, the system (66) does not show any damping. The exponential decay towards the steady state is only due to outflow through the boundary.*

Proof. Transforming to homogeneous inflow BCs via $\tilde{w} := w - w_\infty$ yields:

$$\begin{cases} \tilde{w}_t + T\tilde{w}_x - A(x)\tilde{w} = 0 \\ \tilde{w}^+(0,t) = 0; \quad \tilde{w}^-(L,t) = 0 \\ \tilde{w}(x,z) = w^I - w_\infty \end{cases} \tag{67}$$

- A naive approach in $L^2(0,L)$ would only yield $\frac{d}{dt}\|\tilde{w}\|_{L^2(0,L)} \leq 0$, but no decay estimate.
- Hence, we shall prove the decay in a weighted L^2-norm. To this end we construct a multiplier matrix $\Phi(x) = diag(\varphi_j(x))_{j \in J}$ with:

$$0 < c \leq \varphi_j(x) \leq C \qquad \forall j \in J, \ x \in [0,L]$$

- We take the $L^2(0,L)$-inner product of (67) with $\Phi\tilde{w}$. With the notation

$$\|\tilde{w}\|_\Phi^2 := \int_0^L \tilde{w}^\top \cdot \Phi \cdot \tilde{w} \, dx$$

we have

$$\frac{1}{2}\frac{d}{dt}\|\tilde{w}\|_\Phi^2 - \frac{1}{2}\langle \tilde{w}, T\Phi_x\tilde{w}\rangle_{L^2} - \langle A\tilde{w}, \Phi\tilde{w}\rangle_{L^2} \leq 0.$$

The last inequality is due to outflow at the boundary.
- Now we have to find a matrix function $\Phi(x)$ such that the following inequality holds for some (small) $\varepsilon > 0$:

$$\tfrac{1}{2}\big(f, T\Phi_x(x)f\big) + \big(A(x)f, \Phi(x)f\big) \leq -\varepsilon\big(f, \Phi(x)f\big)$$

$$\forall x \in [0,L], \ \forall f \in \mathbb{R}^{|J|}$$

- This holds true, if we have componentwise (i.e. $\forall j \in J$):

$$\frac{1}{2}v_j\frac{\partial}{\partial_x}\varphi_j + \frac{1}{2}\sum_{i\neq j}|a_{ij}(x)||\varphi_i - \varphi_j| \leq -\varepsilon\varphi_j \qquad \text{on } [0,L].$$

- After constructing these functions φ_j, we obtain:

$$\frac{d}{dt}\|\tilde{w}\|_\Phi \leq -\varepsilon\|\tilde{w}\|_\Phi := -\varepsilon\left\|\sqrt{\varphi_j}\tilde{w}_j(.,t)\right\|_{L^2(0,L)}.$$

$\square$

Open Problems

- Removing the velocity cut-off ($\zeta(v) \longrightarrow v$) makes $\frac{1}{\zeta(v)}\Theta[V]$ an unbounded operator in the forward/backward IVPs involved in (63), (64). In this case it is not yet clear how to construct the propagator $U(0,L)$.

- To find conditions on $f^\pm$ such that $\hat{\varrho}[w] \geq 0$ or at least $n(x) = \int_\mathbb{R} w(x,v)\,dv \geq 0$ on $[0,L]$ seems to be very difficult. Actually, the Wigner framework is probably not appropriate to answer this question.

- Non-linear (self-consistent) extension of (57):

$$\begin{cases} vw_x - \Theta[V]w &= 0, & 0 < x < L, \quad v \in \mathbb{R}, \\ V_{xx} &= D(x) - \int_\mathbb{R} w(x,v)\,dv, & 0 < x < L, \\ w(0,v) &= f^+(v), & v > 0, \\ w(L,v) &= f^-(v), & v < 0. \end{cases}$$

would need estimates on the quantum-repulsive/attractive potential. Here, some strategies from the stationary quantum-classical coupling in [Ben98] might be extendable.

Acknowledgements The author was partly supported be the DFG-project no. AR 277/3-3 and the Wissenschaftskolleg *Differentialgleichungen* of the FWF. He thanks Roberta Bosi for many suggestions to improve the manuscript.

References

[AABES07] Antoine, X., Arnold, A., Besse, C., Ehrhardt, M., and Schädle A.: A Review of Transparent and Artificial Boundary Conditions Techniques for Linear and Nonlinear Schrödinger Equations. Commun. Compound. Phys. 4, Nr. 4, 729–796 (2008)

[ACD02] Arnold, A., Carrillo, J.A., Dhamo, E.: The periodic Wigner-Poisson-Fokker-Planck system. JMAA, **275**, 263–276 (2002)

[ACT02]   Arnold, A., Carrillo, J.A., Tidriri, M.D.: Large-time behavior of discrete ki-
            netic equations with non-symmetric interactions. Math. Models Meth. Appl.
            Sc., **12**, no.11, 1555–1564 (2002)
[ADM07]   Arnold, A., Dhamo, E., Manzini, C.: The Wigner-Poisson-Fokker-Planck sys-
            tem: global-in-time solutions and dispersive effects. Annales de l'IHP (C) -
            Analyse non linéaire **24**, no.4, 645–676 (2007)
[ADM07a]   Arnold, A., Dhamo, E., Manzini, C.: Dispersive effects in quantum kinetic
            equations. Indiana Univ. Math. J., **56**, no.3, 1299–1331 (2007)
[AF01]   Alicki, R., Fannes, M.: Quantum Dynamical Systems. Oxford University Press
            (2001)
[AGGS07]   Arnold, A., Gamba, I.M., Gualdani, M.P., Sparber, C.: The Wigner-Fokker-
            Planck equation: Stationary states and large time behavior. Submitted (2007)
[AJ06]   Arnold, A., Jüngel, A.: Multi-scale modeling of quantum semiconductor de-
            vices., p. 331–363 in: Analysis, Modeling and Simulation of Multiscale Prob-
            lems, A. Mielke (Ed.), Springer, Berlin-Heidelberg (2006)
[AJP06]   Attal, S., Joye, A., Pillet, C.A. (Eds.): *Open Quantum Systems I-III* - Lecture
            Notes in Mathematics, 1880–1882, Springer, Berlin-Heidelberg (2006)
[ALMS04]   Arnold, A., López, J.L., Markowich, P., Soler, J.: An analysis of quan-
            tum Fokker-Planck models: A Wigner function approach. Revista Matem.
            Iberoam., **20**, no.3, 771–814 (2004)
[ALZ00]   Arnold, A., Lange, H., Zweifel, P.F.: A discrete-velocity, stationary Wigner
            equation. J. Math. Phys., **41**, no.11, 7167–7180 (2000)
[AMTU01]   Arnold, A., Markowich, P., Toscani, G., Unterreiter, A.: Comm. PDE, **26**,
            no.1–2, 43–100 (2001)
[AN91]   Arnold, A., Nier, F.: The two-dimensional Wigner-Poisson problem for an
            electron gas in the charge neutral case. Math. Meth. Appl. Sc., **14**, 595–613
            (1991)
[Arn96]   Arnold, A.: Self-consistent relaxation-time models in quantum mechanics.
            Comm. PDE, **21**, no.3&4, 473–506 (1996)
[Arn96a]   Arnold, A.: The relaxation-time von Neumann-Poisson equation. Proceedings
            of ICIAM 95, Hamburg (1995), Mahrenholtz, O., Mennicken, R. (eds.); ZAMM
            76 S2 (1996)
[Arn01]   Arnold, A.: Mathematical concepts of open quantum boundary conditions.
            Transp. Theory Stat. Phys., **30**, no.4–6, 561–584 (2001)
[Arn02]   Arnold, A.: Entropy method and the large-time behavior of parabolic
            equations. Lecture notes for the XXVII Summer school in mathemati-
            cal physics, Ravello, Italy (2002). http://www.anum.tuwien.ac.at/~arnold/
            papers/ravello.pdf
[AS07]   Arnold, A., Schulte, M.: Transparent boundary conditions for quantum-
            waveguide simulations. to appear in Mathematics and Computers in Simu-
            lation (2007), Proceedings of MATHMOD 2006, Vienna, Austria.
[AS04]   Arnold, A., Sparber, C.: Conservative Quantum Dynamical Semigroups for
            mean-field quantum diffusion models. Comm. Math. Phys., **251**, no.1 179–
            207 (2004)
[Ben98]   Ben Abdallah, N.: A Hybrid Kinetic-Quantum Model for Stationary Electron
            Transport. J. Stat. Phys., **90**, no.3–4, 627–662 (1998)
[Bre87]   Brezis, H.: Analyse fonctionelle - Théorie et applications. Masson (1987)
[BM91]   Brezzi, F., Markowich, P.A.: The three-dimensional Wigner-Poisson problem:
            existence, uniqueness and approximation. Math. Methods Appl. Sci., **14**, no.1,
            35–61 (1991)
[BMP05]   Ben Abdallah, N., Méhats, F., Pinaud, O.: On an open transient Schrödinger-
            Poisson system. Math. Models Methods Appl. Sci., **15**, no.5, 667–688 (2005)
[BN88]   Benatti, F., Narnhofer, H.: Entropy Behaviour under Completely Positive
            Maps. Lett. Math. Phys., **15**, 325–334 (1988)

[Boh89]   Bohm, D.: Quantum Theory. Dover (1989); reprint from 1951.
[Cas97]   Castella, F.: L^2–solutions to the Schrödinger-Poisson System: Existence,
             Uniqueness, Time Behaviour, and Smoothing Effects. Math. Mod. Meth. Appl.
             Sci., **7**, no.8, 1051–1083 (1997)
[Cas98]   Castella, F.: The Vlasov-Poisson-Fokker-Planck System with Infinite Kinetic
             Energy. Indiana Univ. Math. J., **47**, no.3, 939–964 (1998)
[Caz96]   Cazenave, T.: An introduction to nonlinear Schrödinger equation. Textos de
             Métodos Matemáticos **26**, Univ. Federal do Rio de Janeiro, (1996)
[CEFM00]   Castella, F., Erdös, L., Frommlet, F., Markowich, P.A.: Fokker-Planck equa-
             tions as Scaling limits of Reversible Quantum Systems. J. Statist. Phys., **100**,
             no.3-4, 543–601 (2000)
[CL83]   Caldeira, A.O., Leggett, A.J.: Path integral approach to quantum Brownian
             motion. Physica A, **121**, 587–616 (1983)
[CLN04]   Cañizo, J.A., López, J.L., Nieto, J.: Global L^1–theory and regularity for the
             3D nonlinear Wigner-Poisson-Fokker-Planck system. J. Diff. Eq., **198**, 356–
             373 (2004)
[CP96]   Castella, F., Perthame, B.: Estimations de Strichartz pour les Equations de
             transport Cinétique. C. R. Acad. Sci. Paris, t. 322, Série I, 535–540 (1996)
[Dav76]   Davies, E.B.: Quantum Theory of Open Systems. Academic Press, London-
             New York (1976)
[Dav77]   Davies, E.B.: Quantum dynamical semigroups and the neutron diffusion equa-
             tion. Rep. Math. Phys., **11**, no.2, 169–188 (1977)
[Deg]   Degond, P.: Introduction à la théorie quantique, DEA–Lecture Notes, UPS
             Toulouse.
[DL88]   Dautray, R., Lions, J.L.: Mathematical Analysis and Numerical Methods of
             Science and Technology; vol 1 Pysical Origins and Classical Methods. Springer
             (1990)
[DL88a]   Dautray, R., Lions, J.L.: Mathematical Analysis and Numerical Methods of
             Science and Technology; vol 5 Evolution Problems I. Springer (1988)
[FR98]   Fagnola, F., Rebolledo, R.: The approach to equilibrium of a class of quan-
             tum dynamical semigroups. Infinite Dim. Analysis, Quantum Prob. and Rel.
             Topics, **1**, no.4, 561–572 (1998)
[Fre90]   Frensley, W.R.: Boundary conditions for open quantum systems driven
             far from equilibrium. Rev. Mod. Phys., **62**, 745–791 (1990). http://www.
             utdallas.edu/~frensley/technical/opensyst/opensyst.html
[Fre94]   Frensley, W.R.: Quantum Transport. In: Frensley, W.R., Einspruch, N.G.
             (eds.) Heterostructures and Quantum Devices. Academic Press, San Diego
             (1994). http://www.utdallas.edu/~frensley/technical/qtrans/qtrans.html
[GvMP87]   Greenberg, W., van der Mee, C., Protopopescu, V.: Boundary Value Problems
             in Abstract Kinetic Theory. Birkhäuser, Basel-Boston-Stuttgart (1997)
[GV94]   Ginibre, J., Velo, G.: The global Cauchy problem for the non linear
             Schrödinger equation revisited. Annales de l'institut Henri Poincaré (C)
             Analyse non linéaire, **2**, no.4, 309–327 (1985)
[HO89]   Hayashi, N., Ozawa, T.: Smoothing Effect for Some Schrödinger Equations.
             J. Funct. Anal, **85**, 307–348 (1989)
[ILZ94]   Illner, R., Lange, H., Zweifel, P.: Global existence, uniqueness, and asymptotic
             behaviour of solutions of the Wigner-Poisson and Schrödinger systems. Math.
             Meth. Appl. Sci., **17**, 349–376 (1994)
[KKFR89]   Kluksdahl, N.C., Kriman, A.M., Ferry, D.K., Ringhofer, C.: Self-consistent
             study of the resonant-tunneling diode. Phys. Rev. B, **39**, 7720–7735 (1989)
[KN06]   Kosina, H., Nedjalkov, M.: Wigner Function-Based Device Modeling, §67.
             In: Handbook of Theoretical and Computational Nanotechnology. Rieth, M.,
             Schommers, W. (eds.) American Scientific Publishers. (2006)

[Lev70]   Levinson, I.B.: Translational invariance in uniform fields and the equation for the density matrix in the Wigner representation. Sov. Phys. JETP, **30**, 362–367 (1970)

[Lin76]   Lindblad, G.: On the generators of quantum mechanical semigroups. Comm. Math. Phys., **48**, 119–130 (1976)

[LK90]   Lent, C.S., Kirkner, D.J.: The quantum transmitting boundary method, J. of Appl. Phys., **67**, no.10, 6353–6359 (1990)

[LL85]   Landau, L.D., Lifschitz, E.M.: Quantenmechanik. Akademie-Verlag, Berlin. (1985)

[LP93]   Lions, P.L., Paul, T.: Sur les measures de Wigner, Rev. Math. Iberoam., **9**, no.3, 553–618 (1993)

[LT76]   Lieb, E.H., Thirring, W.E.: Inequalities for the moments of the eigenvalues of the Schrödinger Hamiltonian and their relation to Sobolev inequalities. In: Lieb, E.H., Simon, B., Wightman, A.S. (eds.) Studies in Mathematical Physics. Essays in honor of Valentine Bargmann. Princeton Univ. Press (1976)

[MRS90]   Markowich, P.A., Ringhofer, C.A., Schmeiser, C.: Semiconductor Equation. Springer-Verlag, Wien-New York. (1990)

[Paz83]   Pazy, A.: Semigroups of Linear Operators and Applications to Partial Differential Equations. Springer, New York (1983)

[Per96]   Perthame, B.: Time decay, propagation of low moments and dispersive effects for kinetic equations. Comm. P.D.E., **21**, no.1&2, 659–686 (1996)

[Rin03]   Ringhofer, C.: Thermodynamic principles in modeling nano-scale transport in semiconductors. Lecture Notes for: Nanolab Spring School. Toulouse (2003). http://math.la.asu.edu/~chris/nano030529.pdf

[RS75]   Reed, M., Simon, B.: Methods of modern mathematical physics. II. Fourier analysis, self-adjointness. Academic Press (1975)

[SA07]   Schulte, M., Arnold, A.: Discrete transparent boundary conditions for the Schrödinger equation – a compact higher order scheme. Kinetic and Related Models 1, no.1, 101–125 (2008)

[SS87]   Săndulescu, A., Scutaru, H.: Open Quantum Systems and the Damping of Collective Modes in Deep Inelastic Collisions. Annals of Phys., **173**, 277–317 (1987)

[SCDM04]   Sparber, C., Carrillo, J.A., Dolbeault, J., Markowich, P.A.: On the Long-Time Behavior of the Quantum Fokker-Planck Equation. Monatsh. Math., **141**, 237–257 (2004)

[Sim79]   Simon, B.: Trace Ideals and Their Applications. Cambridge University Press (1979)

[Spo76]   Spohn, H.: Approach to equilibrium for completely positive dynamical semigroups. Rep. Math. Phys., **10**, no.2, 189–194 (1976)

[Spo78]   Spohn, H.: Entropy production for quantum dynamical semigroups. J. Math. Phys., **19**, no.5, 1227–1230 (1978)

[Tha05]   Thaller, B.: Advanced Visual Quantum Mechanics. Springer (2005)

[W32]   Wigner, E.: On the quantum correction for thermodynamic equilibrium. Phys. Rev., **40**, 749–759 (1932)

Quantum Hydrodynamic and Diffusion Models Derived from the Entropy Principle

Pierre Degond, Samy Gallego, Florian Méhats, Christian Ringhofer

Abstract In these notes, we review the recent theory of quantum hydrodynamic and diffusion models derived from the entropy minimization principle. These models are obtained by taking the moments of a collisional Wigner equation and closing the resulting system of equations by a quantum equilibrium. Such an equilibrium is defined as a minimizer of the quantum entropy subject to local constraints of given moments. We provide a framework to develop this minimization approach and successively apply it to quantum hydrodynamic models and quantum diffusion models. The results of numerical simulations show that these models capture well the various features of quantum transport.

1 General Introduction

The goal of these lecture notes is to give a comprehensive introduction to the theory of quantum hydrodynamic and diffusion models derived from the entropy principle. These lecture notes report on previously published works [21–26].

These models are obtained by taking the moments of a collisional Wigner equation and closing the resulting system of equations by a quantum equilibrium. Such an equilibrium is defined as a minimizer of the quantum entropy

Pierre Degond and Samy Gallego
Institut de Mathématiques, Université de Toulouse, 118 route de Narbonne, 31062 Toulouse Cedex 9, France, e-mail: Pierre.degond@math.univ-toulouse.fr
Florian Méhats
IRMAR (UMR CNRS 6625), Université de Rennes, Campus de Beaulieu, 35042 Rennes Cedex, France, e-mail: florian.mehats@univ-rennes1.fr
Christian Ringhofer
Department of Mathematics, Arizona State University, Tempe, AZ 85284-184, USA, e-mail: ringhofer@asu.edu

N. Ben Abdallah, G. Frosali (eds.), *Quantum Transport*.
Lecture Notes in Mathematics 1946.
© Springer-Verlag Berlin Heidelberg 2008

subject to local constraints of given moments. We provide a framework to develop this minimization approach and successively apply it to quantum hydrodynamic models and quantum diffusion models. We also give some preliminary numerical results.

These lecture notes are organized as follows. In Sect. 2 and particularly in Sect. 2.1, an introduction to quantum kinetic theory is given. Section 2.2 is devoted to a presentation of the N-particle problem in quantum mechanics and the Hartree and Hartree–Fock approximations. With these notions in hand, it is possible to give a very short and incomplete summary of methods that are in use for modeling large quantum particle systems in Sect. 2.3. Current quantum hydrodynamic approaches are then introduced in Sect. 2.4.

Section 3 tackles the core of the matter, the derivation of quantum hydrodynamic models based on the entropy principle. The setting of the problem is recalled in Sect. 3.1. Then, general quantum hydrodynamic models are derived in Sect. 3.2. Finally, a detailed analysis of the isothermal quantum Euler model is developed in Sect. 3.3, and illustrated by some preliminary numerical simulations.

Section 4 turns towards quantum diffusion models. First, the quantum Energy-Transport model is developed in Sect. 4.1. The case of the quantum drift-diffusion model is reviewed in Sect. 4.2. In the latter case, detailed numerical simulations are given which show that the model captures well the various features of diffusive quantum transport.

Finally, conclusions and perspectives are drawn in Sect. 5.

2 Quantum Kinetic Equations: An Introduction

2.1 Quantum Statistical Mechanics of Nonequilibrium Systems

In quantum mechanics, the state of a particle is defined by a wave-function $\psi(x,t) \in \mathbb{C}$, where $x \in \mathbb{R}^d$ is the position and t the time. The quantity $dP_t(x) = |\psi(x,t)|^2 \, dx$ represents the probability of finding the particle in the elementary volume dx at time t. As such, we have the normalization condition

$$\int |\psi(x,t)|^2 \, dx = \int dP_t(x) = 1. \tag{1}$$

Therefore, $\psi(\cdot,t) \in L^2(\mathbb{R}^d)$. We denote by d the dimension of configuration space. The evolution of ψ is ruled by the Schrödinger equation

$$i\hbar \partial_t \psi = \mathcal{H}\psi, \tag{2}$$

where $\hbar$ is the Planck constant and $\mathcal{H}$ is the Hamiltonian operator

$$\mathcal{H}\psi = -\frac{\hbar^2}{2}\Delta\psi + V(x,t)\psi, \tag{3}$$

where $V(x,t)$ is the potential energy.

A measurement of the system gives rise to the operation

$$(\psi, A\psi)_{L^2} = \int \psi \overline{A\psi}\, dx, \tag{4}$$

where A is a Hermitian operator on L^2 which corresponds to the physical quantity to be measured and the bar denotes complex conjugation. For instance, the position operator $X : \psi \to x\psi(x)$ corresponds to the observation of the mean particle position:

$$(\psi, X\psi) = \int x\,|\psi|^2\, dx.$$

Similarly, the momentum operator $P : \psi \to -i\hbar\nabla\psi$ gives rise to the observation

$$(\psi, P\psi) = -\int \psi\, \overline{i\hbar\nabla\psi}\, dx = \int \hbar k\,|\hat{\psi}(k)|^2\, dk,$$

where $\hat{\psi}(k)$ is Fourier transform of ψ:

$$\hat{\psi}(k) = (2\pi)^{-d/2} \int e^{-ik\cdot x}\psi(x)\, dx.$$

Any classical observable $a(x,p)$ gives rise to a quantum observable $A = \mathrm{Op}(a)$ according to the Weyl quantization rule:

$$\mathrm{Op}(a)\psi = \frac{1}{(2\pi)^d} \int a\left(\frac{x+y}{2}, \hbar k\right) \psi(y)e^{ik(x-y)}\, dk\, dy, \tag{5}$$

and a is called the Weyl symbol of $\mathrm{Op}(a)$. An example of such observable is the mechanical energy or Hamiltonian $\mathcal{H}_c(x,p) = |p|^2/2 + V$ which gives rise to the quantum Hamiltonian operator

$$\mathrm{Op}(\mathcal{H}_c) = \mathcal{H} = -(\hbar^2/2)\Delta + V. \tag{6}$$

For an N-particle systems, the wave-function $\psi(x_1, \ldots, x_N)$ depends on all the coordinates of the N particles: here, x_i denotes the coordinate of the i-th particle. In this case, the classical Hamiltonian is written

$$\mathcal{H}_c = \sum_{i=1}^{N} \frac{1}{2}|p_i|^2 + \frac{1}{2}\sum_{i\neq j} \phi_{int}(x_i - x_j) + \sum_{i} \phi_{ext}(x_i), \tag{7}$$

with $\phi_{int}(x - y)$ a binary interaction potential and $\phi_{ext}(x)$ an external potential. Its quantum counterpart is:

$$\mathcal{H} = -\sum_{i=1}^{N} \frac{\hbar^2}{2} \Delta_{x_i} + \frac{1}{2} \sum_{i \neq j} \phi_{int}(x_i - x_j) + \sum_{i} \phi_{ext}(x_i). \tag{8}$$

When the state of an N-particle system is incompletely known, a statistical description is needed. To describe the uncertainty about the state of the system we make use of a complete orthonormal basis of the system $(\phi_s)_{s \in S}$ and ρ_s lists the occupation probability of state s i.e. the probability that a particle lies in the state s. Being probabilities, ρ_s satisfies

$$0 \leq \rho_s \leq 1, \quad \sum_{s \in S} \rho_s = 1. \tag{9}$$

The probability of presence of a particle in the incompletely known system described by the set $(\rho_s)_{s \in S}$ is given by:

$$P(x, t)dx = \sum_{s \in S} \rho_s \, |\phi_s|^2 \, dx. \tag{10}$$

To compute the evolution of an incompletely known system, it is necessary to make a bookkeeping of the states $(\phi_s)_{s \in S}$ and the probabilities $(\rho_s)_{s \in S}$. In order to do so, one introduces an operator on L^2, the so-called density operator ρ, defined by its action on an arbitrary $\psi \in L^2$ by:

$$\rho\psi = \sum_{s \in S} \rho_s(\psi, \phi_s) \, \phi_s. \tag{11}$$

This defines ρ as an Hermitian, positive operator whose eigenvectors are ϕ_s and eigenvalues ρ_s. Then, because of (9), ρ is trace-class and its trace is unity:

$$\text{Tr}\, \rho = \sum_{s \in S} \rho_s = 1. \tag{12}$$

Such an incompletely known system is said to be in a mixed state. The case of a pure state is when all ρ_s are zero but one, say $\rho_{s_0} = 1$. Then $\rho = (\cdot, \phi_{s_0}) \, \phi_{s_0}$ is nothing but the projection on the one-dimensional manifold spanned by ϕ_{s_0}.

Now, to compute the evolution of ρ we suppose that $\phi_s(t)$ is a solution of Schrödinger equation for all s and that the probabilities ρ_s do not evolve with time: $\rho_s(t) = $ constant. Then, the equation for ρ reads

$$i\hbar \partial_t \rho = \mathcal{H}\rho - \rho\mathcal{H} = [\mathcal{H}, \rho], \tag{13}$$

and is called the Quantum Liouville equation. From now on $[\mathcal{A}, \mathcal{B}]$ will denote the commutator of operators $\mathcal{A}$ and $\mathcal{B}$, i.e. $[\mathcal{A}, \mathcal{B}] = \mathcal{A}\mathcal{B} - \mathcal{B}\mathcal{A}$.

Another way of defining ρ is through its distribution kernel $\underline{\rho}(x, x')$ defined by

$$\rho\psi = \int \underline{\rho}(x, x')\psi(x')\,dx'. \tag{14}$$

With (11), we have

$$\underline{\rho}(x, x') = \sum_s \rho_s \phi_s(x)\overline{\phi_s(x')}. \tag{15}$$

The Quantum Liouville equation expressed in terms of $\underline{\rho}(x, x')$ is written

$$i\hbar\partial_t\underline{\rho} = (\mathcal{H}_x - \mathcal{H}_{x'})\underline{\rho}, \tag{16}$$

where $\mathcal{H}_x$ (resp. $\mathcal{H}_{x'}$) denotes the Hamiltonian applied to the x (resp. x') dependence of $\underline{\rho}(x, x')$. We note that, because ρ is Hermitian,

$$\underline{\rho}(x', x) = \overline{\underline{\rho}(x, x')}, \tag{17}$$

and that its trace is given by

$$\mathrm{Tr}\rho = \int \underline{\rho}(x, x)\,dx. \tag{18}$$

Now, we turn to the observation of a statistical quantum system defined by its density operator ρ. Let A be an observable. Then, the observation of the system is

$$\langle A \rangle_\rho = \sum_s \rho_s (A\phi_s, \phi_s) = \mathrm{Tr}\{\rho A\}. \tag{19}$$

For example the probability of presence at x_0 is given by

$$P(x_0) = \sum_{s \in S} \rho_s |\phi_s(x_0)|^2$$
$$= \underline{\rho}(x_0, x_0) = \mathrm{Tr}\{\rho\,\mathrm{Op}(\delta_{x-x_0})\}, \tag{20}$$

and is the observation of the state at $x = x_0$.

The Wigner transform of ρ allows to write the observation associated with observable A in terms of its Weyl symbol $A = \mathrm{Op}(a)$, according to the following formula:

$$\langle \mathrm{Op}(a)\rangle_\rho = \mathrm{Tr}\{\rho\mathrm{Op}(a)\}$$
$$= \frac{1}{(2\pi\hbar)^d} \int W[\rho](x, p)\, a(x, p)\, dx\, dp, \tag{21}$$

and $W[\rho]$ is the so-called Wigner transform of ρ. In other words,

$$W[\rho](x_0, p_0) = (2\pi\hbar)^d \langle \mathrm{Op}(\delta_{x-x_0}\delta_{p-p_0})\rangle_\rho, \tag{22}$$

is up to the factor $(2\pi\hbar)^d$, the observation of the system at (x_0, p_0).

For such a formula to be valid, $W[\rho]$ must be defined by

$$W[\rho](x,p) = \int \underline{\rho}(x - \frac{\eta}{2}, x + \frac{\eta}{2}) e^{\frac{i\eta \cdot p}{\hbar}} d\eta, \qquad (23)$$

and, with ρ defined by (15), we find

$$W[\rho](x,p) = \sum_s \rho_s \int \phi_s(x - \frac{\eta}{2}) \overline{\phi_s(x + \frac{\eta}{2})} e^{\frac{i\eta \cdot p}{\hbar}} d\eta. \qquad (24)$$

Note that $W[\rho]$ is real-valued (thanks to (17)) but is not necessary positive. Therefore, $W[\rho] \, dx \, dp$ is not a probability distribution function (as would a classical distribution function be). However, the Husimi regularization, defined by

$$W_H[\rho] = W[\rho] * G, \quad G = \frac{1}{(\hbar\pi)^3} e^{-(|x|^2 + |p|^2)/\hbar},$$

is actually non negative.

Applying the Wigner transform to the Quantum Liouville equation for ρ (13), we find the Wigner equation for $W[\rho]$:

$$\partial_t W + p \cdot \nabla_x W + \Theta^\hbar[V]W = 0, \qquad (25)$$

where the field operator $\Theta^\hbar[V]W$ is defined by:

$$\Theta^\hbar[V]W = -\frac{i}{(2\pi)^3 \hbar} \int (V(x + \frac{\hbar}{2}\eta) - V(x - \frac{\hbar}{2}\eta)) \times$$
$$\times W(x,q) \, e^{i\eta \cdot (p-q)} \, dq \, d\eta. \quad (26)$$

The Wigner equation resembles the classical collisionless kinetic equation (or Vlasov equation) but for the field term $\Theta^\hbar[V]$. However, as the Planck constant $\hbar$ tends to zero, we have

$$\Theta^\hbar[V]W \xrightarrow{\hbar \to 0} -\nabla_x V \cdot \nabla_p W. \qquad (27)$$

Using the Parseval identity, we have

$$\int W[\rho] \overline{W[\sigma]} \frac{dx \, dp}{(2\pi\hbar)^d} = \mathrm{Tr}\{\rho \, \sigma^\dagger\}, \qquad (28)$$

$$\int a \overline{b} \frac{dx \, dp}{(2\pi\hbar)^d} = \mathrm{Tr}\{\mathrm{Op}(a) \, \mathrm{Op}(b)^\dagger\}, \qquad (29)$$

where the superscript $\dagger$ indicates the Hermitian conjugate operator. The Wigner transformation and the Weyl quantization are inverse operations one to each other and are isometries between the space of $L^2(\mathbb{R}^{2d})$ functions of (x,p) (equipped with the measure $(2\pi\hbar)^{-d} \, dx \, dp$) and the space $\mathcal{L}^2$ of operators σ on $L^2(\mathbb{R}^d)$ such that the product $\sigma\sigma^\dagger$ is trace class (equipped with the norm $|\sigma|_{\mathcal{L}^2}^2 = \mathrm{Tr}\{\sigma\sigma^\dagger\}$):

$$W = \mathrm{Op}^{-1}, \quad \mathrm{Op} = W^{-1}.$$

Note that elements of the space $\mathcal{L}^2$ are the so-called Hilbert-Schmidt operators.

2.2 N-Particle Quantum System

In this section, we review some of the standard techniques to deal with many-particle quantum systems, namely the Hartree and the Hartree–Fock methods.

For an N-particle quantum system, the density operator is an operator ρ^N operating on $L^2(\mathbb{R}^{3N})$ whose distribution kernel is $\underline{\rho}^N(x_1, x_1', \ldots, x_N, x_N')$. Now, we have to enforce the undistinguishability of the particles, which translates to the fact that the density operator is invariant under permutations of the particles:

$$\underline{\rho}^N(x_{\sigma(1)}, x_{\sigma(1)}', \ldots, x_{\sigma(N)}, x_{\sigma(N)}') = \underline{\rho}^N(x_1, x_1', \ldots, x_N, x_N'), \quad (30)$$

for all permutations σ of the set $\{1, 2, \ldots, N\}$. The Quantum Liouville equation is written

$$i\hbar \partial_t \rho^N = [\mathcal{H}^N, \rho^N], \quad (31)$$

$$\mathcal{H}^N = \sum_{i=1}^{N} \frac{1}{2}|p_i|^2 + \frac{1}{2}\sum_{i \neq j} \phi(x_i - x_j), \quad (32)$$

where ϕ describes the potential for binary interactions between the particles.

We now define partial density operators by taking the partial trace with respect to the last $N - j$ variables, i.e.

$$\rho^j = \mathrm{Tr}_{j+1}^{N}\{\rho^N\}, \quad (33)$$
$$\underline{\rho}^j(x_1, x_1', \ldots, x_j, x_j') =$$
$$= \int \underline{\rho}^N(\ldots, x_{j+1}, x_{j+1}, \ldots, x_N, x_N) \, dx_{j+1} \ldots dx_N. \quad (34)$$

Taking the partial trace of the quantum Liouville equation over the last $N - j$ variables, we find the equation satisfied by ρ^j

$$i\hbar \partial_t \rho^j = [\mathcal{H}^j, \rho^j] + Q^j(\rho^{j+1}), \quad (35)$$

$$\mathcal{H}^j = \sum_{i=1}^{j} \frac{1}{2}|p_i|^2 + \frac{1}{2}\sum_{i,k=1, i \neq k}^{j} \phi(x_i - x_k). \quad (36)$$

The equation for ρ^j depends on ρ^{j+1} and so on up to ρ^N. This forms a hierarchy of equations called the quantum BBGKY hierarchy, after the works of Bogoliubov, Born, Green, Kirkwood and Yvon. The interaction operator is written,

$$Q^j(\rho^{j+1}) = (N-j)\sum_{i=1}^{j} \text{Tr}_{j+1}\{[\phi(x_i - x_{j+1}), \rho^{j+1}]\}. \tag{37}$$

The invariance of ρ^N under permutations allows to have only the variable x_{j+1} coming into this expression instead of the whole list $x_{j+1}, \ldots, x_N$. In particular, the equation for ρ^1 is given by

$$i\hbar\partial_t\rho^1 = [\mathcal{H}^1, \rho^1] + Q^1(\rho^2), \tag{38}$$

$$\mathcal{H}^1 = \frac{1}{2}|p_1|^2, \tag{39}$$

$$Q^1(\rho^2) = (N-1)\text{Tr}_2\{[\phi(x_1 - x_2), \rho^2]\}, \tag{40}$$

or,

$$\underline{Q}^1(\rho^2) = (N-1)\int[\phi(x_1 - x_2) - \phi(x_1' - x_2)]\underline{\rho}^2(x_1, x_1', x_2, x_2)\,dx_2. \tag{41}$$

The goal is to find a closed system for ρ^1 only. For that purpose, we need to find a prescription for ρ^2. The most simple one is that of statistical independence, also called propagation of chaos, which states that ρ^2 is a product of a one-particle density:

$$\underline{\rho}^2(x_1, x_1', x_2, x_2') = \underline{\rho}^1(x_1, x_1')\,\underline{\rho}^1(x_2, x_2'). \tag{42}$$

The choice can be rigorously justified in the limit $N \to \infty$, provided that a rescaling of the interaction strength between the particles is made, namely rescaling ϕ into $\frac{1}{N}\phi$. The rationale of this rescaling is keeping the force acting on a single particle finite, which allows to capture the limiting dynamics as $N \to \infty$ without any time rescaling.

Using (42), we find

$$\underline{Q}^1(\rho^2) = (1 - (1/N))\int[\phi(x_1 - x_2) - \phi(x_1' - x_2)]\underline{\rho}^2(x_1, x_1', x_2, x_2)\,dx_2$$

$$\approx \int[\phi(x_1 - x_2) - \phi(x_1' - x_2)]\underline{\rho}^1(x_2, x_2)\,dx_2\underline{\rho}^1(x_1, x_1'). \tag{43}$$

We can write

$$\underline{Q}^1(\rho^2) \approx (V_\rho(x_1) - V_\rho(x_1'))\underline{\rho}^1(x_1, x_1'), \tag{44}$$

or

$$Q^1(\rho^2) \approx [V_\rho, \rho^1] \tag{45}$$

with

$$V_\rho(x) = \int \phi(x-y)\underline{\rho}^1(y,y)\,dy. \tag{46}$$

Finally, we find the density operator formulation of Schrödinger mean-field equations:

$$i\hbar\partial_t\rho = [\mathcal{H}_{mf}, \rho], \tag{47}$$

$$\mathcal{H}_{mf} = \frac{1}{2}|p|^2 + V_\rho, \tag{48}$$

$$V_\rho(x) = \int \phi(x-y)n(y)\,dy, \tag{49}$$

$$n(y) = \rho(y,y). \tag{50}$$

If a pure-state is considered, i.e. if $\rho = (\cdot,\psi)\psi$ is a projector, then ψ satisfies the Schrödinger mean-field equation

$$i\hbar\partial_t\psi = \mathcal{H}_{mf}\psi, \tag{51}$$

$$\mathcal{H}_{mf} = \frac{1}{2}|p|^2 + V_\psi, \tag{52}$$

$$V_\psi(x) = \int \phi(x-y)n(y)\,dy, \tag{53}$$

$$n(y) = |\psi(y)|^2, \tag{54}$$

However, for a more precise mean-field limit, it is important to take into account some more refined statistical properties of the particles. If fermions are considered (which is the case of electrons), the N-particle wave function is antisymmetric under particle permutations, i.e.:

$$\psi(x_{\sigma(1)}, \ldots, x_{\sigma(N)}) = (-1)^{\varepsilon(\sigma)}\psi(x_1, \ldots, x_N), \tag{55}$$

where σ is a permutation of $\{1, \ldots, N\}$ and $\varepsilon(\sigma)$ is the signature of this permutation. Then, the density matrix satisfies

$$\rho(x_1, x'_{\sigma(1)}, \ldots, x_N, x'_{\sigma(N)}) = (-1)^{\varepsilon(\sigma)}\rho(x_1, x'_1, \ldots, x_N, x'_N). \tag{56}$$

Clearly, the Hartree mean-field closure (42) does not satisfy this property. The most simple closure which does satisfy this antisymmetry property is the Slater determinant closure:

$$\underline{\rho}^2(x_1, x'_1, x_2, x'_2) = \underline{\rho}^1(x_1, x'_1)\,\underline{\rho}^1(x_2, x'_2) - \underline{\rho}^1(x_1, x'_2)\,\underline{\rho}^1(x_2, x'_1). \tag{57}$$

Inserting this closure into (41), we find

$$\underline{Q}^1(\rho^2) \approx (V_\rho(x_1) - V_\rho(x'_1))\underline{\rho}^1(x_1, x'_1) - \underline{Q}_{\mathrm{ex}}(\rho^1), \tag{58}$$

with

$$Q_{\underline{ex}}(\rho^1) = \int [\phi(x_1 - x_2) - \phi(x_1' - x_2)]\underline{\rho}^1(x_1, x_2')\underline{\rho}^1(x_2, x_1')\, dx_2 \,. \tag{59}$$

The quantity $Q_{\underline{ex}}$ is called the exchange-correlation potential. We can write it shortly:

$$Q_{ex} = \text{Tr}\{[\phi, (\rho \otimes \rho)_{ex}]\}_2, \tag{60}$$

with

$$(\rho \otimes \rho)_{\underline{ex}} = \underline{\rho}^1(x_1, x_2')\, \underline{\rho}^1(x_2, x_1'), \tag{61}$$

and $\text{Tr}\{\}_2$ is the trace w.r.t the second variable.

Finally, we find the Hartree–Fock mean-field model

$$i\hbar\partial_t \rho = [\mathcal{H}_{mf}, \rho] - Q_{ex}(\rho), \tag{62}$$

$$\mathcal{H}_{mf} = \frac{1}{2}|p|^2 + V_\rho, \tag{63}$$

$$Q_{ex} = \text{Tr}\{[\phi, (\rho \otimes \rho)_{ex}]\}_2, \tag{64}$$

$$V_\rho(x) = \int \phi(x - y)n(y)\, dy, \tag{65}$$

$$n(y) = \underline{\rho}(y, y). \tag{66}$$

2.3 Quantum Methods: A Brief and Incomplete Summary

The mean-field limit has first been rigorously proven in the case of a smooth interaction potential ϕ by Spohn [59] and recently in the case of the Coulomb potential by Bardos et al. [6]. The Hartree–Fock case has been solved by Bardos, Golse, Gottlieb, Mauser in [7,8].

The existence theory for the Schrödinger mean-field equation has first been developed in the stationary case by Nier [53–55] and Kaiser and Rehberg [43]. The semiclassical limit $\hbar \to 0$ of the Wigner formulation of the density-matrix mean-field equations has been developed by Lions and Paul [47] and by Markowich and Mauser [50].

There is no such thing as a BBGKY hierarchy for Hard-Spheres dynamics in quantum mechanics and so no quantum equivalent of the Boltzmann equation, at least up to our knowledge.

We refer to [44] for a complete monography of up-to-date techniques. Numerical simulations of "Small" systems (such as atoms, molecules up to a few tens of electrons) have first focused on the eigenvalue problem for finding the minimal energy (lowest eigenvalue of the Hamiltonian) or the excited states (i.e. the lower part of the spectrum of the Hamiltonian). For this purpose, the techniques have primarily concentrated upon the Hartree–Fock (i.e. assuming that the wave-function assumes the form of a Slater determinant) or multi-configuration techniques (i.e. the wave-function is a combination of a finite number of Slater determinants). Usually, for quantum chemistry computations, the Born–Oppenheimer approximation is made. It consists in decoupling the electron and nuclei dynamics based on the small mass ratio of the electrons to the ions and assuming that the ions move classically on quantum mechanically computed electron energy surfaces. To reach convergence of the electron minimization problem for a given configuration of nuclei positions is sometimes very time consuming. This is why Car and Parinello proposed to optimize the nuclei position on the fly, without waiting for the convergence of the electron minimization problem [18].

The dynamical study of small systems is also very important for applications such as chemical reactions dynamics particularly when energy surface crossings are involved or the determination of chemical intermediates is needed. Modern applications of such dynamical problems involve the control of chemical reactions by lasers. Such dynamical studies make use either of direct computations of the time-dependent Schrödinger equation or that of the time-dependent Hartree–Fock model.

In the case of large systems such as large molecules, crystals, nano-objects or those systems involved in molecular dynamics computations of phase changes, the direct resolution of the Schrödinger equation or time-dependent Hartree–Fock equation is computationally too demanding. Then, Density Functional Theory (DFT) is used. DFT reduces the problem of finding the minimal energy of such a system to the resolution of a one-particle Schrödinger equation in a nonlinear potential which is a functional of the density. This observation is originally due to Hohenberg and Kohn in [39]. Although the method is exact, the functional of the density is not known and various kinds of approximations are needed among which the most popular ones are the Thomas–Fermi and Kohn–Sham models (see, e.g. [28]). However, the validity of these approximations is still under scrutiny.

The modeling of open systems such as electrons in a semiconductor, molecules in a solvent, proteins in a biological cell is even more complex. The main question is how to account for the environment of these molecules without having to compute their environment in full detail. Typically, a model for an open system would start with a density matrix formulation of the problem in terms of

$$\underline{\rho}(x_1, x_1', \ldots, x_N, x_N', y_1, y_1', \ldots, y_P, y_P')$$

where $x_1, \ldots, x_N$ denote the variables of the system of interest and $y_1, \ldots, y_P$ denote the environment variables. Then, taking the partial trace of the Quantum Liouville equation for ρ over the y variables we find an equation for the partial density matrix corresponding to the variables of interest. However two problems with this approach need to be solved. First, in general, the environment variables, their dynamics and their interaction with the system are poorly known. Second, a closure assumption for the environment variables is needed. Such closures are hard to find and usually rely on a thermodynamical equilibrium hypothesis. Still, this approach has been used in a number of cases, such as the electron–phonon in semiconductors where the partial trace over phonon variables is realized, see e.g. the work of Argyres [4]. In this case, this leads to a quite complex "collision operator" which exhibits nonlocality in space and time and which is very difficult to deal with numerically. Furthermore, the precise domain of validity of the closure is still to be determined. Quantum kinetic models with collisions have been investigated in [5, 30, 31].

A somewhat different route to model such open systems is via hydrodynamic models. Hydrodynamic models are expected to be valid at a mesoscale, i.e. in conditions such that the system is large enough so that a notion of thermodynamic limit is valid while being not too large in order to prevent quantum decoherence to occur. They rely on a scale separation hypothesis, namely that small scale phenomena are clearly separated from large scale ones; small scales are quickly dissipated towards a local thermodynamical equilibrium while large scales follow the macroscopic hydrodynamic evolution. We are now going to review the existing attempts in this direction.

2.4 Hydrodynamic Limits: A Review

The difficulty with deriving quantum hydrodynamic models is that there does not exist a kinetic model which would clearly play the role which the Boltzmann equation plays for the derivation of classical hydrodynamic models.

It is known since Madelung that the Schrödinger equation has a formulation in the form of a pressureless gas dynamics equation perturbed by a quantum term. As such, it would describe a single-particle hydrodynamics. It has given rise to the concept of quantum trajectories also known as Bohmian mechanics because it led Bohm to propose a sort of "deviant" interpretation of quantum mechanics. Now the concept of Bohmian trajectories is routinely used as a numerical integration tool for solving the Schrödinger equation [48], [60].

In the follow-up of this document, we shall propose an extension of this concept to many-particle hydrodynamics by using the entropy minimization principle "à la Levermore" [45].

In the classical case, the single particle hydrodynamics follows from considering the Free Transport equation (where $f(x, v, t)$ is the distribution function, x being the position, v, the velocity and t the time):

$$\frac{\partial f}{\partial t} + v \cdot \nabla_x f - \nabla_x V \cdot \nabla_v f = 0$$

and looking for monophase solutions of the form

$$f = n(x,t)\,\delta(v - u(x,t)).$$

Then, n and u satisfy exactly the Pressureless gas dynamics:

$$\partial_t n + \nabla_x \cdot nu = 0, \tag{67}$$

$$\partial_t u + u \cdot \nabla_x u = -\nabla_x V. \tag{68}$$

This system has some unpleasant features, such as being non strictly hyperbolic. It has been investigated and used by Brenier and Grenier [12], Bouchut [13], E and coauthors [29].

The quantum case follows a similar methodology. We consider a single state ψ and the associated Schrödinger equation

$$i\hbar\partial_t \psi = -\frac{\hbar^2}{2}\Delta\psi + V(x,t)\psi, \tag{69}$$

and we decompose the wave function into its amplitude $\sqrt{n}$ and its phase S as follows:

$$\psi = \sqrt{n}e^{iS/\hbar}. \tag{70}$$

Inserting this Ansatz into (69) and taking real and imaginary parts, we get

$$\partial_t n + \nabla_x \cdot nu = 0, \tag{71}$$

$$\partial_t S + \frac{1}{2}|\nabla S|^2 + V - \frac{\hbar^2}{2}\frac{1}{\sqrt{n}}\Delta\sqrt{n} = 0. \tag{72}$$

Defining $u = \nabla_x S$ and taking the gradient of the phase equation, we get:

$$\partial_t n + \nabla_x \cdot nu = 0, \tag{73}$$

$$\partial_t u + u \cdot \nabla_x u = -\nabla_x(V + V_B), \tag{74}$$

$$V_B = -\frac{\hbar^2}{2}\frac{1}{\sqrt{n}}\Delta\sqrt{n}. \tag{75}$$

This system appears as the Pressureless Gas Dynamics equations with an additional potential V_B called the Bohm potential which is an order $O(\hbar^2)$ term. If this term is neglected, the phase equation leads to the classical Hamilton–Jacobi equation. The Bohm potential involves dispersive term which adds high frequency oscillations and leads to delicate numerics. This model has been mathematically investigated by Markowich and coauthors [37].

Of course, the natural question is whether it is possible to add an energy balance equation which would allow to describe a finite temperature many-particle system. To do so, a straightforward idea is to start for a mixed-state (in the density operator or Wigner distribution formulations), to compute the mass and momentum balance equation for each of the states involved in the mixture and to average over the statistics of the states. But clearly, we are facing a closure problem, since the average in the quantities involved in the mixture are unlikely to be expressed in terms of total mass, momentum and energy only. This closure problem has been solved by various methods in the literature: Gardner [34] has used a classical Fourier law for the heat flux. Gasser et al. [38] have proposed a small temperature asymptotics. Gardner and Ringhofer [35, 36] have developed a Chapman–Enskog expansion for a phenomenological BGK-type collision term. Related approaches can be found in [14–16, 49].

In these notes, we propose a different approach, based on an entropy minimization principle "à la Levermore" [45]. This method has been developed in [25].

3 Quantum Hydrodynamic Models Derived from the Entropy Principle

3.1 Quantum Setting

In this section, we consider a collisional Quantum Liouville equation

$$i\hbar\partial_t\rho = [\mathcal{H}, \rho] + i\hbar\mathcal{Q}(\rho), \tag{76}$$

where $\mathcal{H}$ is the Hamiltonian:

$$\mathcal{H}\psi = -\frac{\hbar^2}{2}\Delta\psi + V(x, t)\psi, \tag{77}$$

and $\mathcal{Q}(\rho)$ is an unspecified collision operator which describes the interaction of the particles with themselves and with their environment and accounts for dissipation mechanisms. The only assumption that will be used is that this operator dissipates entropy (see below).

Let $W[\rho](x, p)$ denote the Wigner transform of ρ:

$$W[\rho](x, p) = \int \underline{\rho}(x - \frac{1}{2}\xi, x + \frac{1}{2}\xi)\, e^{i\frac{\xi \cdot p}{\hbar}}\, d\xi, \tag{78}$$

where $\underline{\rho}(x, x')$ is the distribution kernel of ρ:

$$\rho\psi = \int \underline{\rho}(x, x')\psi(x')\, dx'.$$

Note that we use the momentum p instead of the velocity v used in the classical setting. We make $m = 1$ so that $v = p$. We recall that the inverse Wigner transform (or Weyl quantization) is given by the following formula:

$$W^{-1}(w)\psi = \frac{1}{(2\pi)^d}\int w(\frac{x + y}{2}, \hbar k)\, \psi(y)e^{ik(x-y)}\, dk\, dy, \tag{79}$$

and defines $W^{-1}(w)$ as an operator acting on the element ψ of L^2. The function w is also called the Weyl symbol of ρ. Again, W and W^{-1} are Isometries between $\mathcal{L}^2$ (the space of operators such that the product $\rho\rho^\dagger$ is trace-class, where $\rho^\dagger$ is the Hermitian conjugate of ρ) and $L^2(\mathbb{R}^{2d})$:

$$\mathrm{Tr}\{\rho\sigma^\dagger\} = \int W[\rho](x,p)\overline{W[\sigma](x,p)}\,\frac{dx\,dp}{(2\pi\hbar)^d}. \tag{80}$$

Taking the Wigner transform of (76), we get the following collisional Wigner equation for $w = W[\rho]$:

$$\partial_t w + p \cdot \nabla_x w + \Theta^\hbar[V]w = Q(w). \tag{81}$$

with

$$\Theta^\hbar[V]w = -\frac{i}{(2\pi)^d\hbar} \int (V(x + \frac{\hbar}{2}\eta) - V(x - \frac{\hbar}{2}\eta)) \times$$
$$\times w(x,q)\,e^{i\eta\cdot(p-q)}\,dq\,d\eta. \tag{82}$$

and $Q(w)$ is the Wigner transform of $\mathcal{Q}(\rho)$.

In the next section, we are going to make use of the entropy dissipation properties of Q to derive quantum hydrodynamic models.

3.2 QHD via Entropy Minimization

This section is a summary of [25, 26]. A related approach can be found in [46, 52, 61]. We are following Levermore's approach for the classical Boltzmann equation [45]. We take the moments of (76) and close the resulting system by the assumption that $\mathcal{Q}(\rho)$ relaxes the system to an equilibrium ρ_α defined as an entropy minimizer constrained to have the same prescribed moments as ρ. The question is how such an equilibrium is defined.

First, let us define the moments of a quantum system. They are defined like in classical mechanics as the moments of the Wigner distribution function. Let $\mu_i(p)$, $i = 0, \ldots, N$ denote a list of monomial functions of the momentum p such as e.g the list of hydrodynamic monomials $(1, p, |p|^2)$. We denote by $\mu(p) = (\mu_i(p))_{i=0}^N$ the vector listing all these monomials. We construct a list of moments $m[w] = (m_i[w])_{i=0}^N$ of moments of w by setting

$$m_i[w] = \int w(x,p)\,\mu_i(p)\,\widetilde{dp}, \quad \widetilde{dp} := \frac{dp}{(2\pi\hbar)^d}. \tag{83}$$

For instance, $m = (n, q, \mathcal{W})$ (where n is the local particle density, q the momentum density and $\mathcal{W}$ the energy density) in the case of the list of hydrodynamic monomials $(1, p, |p|^2)$:

$$\begin{pmatrix} n \\ q \\ 2\mathcal{W} \end{pmatrix} = \int W[\rho] \begin{pmatrix} 1 \\ p \\ |p|^2 \end{pmatrix} \widetilde{dp}. \tag{84}$$

Note that

$$m_i[\rho](y) = \mathrm{Tr}\{\rho\, W^{-1}(\mu_i(p)\delta(x-y))\},$$

i.e. $m_i[\rho](y)$ is the observation of the observable $\mu_i(p)$ locally at point y. This definition of the moments is therefore consistent with the quantum definition of an observable.

The moment method now consists in taking the moments of the Wigner equation:

$$\partial_t m[w] + \nabla_x \cdot \int w\, \mu p\, \widetilde{dp} + \int \Theta[V]w\, \mu\, \widetilde{dp} = \int Q(w)\, \mu\, \widetilde{dp}. \tag{85}$$

In general $\int Q(w)\, \mu\, \widetilde{dp} \neq 0$ except for those moments conserved by the collision operator (e.g mass, momentum and energy) in the case of a "Boltzmann-like" collision operator.

Now, the closure problem reads as follows: find an expression of the integrals by setting w to be a local thermodynamical equilibrium, i.e. a solution of the entropy minimization problem. Now, we turn to the definition of the quantum entropy.

Recall that the density operator ρ in its eigenfunction basis ϕ_s reads

$$\rho\psi = \sum_{s\in S} \rho_s(\psi, \phi_s)\phi_s, \tag{86}$$

with

$$0 \leq \rho_s \leq 1, \quad \sum_{s\in S} \rho_s = 1. \tag{87}$$

In this setting, the quantum Boltzmann entropy reads

$$H[\rho] = \sum_{s\in S} \rho_s(\ln \rho_s - 1). \tag{88}$$

It should be noted that the theory can be developed for any entropy, such as the Fermi–Dirac or Bose–Einstein entropy but we restrict ourselves to the Boltzmann entropy for simplicity and refer to [25] for the general case. Here, the entropy is nothing but the information entropy of a system which has discrete states ρ_s.

Functional calculus tells us how to compute a function of an operator. Let $h : \mathbb{R} \to \mathbb{R}$ be an arbitrary function (say continuous to fix the ideas). Then: $h(\rho)$ is defined by

$$h(\rho)\psi = \sum_{s\in S} h(\rho_s)(\psi, \phi_s)\phi_s.$$

Then, the quantum entropy reads

$$H[\rho] = \text{Tr}\{\rho(\ln\rho - 1)\}. \tag{89}$$

Note that we do not take into account that $\text{Tr}\,\rho = 1$ for the time being. We also note that this definition of the entropy is opposite to the usual physical definition. We choose the mathematical convention to prefer convex to concave functions but of course, this unusual convention will not change anything to the results.

Now, the entropy minimization principle reads as follows: Given a set of moments $m = (m_i(x))_{i=0}^{N}$, minimize $H(\rho)$ subject to the constraint that

$$\int W[\rho](x,p)\,\mu(p)\,\widetilde{dp} = m(x) \quad \forall x \in \mathbb{R}^d.$$

As it appears in the formulation of the problem, the entropy is naturally defined in terms of the density operator while the moments are naturally defined in terms of Wigner functions. We now need to reconcile these two representations by invoking the Wigner or inverse Wigner transformations. Because these transformations are non local transformations, the entropy minimization problem must be stated globally (in space) and not locally like in classical mechanics. In other words, we expect that the value of the quantum local thermodynamical equilibrium at one point be dependent on the value of the moments at all points and not specifically at that point like in classical mechanics. This is a signature of the non-local character of quantum mechanics.

We choose to express the moment constraints in terms of the density operator ρ. For this purpose, we dualize the constraint: Let $\lambda(x) = (\lambda_i(x))_{i=0}^{N}$ be an arbitrary (vector) test function. We compute

$$\int w(x,p)\,\mu(p) \cdot \lambda(x)\,dx\,\widetilde{dp} = \int m(x) \cdot \lambda(x)\,dx, \tag{90}$$

and use (80) to express the left-hand side of this equation in terms of ρ:

$$\text{Tr}\{\rho\,W^{-1}[\mu(p) \cdot \lambda(x)]\} = \int m(x) \cdot \lambda(x)\,dx. \tag{91}$$

Here, $\mu(p) \cdot \lambda(x)$ for instance denotes $\sum_{i=0}^{N}\mu_i(p)\lambda_i(x)$.

Therefore, the entropy minimization principle has the following expression: given a set of (physically admissible) moments $m = (m_i(x))_{i=0}^{N}$, solve

$$\min\{H[\rho] = \text{Tr}\{\rho(\ln\rho - 1)\} \quad \text{subject to:}$$
$$\text{Tr}\{\rho\,W^{-1}[\mu(p) \cdot \lambda(x)]\} = \int m \cdot \lambda\,dx, \quad \forall \lambda = (\lambda_i(x))_{i=0}^{N}\}. \tag{92}$$

To solve this problem, we need to compute the Gâteaux derivative of H. We have:

$$\frac{\delta H}{\delta \rho} \delta \rho \stackrel{\text{def}}{=} \lim_{t \to 0} \frac{1}{t} (H[\rho + t\delta\rho] - H[\rho]) = \text{Tr}\{\ln \rho \, \delta\rho\}. \tag{93}$$

This result has first been given by Nier [53–55]. An elementary proof can be found in [25].

Now, using this result and the classical Lagrange multiplier technique, we can assert that there exists a set of Lagrange multipliers, which, here, are functions of x, $\alpha(x) = (\alpha_i(x))_{i=0}^N$, such that:

$$\text{Tr}\{\ln \rho \, \delta\rho\} = \text{Tr}\{\delta\rho \, W^{-1}[\mu(p) \cdot \alpha(x)]\}. \tag{94}$$

We deduce that

$$\ln \rho = W^{-1}[\mu(p) \cdot \alpha(x)]. \tag{95}$$

Therefore, the solution of the entropy problem is ρ_α such that,

$$\rho_\alpha = \exp(W^{-1}[\alpha(x) \cdot \mu(p)]), \tag{96}$$

where $\alpha = (\alpha_i(x))_{i=0}^N$ is determined such that the moments of ρ_α are given by m i.e. $m[\rho_\alpha] = m$.

Introducing $\mathcal{M}_\alpha = W[\rho_\alpha]$ the Wigner transform of ρ_α, we have

$$\mathcal{M}_\alpha = \mathcal{E}\text{xp}(\alpha(x) \cdot \mu(p)), \quad \mathcal{E}\text{xp} \cdot = W[\exp(W^{-1}(\cdot))]. \tag{97}$$

The term $\mathcal{E}\text{xp}$ denotes the exponential in the operator sense, i.e. taking a symbol w, make it an operator using the Weyl quantization, taking the operator exponential and Wigner transforming the resulting operator. It is a highly nonlinear and non-local operator acting on the function w and will be referred to as the "Quantum exponential". From (97) the analogy between the quantum Local Thermodynamical Equilibrium (LTE) and the classical one given by $\mathcal{M}_\alpha = \exp(\alpha \cdot \mu)$ is clear. The only (but tremendous) difference is the replacement of the classical exponentiation by this new operator exponentiation.

Now, we can close the moment system (85) with the quantum Maxwellian and get

$$\partial_t \int \mathcal{E}\text{xp}(\alpha \cdot \mu) \, \mu \, dp + \nabla_x \cdot \int \mathcal{E}\text{xp}(\alpha \cdot \mu) \, \mu \, p \, dp$$

$$+ \int \Theta[V] \mathcal{E}\text{xp}(\alpha \cdot \mu) \, \mu \, dp = \int Q(\mathcal{E}\text{xp}(\alpha \cdot \mu)) \, \mu \, dp. \tag{98}$$

This gives an evolution system for the vector function $\alpha(x, t)$ which will be referred to in the sequel as the Quantum Moment Model (QMM). Note that the right-hand side would be identically zero for the hydrodynamic moments (mass, momentum and energy) if the quantum collision operator conserves these quantities.

It is often more informative to use the density operator formulation in (QMM). For this purpose, we transform (QMM) into density operator formalism using (91). We start from the quantum Liouville equation (76), take moments and close with the equilibrium $\rho = \rho_\alpha$. This gives

$$\partial_t \mathrm{Tr}\{\rho_\alpha W^{-1}(\lambda \cdot \mu)\} = -\frac{i}{\hbar}\mathrm{Tr}\{[\mathcal{H}, \rho_\alpha]W^{-1}(\lambda \cdot \mu)\}$$
$$+\mathrm{Tr}\{\mathcal{Q}(\rho_\alpha)W^{-1}(\lambda \cdot \mu)\}, \quad \forall \text{ test fct } \lambda(x) = (\lambda_i(x))_{i=0}^N. \quad (99)$$

Then, using the cyclic property of the trace, we get

$$\partial_t \int m[\rho_\alpha]\lambda \, dx = -\frac{i}{\hbar}\mathrm{Tr}\{\rho_\alpha[W^{-1}(\lambda \cdot \mu), \mathcal{H}]\} +$$
$$+\mathrm{Tr}\{\mathcal{Q}(\rho_\alpha)W^{-1}(\lambda \cdot \mu)\}, \quad \forall \text{ test fct } \lambda(x) = (\lambda_i(x))_{i=0}^N, \quad (100)$$

which appears as the weak formulation of (QMM) using density operator formulation.

We now return to the entropy concept. First, we note that the kinetic entropy $H[\rho]$ in terms of $w = W[\rho]$ is written

$$H[\rho] = \mathrm{Tr}\{\rho(\ln \rho - 1)\} = \int w(\mathcal{L}\mathrm{n}\, w - 1) \, dx \, \widetilde{dp}, \quad (101)$$

where the quantum logarithm is defined analogously as the quantum exponential $\mathcal{L}\mathrm{n}\, w = W[\ln(W^{-1}(w))]$.

Now, the fluid entropy is defined as the following function $S(m)$ of the moments m:

$$S(m) = H[\rho_\alpha] = \int \mathcal{E}\mathrm{xp}(\alpha \cdot \mu)((\alpha \cdot \mu) - 1) \, dx \, \widetilde{dp}, \quad (102)$$

where α is such that the moments of $\mathcal{E}\mathrm{xp}(\alpha \cdot \mu)$ are m, i.e.

$$m[\alpha] := \int \mathcal{E}\mathrm{xp}(\alpha \cdot \mu)\,\mu \, dp = m. \quad (103)$$

The fluid entropy is therefore the kinetic entropy evaluated for the equilibrium and considered as a function of the moments. We can show that $S(m)$ convex. The proof is given in [25] and is a consequence of the convexity of the Boltzmann entropy function $h(\rho) = \rho(\ln \rho - 1)$

Now, with (103), we can write

$$S(m) = \int \alpha \cdot m \, dx - \Sigma(\alpha), \quad (104)$$

with $\Sigma(\alpha)$ the Legendre dual of S (also known as the Massieu–Planck potential):

$$\Sigma(\alpha) = \int \mathcal{E}\mathrm{xp}(\alpha \cdot \mu)dx\,\widetilde{dp}. \tag{105}$$

$\Sigma(\alpha)$ is also convex and we have the following formulae for the inversion of the mapping $\alpha \to m$:

$$\frac{\delta S}{\delta m} = \alpha, \qquad \frac{\delta \Sigma}{\delta \alpha} = m, \tag{106}$$

where the δ's denote Gâteaux derivatives.

We sketch the proof of (106). Using (93), we have

$$
\begin{aligned}
\delta\Sigma &= \mathrm{Tr}\{\exp(W^{-1}(\alpha \cdot \mu))\,(W^{-1}(\delta\alpha \cdot \mu))\} \\
&= \int \mathcal{E}\mathrm{xp}(\alpha \cdot \mu)\,(\delta\alpha \cdot \mu)dx\,\widetilde{dp} \\
&= \int \delta\alpha \cdot m\,dx,
\end{aligned} \tag{107}
$$

which proves the second formula (106).

Now, we have

$$\delta S = \int (\delta\alpha \cdot m + \alpha \cdot \delta m)\,dx - \delta\Sigma, \tag{108}$$

and with (107) we deduce that

$$\delta S = \int \alpha \cdot \delta m\,dx, \tag{109}$$

which proves the first formula (106).

If we assume that the collision operator is entropy dissipative, then we can prove that the moment models are compatible with the entropy dissipation, i.e. satisfy:

$$\partial_t S(m(t)) \leq 0, \tag{110}$$

for any solution $m(t)$ of (QMM).

To prove this result, we use the density matrix formulation of (QMM) (100) or (99) and choose $\lambda = \alpha$ as a test function:

$$
\begin{aligned}
\partial_t \int m[\rho_\alpha]\alpha\,dx = -\frac{i}{\hbar}\mathrm{Tr}\{\rho_\alpha[W^{-1}(\alpha \cdot \mu), \mathcal{H}]\} \\
+ \mathrm{Tr}\{\mathcal{Q}(\rho_\alpha)W^{-1}(\alpha \cdot \mu)\}.
\end{aligned} \tag{111}
$$

The first term of the left-hand side of (111) is the time derivative of the entropy. Indeed, we have

$$\int m[\rho_\alpha]\alpha\,dx = \mathrm{Tr}\{\rho_\alpha W^{-1}(\alpha \cdot \mu)\} = \mathrm{Tr}\{\rho_\alpha \ln \rho_\alpha\}, \tag{112}$$

and taking the time derivative leads to

$$\partial_t \int m[\rho_\alpha]\alpha\, dx = \partial_t(\mathrm{Tr}\{\rho_\alpha \ln \rho_\alpha\}) = \partial_t(\mathrm{Tr}\{\rho_\alpha(\ln \rho_\alpha - 1)\}) = \partial_t S(m). \quad (113)$$

Here, we have used the fact that $\mathrm{Tr}\{\rho_\alpha\} = 1$. Indeed, in this theory, we always normalize the moments by the total mass in order to keep total mass equal to unity (i.e. $\int n\, dx = 1$) and keep the trace of the density matrix equal to unity. The first term at the right-hand side of (111) is transformed using the cyclic invariance of the trace and the fact that two operators such that one is a function of the other one commute (they have the same eigenfunctions):

$$\mathrm{Tr}\{\rho_\alpha[W^{-1}(\alpha \cdot \mu), \mathcal{H}]\} = \mathrm{Tr}\{[\rho_\alpha, \ln \rho_\alpha]\mathcal{H}]\} = 0. \quad (114)$$

Finally, we express that $\mathcal{Q}$ is entropy dissipative:

$$\mathrm{Tr}\{\mathcal{Q}(\rho_\alpha)W^{-1}(\alpha \cdot \mu)\} = \mathrm{Tr}\{\mathcal{Q}(\rho_\alpha) \ln \rho_\alpha\} \leq 0. \quad (115)$$

Collecting (111) to (115) leads to (110) and proves entropy dissipation.

A remarkable special case of (QMM) is the Quantum Hydrodynamic Model (QHD) which is obtained by choosing the set of hydrodynamic monomials $\mu = \{1, p, |p|^2\}$ to generate an evolution system for the hydrodynamic moments. In this setting, using that the collision operator is supposed to preserve mass, momentum and energy, we find the following balance equations:

$$\partial_t n + \nabla_x \cdot nu = 0, \quad (116)$$
$$\partial_t nu + \nabla_x \Pi = -n\nabla_x V, \quad (117)$$
$$\partial_t W + \nabla_x \cdot \Phi = -nu \cdot \nabla_x V, \quad (118)$$

with the pressure tensor Π and the energy flux Φ given by

$$\Pi = \int \mathcal{E}\mathrm{xp}(\alpha \cdot \mu)\, p \otimes p\, \widetilde{dp}, \quad (119)$$

$$2\Phi = \int \mathcal{E}\mathrm{xp}(\alpha \cdot \mu)\, |p|^2\, \widetilde{dp}, \quad (120)$$

and

$$\alpha \cdot \mu = A(x) + B(x) \cdot p + C(x)|p|^2, \quad (121)$$

such that

$$\int \mathcal{E}\mathrm{xp}(\alpha \cdot \mu) \begin{pmatrix} 1 \\ p \\ |p|^2 \end{pmatrix} \widetilde{dp} = \begin{pmatrix} n \\ q \\ 2\mathcal{W} \end{pmatrix}. \quad (122)$$

$A(x)$ and $C(x)$ are scalar functions of x while $B(x)$ is a vector valued function with values in $\mathbb{R}^d$.

By a "Quantum Maxwellian" we now refer to an equilibrium associated with the prescription of the hydrodynamic moments, i.e.

$$\mathcal{M}_\alpha = \mathcal{E}\mathrm{xp}(\alpha \cdot \mu) = W(\exp(W^{-1}(\alpha \cdot \mu))), \tag{123}$$

with $\alpha \cdot \mu$ given by (121) where $\alpha = (A, B, C)$ are related with $(n, nu, \mathcal{W})$ in a non-local way through (122). We note that $u \neq B/2C$ in general while this is true in the classical case.

The operator $W^{-1}(\alpha \cdot \mu)$ is a second order differential operator:

$$W^{-1}(\alpha \cdot \mu)\psi = -\hbar^2 \nabla \cdot (C\nabla \psi)$$
$$- i\hbar(B \cdot \nabla \psi + (1/2)(\nabla \cdot B)\psi) + (A - (\hbar^2/4)\Delta C)\psi. \tag{124}$$

To prove this result, we note the following identities:

$$W^{-1}(A) = A, \tag{125}$$

$$W^{-1}(B \cdot p) = -i\hbar(B \cdot \nabla + \frac{1}{2}(\nabla \cdot B)), \tag{126}$$

$$W^{-1}(C|p|^2) = -\hbar^2(C\Delta + \nabla C \cdot \nabla + \frac{1}{4}\Delta C). \tag{127}$$

For instance, to show (126), we use (79) and compute

$$W^{-1}(B \cdot p)\,\psi = \int B(\frac{x+y}{2}) \cdot p\,\psi(y)e^{\frac{ip(x-y)}{\hbar}}\,\widetilde{dp}\,dy. \tag{128}$$

Elementary manipulations of Fourier transform lead to

$$\int p\,e^{\frac{ip(x-y)}{\hbar}}\,\widetilde{dp} = i\hbar\nabla\delta(y-x). \tag{129}$$

Then

$$W^{-1}(B \cdot p)\,\psi = i\hbar \int B(\frac{x+y}{2}) \cdot \nabla\delta(y-x)\,\psi(y)\,dy$$
$$= -i\hbar \int \nabla_y \cdot (B(\frac{x+y}{2})\psi(y))\,\delta(y-x)\,dy$$
$$= -i\hbar(\frac{1}{2}\nabla \cdot B(x)\psi(x) + B(x) \cdot \nabla\psi(x)), \tag{130}$$

which is the result to be proved. Combining identities (125) to (127), we find (124).

Now, suppose that $W^{-1}(\alpha \cdot \mu)$ has point spectrum only, i.e. has eigenvalues $a_s[\alpha]$, and eigenvectors $\phi_s[\alpha]$:

$$W^{-1}(\alpha \cdot \mu) = \sum_s a_s(\cdot, \phi_s)\phi_s, \tag{131}$$

$$\rho_\alpha = \exp(W^{-1}(\alpha \cdot \mu)) = \sum_s e^{a_s}(\cdot, \phi_s)\phi_s. \tag{132}$$

Then, from the fact that $\mathrm{Tr}\rho_\alpha = 1$, we deduce that $\sum_s e^{a_s} = 1$. It follows that $a_s < 0$ and $a_s \overset{s\to\infty}{\longrightarrow} -\infty$. We deduce that $-W^{-1}(\alpha \cdot \mu)$ must be an elliptic operator which indicates that $C(x)$ is likely to be non-positive, at least in an important part of the domain. This is to be related to the positivity of the temperature for the classical maxwellian case. Of course, this is not a rigorous proof and the study of operator $-W^{-1}(\alpha \cdot \mu)$ would require further investigations.

In practice, it is necessary to compute the mapping between the so-called "entropic variables" (A, B, C) and the associated "conservative variables" $(n, u, \mathcal{W})$. This can be done by a minimization problem thanks to the first formula of (106) and the fact that Σ is convex. This formula can indeed be put in a minimization form

$$\min_\alpha \{\Sigma(\alpha) - \int \alpha \cdot m \, dx\}, \tag{133}$$

i.e.

$$\min_\alpha \{\sum_s e^{a_s[\alpha]} - \int \alpha \cdot m \, dx\}. \tag{134}$$

This idea is used in practical computations (see, e.g. [32, 33]).

3.3 Quantum Isothermal Euler Model

This part is a summary of [22]. We want to develop a quantum counterpart of the classical isothermal Euler equations. In this setting, the temperature of the medium is fixed and uniform: $T = $ Constant. To deal with a constant temperature background, it is necessary to change the entropy into the Free Energy

$$G(\rho) = \mathrm{Tr}\{Th(\rho) + \mathcal{H}\rho\}, \tag{135}$$

$$h(\rho) = \rho(\ln \rho - 1), \quad \mathcal{H} = W^{-1}(\frac{|p|^2}{2} + V). \tag{136}$$

From now on, we shall omit the W^{-1}. Any function $\phi(x, p)$ will be identified to the operator $W^{-1}(\phi(x, p))$. Two moments are considered: the density n and the momentum nu. The entropy minimization problem reads: To find

$$\min G(\rho) = \min(\mathrm{Tr}\{T\rho(\ln \rho - 1) + \mathcal{H}\rho\}), \tag{137}$$

subject to the moment constraints

$$\mathrm{Tr}\{\rho\phi\} = \int n\phi \, dx, \tag{138}$$

$$\mathrm{Tr}\{\rho W^{-1}(p \cdot \Phi)\} = \int nu \cdot \Phi \, dx, \tag{139}$$

for all (scalar and vector respectively) test functions ϕ and Φ.

The solution of the entropy minimization problem must satisfy

$$T \ln \rho + \mathcal{H} = \tilde{A} + \tilde{B} \cdot p. \tag{140}$$

After rearrangement, this is equivalent to

$$\ln \rho = -\frac{H(A, B)}{T}, \quad H(A, B) = \frac{|p - B|^2}{2} + A, \tag{141}$$

with

$$A = V - \tilde{A} - |\tilde{B}|^2/2 \,, \qquad B = \tilde{B}. \tag{142}$$

We shall refer to $H(A, B)$ as a modified Hamiltonian.

The quantum Maxellian in density operator formulation reads

$$\rho_{n,nu} = \exp(-\frac{H(A, B)}{T}), \tag{143}$$

and in Wigner formulation

$$\mathcal{M}_{n,nu} = \mathcal{E}\mathrm{xp}(-\frac{H(A, B)}{T}), \tag{144}$$

with (A, B) related with (n, nu) by the moment conditions (138), (139). We shall suppose that $T = 1$ for the sake of simplicity from now on.

Let us consider the moment reconstruction problem and suppose that $H(A, B)$ has a discrete spectrum with eigenvalues $\lambda_p(A, B)$ and eigenfunctions $\psi_p(A, B)$ for $p = 1, \ldots, \infty$. Then we have

$$n(A, B)(x) = \sum_{p=1}^{\infty} \exp(-\lambda_p(A, B)) \, |\psi_p(A, B)(x)|^2, \tag{145}$$

$$nu(A, B)(x) = \sum_{p=1}^{\infty} \exp(-\lambda_p(A, B)) \, \mathrm{Im}(\hbar \overline{\psi_p(A, B)(x)} \nabla \psi_p(A, B)(x)). \tag{146}$$

Indeed, by construction, we have

$$\rho_{n,nu} \cdot = \sum_{p=1}^{\infty} \exp(-\lambda_p(A, B))(\cdot, \psi_p)\psi_p \,. \tag{147}$$

Of course, the operator ρ is diagonal in the basis (ψ_p), with diagonal element equal to $\exp(-\lambda_p(A, B))$. On the other hand, the multiplication operator by ϕ has matrix element in this basis

$$\phi_{p,p'} = \int \phi \psi_p \overline{\psi_{p'}} \, dx. \tag{148}$$

Since taking the trace of $\rho\phi$ amounts to summing up the products of the diagonal elements of the operators ρ and ϕ respectively (since ρ is diagonal), we get

$$\mathrm{Tr}\{\rho\phi\} = \sum_{p=1}^{\infty} \exp(-\lambda_p(A, B)) \int \phi |\psi_p|^2 \, dx. \tag{149}$$

Finally

$$n(x_0) = \mathrm{Tr}\{\rho\delta(x - x_0)\}$$
$$= \sum_{p=1}^{\infty} \exp(-\lambda_p(A, B)) |\psi_p(x_0)|^2, \tag{150}$$

which is (145). A similar computation for nu would lead to (146).

Now, we can state our first expression of the Quantum Isothermal Euler model. It comes as a special case of the (QHD) model (116)–(118) with the energy equation replaced by the assumption of constant temperature. Therefore, it reads

$$\partial_t n + \nabla \cdot nu = 0, \tag{151}$$
$$\partial_t nu + \nabla \Pi = -n\nabla V, \tag{152}$$

with the pressure tensor Π given by

$$\Pi = \int \mathcal{E}\mathrm{xp}(-H(A, B)) \, p \otimes p \, \widetilde{dp}, \tag{153}$$

where the modified Hamiltonian $H(A, B)$ is given by (141) and (A, B) is related with (n, nu) by the moment conditions (145), (146).

Now, our next task is to give a more tractable expression of the pressure tensor Π in terms of n, nu, A, B. Given a scalar test function ϕ, we can write

$$\int (\nabla \Pi) \, \phi \, dx = - \int \Pi \, \nabla \phi \, dx$$
$$-- \int \mathcal{E}\mathrm{xp}(-H(A, B)) \, (p \cdot \nabla \phi) p \, dx \, \widetilde{dp}$$
$$= -\mathrm{Tr}\{\exp(-H(A, B)) \, W^{-1}((p \cdot \nabla \phi)p)\}. \tag{154}$$

The idea is to use the commutation with $H(A, B)$ in order to reduce the degree of the p-monomial. More specifically, if we succeed in writing

$$(p \cdot \nabla \phi)p = [H(A, B), \mathcal{L}] + \mathcal{N}, \tag{155}$$

where $\mathcal{N}$ is a polynomial in p of degree strictly less than 2, then

$$\begin{aligned}
\mathrm{Tr}\{\exp(-H(A, B)) \, (p \cdot \nabla \phi)p\} &= \\
&= \mathrm{Tr}\{\exp(-H(A, B)) \, [H(A, B), \mathcal{L}]\} \\
&\quad + \mathrm{Tr}\{\exp(-H(A, B)) \, \mathcal{N}\}.
\end{aligned} \tag{156}$$

Using the cyclic property of the trace, we have

$$\begin{aligned}
\mathrm{Tr}\{\exp(-H(A, B)) \, [H(A, B), \mathcal{L}]\} &= \\
&= \mathrm{Tr}\{[\exp(-H(A, B)), \, H(A, B)]\mathcal{L}\} = 0,
\end{aligned} \tag{157}$$

so that

$$\mathrm{Tr}\{\exp(-H(A, B)) \, (p \cdot \nabla \phi)p\} = \mathrm{Tr}\{\exp(-H(A, B)) \, \mathcal{N}\}, \tag{158}$$

and the degree in p of the trace to be evaluated has decreased. It remains to find the convenient operators $\mathcal{L}$ and $\mathcal{N}$. In the present case, $\mathcal{L}$ will be proportional to $p\phi$. Therefore, we turn to the computation of $[H(A, B), p\phi]$.

We first note the following commutation relations

$$[\phi, \psi] = 0, \tag{159}$$

$$[p \cdot \Phi, \psi] = -i\hbar(\Phi \cdot \nabla \psi), \tag{160}$$

$$[p \cdot \Phi, p \cdot \Psi] = -i\hbar((\Phi \cdot \nabla)\Psi - (\Psi \cdot \nabla)\Phi) \cdot p, \tag{161}$$

$$[|p|^2/2, \phi] = -i\hbar \nabla \phi \cdot p, \tag{162}$$

$$[|p|^2/2, p\phi] = -i\hbar(\nabla \phi \cdot p)p. \tag{163}$$

We note that commutation decreases the degree in p.

For instance, we show (160). We first note, after (126) that

$$p \cdot \Phi = -i\hbar \left(\Phi \cdot \nabla + (\nabla \cdot \Phi)/2 \right). \tag{164}$$

Using that two functions of x commute, we notice that $[(\nabla \cdot \Phi), \psi] = 0$ and we are left with

$$\begin{aligned}
[\Phi \cdot \nabla, \, \psi]f &= \Phi \cdot \nabla(\psi f) - \psi \Phi \cdot \nabla f \\
&= (\Phi \cdot \nabla \psi)f,
\end{aligned} \tag{165}$$

for an arbitrary test function $f \in L^2$.

Now, using formulas (159)–(163), the computation of $[H(A, B), p\phi]$ is easy and leads to

$$\begin{aligned}
[H(A, B), p\phi] = -i\hbar\{(\nabla \phi \cdot p)p - (B \cdot \nabla \phi)p + \\
+ \phi(\nabla B)p - \phi \nabla(A + |B|^2/2)\}.
\end{aligned} \tag{166}$$

Therefore
$$(p \cdot \nabla \phi)p = [H(A, B), \mathcal{L}] + \mathcal{N}, \tag{167}$$

with

$$\mathcal{L} = (i/\hbar)p\phi, \tag{168}$$
$$\mathcal{N} = (B \cdot \nabla \phi)p - \phi(\nabla B)p + \phi\nabla(A + |B|^2/2). \tag{169}$$

Then

$$\begin{aligned}
\mathrm{Tr}\{\exp(-H(A, B))\,(p \cdot \nabla \phi)p\} &= \\
&= \mathrm{Tr}\{\exp(-H(A, B))\,\mathcal{N}\} \\
&= \mathrm{Tr}\{\exp(-H(A, B))\,((B \cdot \nabla \phi)p - \phi(\nabla B)p + \phi\nabla(A + |B|^2/2))\}. \quad (170)
\end{aligned}$$

Using that $\exp(-H(A, B))$ satisfies the moment reconstruction problem (138), (139), we obtain

$$\begin{aligned}
\mathrm{Tr}\{\exp(-H(A, B))\,(p \cdot \nabla \phi)p\} &= \\
&= \int ((B \cdot \nabla \phi)nu - \phi(\nabla B)nu + n\phi\nabla(A + |B|^2/2))\, dx. \quad (171)
\end{aligned}$$

We carry all derivatives outside the test function ϕ by using Green's formula and get:

$$\begin{aligned}
\mathrm{Tr}\{\exp(-H(A, B))\,(p \cdot \nabla \phi)p\} &= \\
&= \int (-\nabla(nu \otimes B) - (\nabla B)nu + n\nabla(A + |B|^2/2))\phi\, dx. \quad (172)
\end{aligned}$$

Finally, according to (154), all this expression equals $-\int (\nabla \varPi)\,\phi\, dx$. The following final expression of $\varPi$ is deduced:

$$\nabla \varPi = \nabla(nu \otimes B) + (\nabla B)nu - n\nabla(A + |B|^2/2). \tag{173}$$

We are now ready to provide a second expression of the Quantum Isothermal Euler model:

$$\partial_t n + \nabla \cdot nu = 0, \tag{174}$$
$$\partial_t nu + \nabla(nu \otimes B) + (\nabla B)nu - n\nabla(A + |B|^2/2) = -n\nabla V, \tag{175}$$

where (A, B) are related with (n, nu) by the moment reconstruction problem (145), (146).

We now consider the Free energy again. The fluid free energy $\mathcal{G}(n, nu)$ is defined as the kinetic free energy evaluated on the equilibrium density matrix:

$$\mathcal{G}(n, nu) = G(\rho_{n,nu}). \tag{176}$$

Using (137), we find:

$$\mathcal{G}(n, nu) = \mathrm{Tr}\{\exp(-H(A,B))(-H(A,B) - 1 + \mathcal{H})\}. \tag{177}$$

Now the expression (141) of $H(A,B)$ leads to

$$\mathcal{G}(n, nu) = \mathrm{Tr}\{\exp(-H(A,B))(B \cdot p - A - |B|^2/2 - 1 + V)\}. \tag{178}$$

Finally, using that $\exp(-H(A,B))$ satisfies the moment reconstruction problem (138), (139), we obtain

$$\mathcal{G}(n, nu) = \int (nu \cdot B + n(V - A - |B|^2/2 - 1)) \, dx. \tag{179}$$

By construction, if V is independent of time, the free energy decays in time

$$\frac{d\mathcal{G}}{dt} \leq 0, \tag{180}$$

with equality for smooth solutions. The proof follows the same steps as for the general (QMM) model. If V solves Poisson equation $-\Delta V = n$ then, again the entropy inequality (180) holds with a modification of the entropy (179), multiplying the term nV by a factor $1/2$ (a usual procedure when computing self-energies). This shows that the Quantum Isothermal Euler system is compatible with free-energy dissipation.

We now turn to a property of the Quantum Isothermal Euler which is a form of gauge invariance. Let $S(x)$ be a smooth function. Then

$$\exp(\frac{iS}{\hbar}) \, H(A,B) \, \exp(-\frac{iS}{\hbar}) = H(A, B + \nabla S). \tag{181}$$

To prove this identity, we write that

$$\exp(iS/\hbar)H(A,B)\exp(-iS/\hbar) - H(A,B) =$$
$$= \exp(iS/\hbar)[H(A,B), \exp(-iS/\hbar)],$$

and we use the commutation relations (159)–(163).

As a consequence, the eigenvalues of $H(A,B)$ and $H(A, B + \nabla S)$ are the same. Also, if two operators are conjugate, any function of these two operators is also conjugate by the same conjugation operator. This implies that

$$\exp(\frac{iS}{\hbar}) \, \exp(-H(A,B)) \, \exp(-\frac{iS}{\hbar}) = \exp(-H(A, B + \nabla S)). \tag{182}$$

Thus, the equilibrium density operators are conjugate. Therefore, the eigenvalues of $\exp(-H(A,B))$ and $\exp(-H(A, B + \nabla S))$ are also the same.

Going back to the fluid Free energy (176), formula (179) implies that

$$\frac{\delta \mathcal{G}}{\delta n} = V - A - |B|^2/2 = \tilde{A}, \tag{183}$$

$$\frac{\delta \mathcal{G}}{\delta (nu)} = B. \tag{184}$$

These formulas are direct consequences of (106). The Legendre dual of the entropy is given by

$$\tilde{\Sigma}(\tilde{A}, B) = \int n \, dx = \mathrm{Tr}\{\exp(-H(A, B))\} := \Sigma(A, B). \tag{185}$$

The inversion formula (106) and the chain rule leads to

$$n(A, B) = \frac{\delta \tilde{\Sigma}}{\delta \tilde{A}} = -\frac{\delta \Sigma}{\delta A}, \tag{186}$$

$$(nu)(A, B) = \frac{\delta \tilde{\Sigma}}{\delta B} = \frac{\delta \Sigma}{\delta B} - B \frac{\delta \Sigma}{\delta A}. \tag{187}$$

It results:

$$\frac{\delta \Sigma}{\delta A} = -n(A, B), \tag{188}$$

$$\frac{\delta \Sigma}{\delta B} = (nu)(A, B) - n(A, B) B. \tag{189}$$

Now, by the gauge invariance, the eigenvalues of $\exp(-H(A, B))$ and $\exp(-H(A, B + \nabla S))$ are the same:

$$\Sigma(A, B) = \mathrm{Tr}\{\exp(-H(A, B))\} =$$
$$= \mathrm{Tr}\{\exp(-H(A, B + \nabla S))\} = \Sigma(A, B + \nabla S). \tag{190}$$

This implies

$$\frac{\delta \Sigma}{\delta A}(A, B + \nabla S) = \frac{\delta \Sigma}{\delta A}(A, B),$$

$$\frac{\delta \Sigma}{\delta B}(A, B + \nabla S) = \frac{\delta \Sigma}{\delta B}(A, B).$$

With (188), (189), this leads to

$$n(A, B + \nabla S) = n(A, B), \tag{191}$$

$$(nu)(A, B + \nabla S) = nu(A, B) + n(A, B)\nabla S, \tag{192}$$

and relates the density and velocity of the Quantum Isothermal Euler for two values of B differing by a gradient.

Another consequence is that for all test functions $S(x)$, we have

$$\lim_{t\downarrow 0} t^{-1}(\varSigma(A, B + t\nabla S) - \varSigma(A, B)) = 0 =$$

$$= \int \frac{\delta\varSigma}{\delta B} \cdot \nabla S\,dx = \int (nu - nB) \cdot \nabla S\,dx, \qquad (193)$$

meaning that

$$\nabla \cdot (n(u - B)) = 0. \qquad (194)$$

This expresses the very strong result that u and B are not equal but differ by a vector field which is a curl divided by the density.

Now, we are ready to propose various equivalent formulations of the momentum equation originally given by (175). First using that $\nabla|B|^2/2 = (\nabla B)B$, we can transform (175) into

$$\partial_t nu + \nabla(nu \otimes B) + n(\nabla B)(u - B) + n\nabla(V - A) = 0. \qquad (195)$$

Now, using the constraint (194), we can transform (195) into

$$\partial_t nu + \nabla(nu \otimes u) + n(\nabla \times u) \times (B - u) +$$
$$+ n\nabla(V - A - |B - u|^2/2) = 0. \qquad (196)$$

Finally, the continuity equation allows us to transform (195) into

$$\partial_t u + (\nabla \times u) \times B + \nabla(u \cdot B - |B|^2/2 + V - A) = 0. \qquad (197)$$

The case of irrotational flows deserves a special mention. We define the vorticity by $\omega = \nabla \times u$. By taking the curl of Form (197), ω satisfies

$$\partial_t \omega + \nabla \times (\omega \times B) = 0. \qquad (198)$$

If $\omega|_{t=0} = 0$, then $\omega \equiv 0$ for all times. Therefore, if a flow is irrotational at time $t = 0$, it stays irrotational for all time.

For an irrotational flow, there exists a scalar function $S(x, t)$ such that $u = \nabla S$. Then, it can be proven that

$$u = B = \nabla S. \qquad (199)$$

To show this property, we first note that $nu(A, 0) = 0$. Indeed,

$$nu(A, 0) = \int \mathcal{E}\mathrm{xp}(-H(A, 0))p\,\widetilde{dp}. \qquad (200)$$

But $H(A, 0) = |p|^2/2 + A$ is even with respect to p. It follows that $\mathcal{E}\mathrm{xp}(-H(A, 0))$ is also even with respect to p. This is not obvious (since $\mathcal{E}\mathrm{xp}$ is not the exponential in the usual sense). To prove it, we can write the series expansion of the operator exponential, prove the evenness property for powers of the operator (using Wigner transform for instance) and then extend it by means of the series expansion, to the exponential. It follows that $nu(A, 0) = 0$ by symmetry.

Now, using the Gauge transformation, we have

$$
\begin{aligned}
nu(A, \nabla S) &= nu(A, 0) + n(A, 0)\nabla S \\
&= \quad 0 \quad + n(A, \nabla S)\nabla S,
\end{aligned}
$$

which shows that the solution (A, B) of the moment problem is given by A solving $n(A, 0) = n$ and $B = \nabla S = u$. Of course, this argument relies on the assumption that the pair (A, B) is unique (given (n, nu)), which has not been rigorously proven yet. Therefore, at this point this argument is only formal.

Nevertheless, applying (199) considerably simplifies the formulation of the Quantum Isothermal Euler model in the case of irrotational flows. This system reads (take $B = u$ in (196)):

$$
\partial_t n + \nabla \cdot nu = 0, \tag{201}
$$

$$
\partial_t nu + \nabla(nu \otimes u) + n\nabla(V - A) = 0, \tag{202}
$$

$$
\nabla \times u = 0, \tag{203}
$$

where A is related with n by (145). In this case, only one quantity A is to be determined from the spectral problem. An important sub-case of irrotational flows is of course one-dimensional flows.

Now, we turn to the semiclassical asymptotics of the Quantum Isentropic Euler model. When $\hbar \to 0$, we recover the classical isothermal Euler equations. If we retain terms of order $\hbar^2$, we obtain:

$$
\partial_t n + \nabla \cdot (nu) = 0, \tag{204}
$$

$$
\partial_t(nu) + \nabla(nu \otimes u) + \nabla n + n\nabla V - \frac{\hbar^2}{6}n\nabla\left(\frac{\Delta\sqrt{n}}{\sqrt{n}}\right) +
$$

$$
+\frac{\hbar^2}{12}\omega \times (\nabla \times (n\omega)) + \frac{\hbar^2}{24}n\nabla(|\omega|^2) = 0, \tag{205}
$$

$$
\omega = \nabla \times u. \tag{206}
$$

This expansion has been given by Jüngel and Matthes in [40]. The computations leading to these formula are quite involved and we refer the reader to [22].

In the case of irrotational flows $\omega = 0$, the semiclassical asymptotics leads to

$$
\partial_t n + \nabla \cdot nu = 0, \tag{207}
$$

$$
\partial_t nu + \nabla(nu \otimes u) + \nabla n + n\nabla V - \frac{\hbar^2}{6}n\nabla\left(\frac{\Delta\sqrt{n}}{\sqrt{n}}\right) = 0 \tag{208}
$$

This model can be found in the literature under the name of "Quantum Hydrodynamic Model" [34]. It is a classical isothermal Euler model with the addition of the Bohm potential in the momentum equation. However, in the literature, it is derived on the basis of phenomenological considerations, and is used in the irrotational as well as the non-irrotational cases. Here, this "Quantum Hydrodynamic Model" is derived on the basis of first principles and it appears to be restricted to irrotational flows, the completely general form valid for non-irrotational flows being (204)–(206).

We now report on preliminary numerical results using this model. Our test problem is a one-dimensional model where the hydrodynamic quantities are coupled with Poisson's equation. A relaxation term has been also added to the momentum relaxation term. We consider a double barrier structure with boundary conditions of Dirichlet type for the wave-function (and consequently for the density) and a zero flux boundary condition for the momentum. We observe the dynamics of electrons which are initially close to the left boundary and which move to eventually fill in the whole domain.

The values of the parameters are given in Table 1. The initial density and velocity are depicted in Fig. 1. Figures 2, 3 and 4 show the evolution of the electron density on the left, and the velocity on the right after 20, 100 and 200 time steps. We can see electrons going through the barriers by tunneling effect. At time step 200 the system seems to achieve an equilibrium. This

Table 1 Values of the parameters for the numerical simulation (Δx is the space step, Δt the time step, $h^2/2T$ the scaled Planck constant, α the scaled Debye length, τ the scaled momentum relaxation time, see [22] for the meaning of these parameters)

Δx	Δt	$h^2/2T$	α^2	τ
0.01	0.005	0.02	0.1	0.1

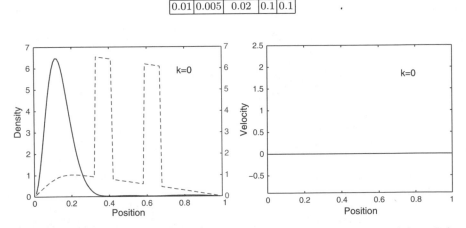

Fig. 1 Numerical solution of the quantum Euler model with relaxation: initial data. *Left*: initial density (*solid line*) and total electrical potential (*dashed line*) as functions of the position x. *Right*: initial velocity as a function of the position x

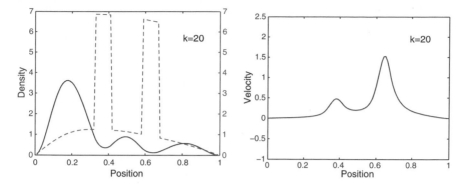

Fig. 2 Numerical solution of the quantum Euler model with relaxation after 20 iterations. *Left*: density (*solid line*) and total electrical potential (*dashed line*) as functions of the position x. *Right*: velocity as a function of the position x

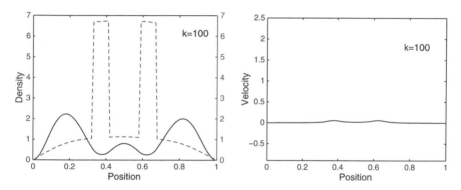

Fig. 3 Numerical solution of the quantum Euler model with relaxation after 100 iterations. *Left*: density (*solid line*) and total electrical potential (*dashed line*) as functions of the position x. *Right*: velocity as a function of the position x

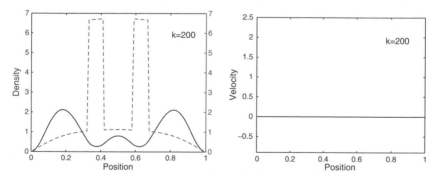

Fig. 4 Numerical solution of the quantum Euler model with relaxation after 200 iterations. *Left*: density (*solid line*) and total electrical potential (*dashed line*) as functions of the position x. *Right*: velocity as a function of the position x

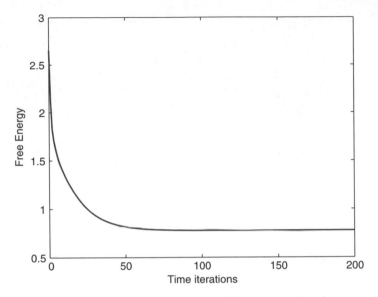

Fig. 5 Evolution of the free energy as a function of the time iteration k

is confirmed by Fig. 5 which shows that the free energy does not evolve any more. On this last graph, we can see that the free energy is a decreasing function of time, as expected.

4 Quantum Diffusion Models

4.1 Quantum Energy-Transport Model

In this section, we report on the work [23, 24].

We notice that the derivation of the Quantum Moment Models, or Quantum Hydrodynamic models did not require any knowledge of the exact form of the collision operator. The only properties which were used were the entropy dissipation and the conservation properties of mass, momentum and energy.

For deriving a diffusion model from a kinetic equation however, the exact form of the collision operator matters and the coefficients of the diffusion model itself depend on this collision operator.

Therefore, in order to derive Quantum Diffusion model, we need to specify the collisions operator $\mathcal{Q}$ in Quantum Liouville equation

$$i\hbar\partial_t\rho = [\mathcal{H}, \rho] + i\hbar\mathcal{Q}(\rho), \qquad (209)$$

or in the Wigner equation

$$\partial_t w + p \cdot \nabla_x w + \Theta^\hbar[V]w = Q(w). \tag{210}$$

In the absence of a precise definition of the physical collision mechanism, the most simple choice is a relaxation operator also called BGK operator. The collision operator expresses the relaxation of the collision operator to the Local Thermodynamical Equilibrium, in our case, the quantum Maxwellian. We want to investigate a case where this collision operator is written

$$Q(w)(p) = -\nu(w - \mathcal{E}\text{xp}(A + C|p|^2/2)), \tag{211}$$

where we recall that $\mathcal{E}\text{xp}\, w = W\,(\exp\,(W^{-1}w))$. The functions $A(x)$ and $C(x)$ are such that the operator Q locally conserves mass and energy. More precisely, let us write

$$\mathcal{M}_{n,\mathcal{W}} = \mathcal{E}\text{xp}(A + C|p|^2/2), \tag{212}$$

the Quantum Maxwellian whose local mass at point x is $n(x)$ and local energy is $\mathcal{W}(x)$. Then, (A, C) is such that

$$\int \mathcal{E}\text{xp}(A + C|p|^2/2) \begin{pmatrix} 1 \\ |p|^2/2 \end{pmatrix} \widetilde{dp} = \begin{pmatrix} n \\ \mathcal{W} \end{pmatrix}. \tag{213}$$

In density operator form, the Quantum Maxwellian is written

$$\rho_{n,\mathcal{W}} = W^{-1}(\mathcal{M}_{n,\mathcal{W}}) = \exp(W^{-1}(A + C|p|^2/2)), \tag{214}$$

with, for all test functions ϕ:

$$\text{Tr}\{\rho_{n,\mathcal{W}}\, \phi\} = \int n\phi\, dx\,, \quad \text{Tr}\{\rho_{n,\mathcal{W}}\, \phi|p|^2/2\} = \int \mathcal{W}\phi\, dx. \tag{215}$$

Again, we recall that the Quantum Maxwellian $\rho_{n,\mathcal{W}} = \exp(W^{-1}(A + C|p|^2/2))$ is a solution of the entropy minimization principle: to find

$$\min\{H[\rho] = \text{Tr}\{\rho(\ln\rho - 1)\}\}, \quad \text{subject to:}$$
$$\text{Tr}\{\rho_{n,\mathcal{W}}\, \phi\} = \int n\phi\, dx\,, \quad \text{Tr}\{\rho_{n,\mathcal{W}}\, \phi|p|^2/2\} = \int \mathcal{W}\phi\, dx\}. \tag{216}$$

In Wigner form, the quantum entropy $H[\rho]$ has the expression:

$$H[\rho] = \text{Tr}\{\rho(\ln\rho - 1)\} = \int w(\mathcal{L}\text{n}\, w - 1)\, dx\, \widetilde{dp}, \tag{217}$$

where the quantum logarithm is defined according to $\mathcal{L}\text{n}\, w = W[\ln(W^{-1}(w))]$.

For a given Wigner distribution w, let us denote by $\mathcal{M}_w := \mathcal{M}_{n,\mathcal{W}}$ the Quantum Maxwellian which possesses the same density n and energy $\mathcal{W}$ as w:

$$\int \mathcal{M}_w \begin{pmatrix} 1 \\ |p|^2/2 \end{pmatrix} dp = \int w \begin{pmatrix} 1 \\ |p|^2/2 \end{pmatrix} dp. \qquad (218)$$

Then, the Quantum BGK operator is written

$$Q(w) = -\nu(w - \mathcal{M}_w). \qquad (219)$$

In density operator form, we shall denote the quantum Maxwellian which has the same mass and energy as ρ by $\mathcal{M}_\rho$. Then the Quantum BGK operator is written

$$\mathcal{Q}(\rho) = -\nu(\rho - \mathcal{M}_\rho). \qquad (220)$$

The physical situation modeled by $Q(w)$ is typically when the energy exchanges among the particles themselves are more efficient than with the surrounding and that a different temperature than that of the background is possible. In short channel transistors, the electron typical energy exceeds the phonon energy by almost two orders of magnitude. Then, the phonon collisions can be viewed as quasi-elastic and most of the energy exchanges are with the other electrons via Coulomb interaction. In plasmas, a similar situation arises between electrons and ions because of the very small electron to ion mass ratio.

Again, we observe that we need the two sets of variables: the conservative variables $(n, \mathcal{W})$ and the entropic variables (A, C). The passage between $(n, \mathcal{W})$ and (A, C) is a functional change of variable which is done through the entropy and its Legendre dual as seen in the previous sections.

Let us now summarize the properties of Q:

(a) Mass and energy conservation

$$\int Q(w) \begin{pmatrix} 1 \\ |p|^2 \end{pmatrix} dp = 0, \qquad (221)$$

(b) Null set of Q (equilibria):

$$Q(w) = 0 \iff \exists (A, C) \text{ such that } w = \mathcal{E}\mathrm{xp}(A + C|p|^2/2), \qquad (222)$$

(c) Entropy decay:

$$\int Q(w) \, \mathcal{L}nw \, dx \, \widetilde{dp} = \mathrm{Tr}\{\mathcal{Q}(\rho) \ln \rho\} \leq 0. \qquad (223)$$

Properties (a) and (b) are obvious from definition (219) and the conservation relations (213). The only delicate point is entropy decay (c). In the classical case, the proof uses that the logarithm is an increasing function. This is no more true here in the case of the quantum logarithm. Indeed, because the dependence between w and $\mathcal{L}n(w)$ is functional, the statement that $\mathcal{L}n(w)$ is increasing w.r.t w is meaningless. So another proof must be developed. It uses convexity argument. Indeed, because of the convexity of the entropy functional H, the function Λ:

$$\lambda \in [0,1] \to \Lambda(\lambda) = H((1-\lambda)\mathcal{M}_\rho + \lambda\rho), \tag{224}$$

is convex. Therefore,

$$\frac{d\Lambda}{d\lambda}(1) \geq \Lambda(1) - \Lambda(0), \tag{225}$$

follows. From (93), we deduce

$$\frac{d\Lambda}{d\lambda}(\lambda) = \mathrm{Tr}\{\ln((1-\lambda)\mathcal{M}_\rho + \lambda\rho)\,(\rho - \mathcal{M}_\rho)\}, \tag{226}$$

and in particular that

$$\frac{d\Lambda}{d\lambda}(1) = \mathrm{Tr}\{\ln\rho\,(\rho - \mathcal{M}_\rho)\} \geq H(\rho) - H(\mathcal{M}_\rho). \tag{227}$$

But, the fact that $\mathcal{M}_\rho$ solves the entropy minimization principle provides us with

$$H(\rho) - H(\mathcal{M}_\rho) \geq 0, \tag{228}$$

out of which we obtain that

$$\mathrm{Tr}\{\mathcal{Q}(\rho)\ln\rho\} = -\nu\mathrm{Tr}\{\ln\rho\,(\rho - \mathcal{M}_\rho)\} \leq 0, \tag{229}$$

which is the result to be proven.

Now, we consider a diffusion scaling of the collisional Wigner equation:

$$\eta^2 \frac{\partial w^\eta}{\partial t} + \eta(v \cdot \nabla_x w^\eta - \Theta(w^\eta)) = Q(w^\eta), \tag{230}$$

This scaling is obtained through the change $t \to t/\eta$ and $Q \to Q/\eta$ which means that the collision operator is large and that we are looking at long time scales.

The limit $\eta \to 0$ of (230) is the so-called Quantum Energy-Transport model. Indeed, as $\eta \to 0$, $w^\eta \longrightarrow \mathcal{E}\mathrm{xp}(A + C|p|^2/2)$ where (A, C) satisfy the Energy-Transport model which consists of the mass and energy conservation equations

$$\frac{\partial n}{\partial t} + \nabla_x \cdot j_n = 0, \tag{231}$$

$$\frac{\partial \mathcal{W}}{\partial t} + \nabla_x \cdot j_{\mathcal{W}} + \nabla_x V \cdot j_n = 0, \tag{232}$$

where (n, W) is related with (A, C) through

$$\int \mathcal{E}\mathrm{xp}(A + C|p|^2/2) \begin{pmatrix} 1 \\ |p|^2/2 \end{pmatrix} \widetilde{dp} = \begin{pmatrix} n \\ W \end{pmatrix}, \tag{233}$$

and the fluxes (j_n, j_W) are given by

$$j_n = -\nu^{-1}[\nabla \Pi + n\nabla V], \tag{234}$$

$$j_W = -\nu^{-1}[\nabla \mathbb{Q} + (W\,\mathrm{Id} + \Pi)\nabla V - \frac{\hbar^2}{8}n\nabla(\Delta V)], \tag{235}$$

with the tensors $\Pi(A, C)$ and $\mathbb{Q}(A, C)$ given by

$$\Pi(\Lambda, C) = \int \mathcal{E}\mathrm{xp}(A + C|p|^2/2)\, p \otimes p \, \widetilde{dp}, \tag{236}$$

$$\mathbb{Q}(A, C) = \int \mathcal{E}\mathrm{xp}(A + C|p|^2/2)\, p \otimes p\, |p|^2/2\, \widetilde{dp}. \tag{237}$$

Like in the classical case (see, e.g. [9, 10, 20]), the system consists of balance equations for the conservative variables (n, W), the fluxes of which are expressed in terms of the gradients of the entropic variables (A, C). The passage (n, W) to (A, C) can be done through the use of the entropy functional or its Legendre dual. However, by contrast with the classical case, there is no clear symmetric positive-definite matrix structure relation between the fluxes and the gradients of the entropic variables.

Let us now consider entropy decay. The fluid entropy is given by the kinetic entropy evaluated for the equilibrium: $S(n, W) = H(\mathcal{M}_{n,W})$ and has the following expressions:

$$S(n, W) = \int \mathcal{M}_{n,W} (\mathcal{L}\mathrm{n}\mathcal{M}_{n,W} - 1)\, dx\, \widetilde{dp}$$

$$= \int \mathcal{E}\mathrm{xp}(A + C|p|^2/2)(A + C|p|^2/2 - 1)\, dx\, \widetilde{dp}$$

$$= \int (n(A - 1) + CW)\, dx. \tag{238}$$

The Quantum Energy-Transport model decreases the entropy:

$$\frac{d}{dt}S(n, W) \le 0. \tag{239}$$

The proof follows exactly the same arguments as for the hydrodynamic model and is omitted.

We now sketch how we prove (formally) that the collisional Wigner equation converges in the diffusive limit $\eta \to 0$ towards the Quantum Energy-Transport model. The (formal) proof follows three steps:

(1) Step 1: Show that w^η converges to a Quantum Maxwellian $\mathcal{E}\mathrm{xp}(A + C|p|^2/2)$ where $(A, C) = (A, C)(x, t)$. Use a Chapman–Enskog expansion to define the first order corrector w_1

(2) Step 2: Write mass and energy conservation equations

(3) Step 3: Compute the fluxes taking the appropriate moment of w_1

We now give more detail about these three points:

(1) Step 1: Convergence to equilibrium. Let us suppose that $w^\eta \to w$ smoothly. Then, using the Wigner-BGK equation (230) we get $Q(w^\eta) = O(\eta)$ and consequently $Q(w) = 0$, which, thanks to (222), gives $w = \mathcal{E}\mathrm{xp}(A + C|p|^2/2)$. Now, we can write a Chapman–Enskog like expansion

$$w^\eta = \mathcal{M}_{w^\eta} + \eta w_1^\eta. \tag{240}$$

This expression defines w_1^η and the equal sign is exact. Then:

$$\frac{1}{\eta}Q(w^\eta) = -\nu w_1^\eta = \mathcal{T}w^\eta + \eta \partial_t w^\eta, \tag{241}$$

with

$$\mathcal{T}w = v \cdot \nabla_x w - \Theta^h[V]w, \tag{242}$$

is the transport operator. Therefore, as $\eta \to 0$:

$$w_1^\eta \to w_1 = -\nu^{-1}\mathcal{T}w, \tag{243}$$

showing that w_1^η has a finite limit.

(2) Step 2: Mass and energy balance equations. We take the moments of the Wigner-BGK equation against 1 and $|p|^2/2$, use that Q preserves mass and energy and get

$$\frac{\partial n^\eta}{\partial t} + \nabla_x \cdot j_n^\eta = 0, \tag{244}$$

$$\frac{\partial W^\eta}{\partial t} + \nabla_x \cdot j_W^\eta + \nabla_x V \cdot j_n^\eta = 0, \tag{245}$$

with

$$j_n^\eta = \eta^{-1} \int w^\eta p \widetilde{dp} = \int w_1^\eta p \widetilde{dp}, \tag{246}$$

$$j_W^\eta = \eta^{-1} \int w^\eta p \, |p|^2/2 \, \widetilde{dp} = \int w_1^\eta p \, |p|^2/2 \, \widetilde{dp}. \tag{247}$$

As $\eta \to 0$

$$j_n^\eta \to j_n = \int w_1 p \widetilde{dp}, \tag{248}$$

$$j_W^\eta \to j_W = \int w_1 p \, |p|^2/2 \, \widetilde{dp}. \tag{249}$$

Therefore, the mass and energy balance equations (244) and (245) are valid in the limit $\eta \to 0$ and lead to (231) and (232).

(3) Step 3: Equations for the fluxes. We start with the equation for j_n. From (243) and (240), we compute:

$$w_1 = -\nu^{-1}[\nabla_x \cdot (p\mathcal{E}\mathrm{xp}(A + C|p|^2/2))$$
$$-\Theta^\hbar[V]\mathcal{E}\mathrm{xp}(A + C|p|^2/2)]. \tag{250}$$

Then

$$j_n = \int w_1 p \widetilde{dp}$$
$$= -\nu^{-1}[\nabla(\int \mathcal{E}\mathrm{xp}(A + C|p|^2/2)p \otimes p\,\widetilde{dp})$$
$$- \int \Theta^\hbar[V](\mathcal{E}\mathrm{xp}(A + C|p|^2/2))\,p\,\widetilde{dp}]. \tag{251}$$

For the moments of the field operator $\Theta^\hbar[V]$, easy computations show that

$$\int \Theta^\hbar[V]w \begin{pmatrix} 1 \\ p \\ |p|^2/2 \end{pmatrix} \widetilde{dp} = \begin{pmatrix} 0 \\ -n\nabla V \\ -nu \cdot \nabla V \end{pmatrix}, \tag{252}$$

and

$$\int \Theta^\hbar[V]w|p|^2/2p\,\widetilde{dp} = -(\mathcal{W}\,\mathrm{Id} + \Pi)\nabla V + \frac{\hbar^2}{8}n\,\nabla(\Delta V). \tag{253}$$

Using these formulas, we finally find

$$j_n = -\nu^{-1}[\nabla \Pi + n\nabla V], \tag{254}$$

which is the formula to be proved.

Now, we turn to the computation of $j_\mathcal{W}$. Similar computation give:

$$j_\mathcal{W} = \int w_1 p\,|p|^2/2\,dp$$
$$= -\nu^{-1}[\nabla(\int \mathcal{E}\mathrm{xp}(A + C|p|^2/2)p \otimes p\,|p|^2/2\widetilde{dp})$$
$$- \int \Theta^\hbar[V](\mathcal{E}\mathrm{xp}(A + C|p|^2/2))\,p\,|p|^2/2\,\widetilde{dp}]. \tag{255}$$

Using (252) and (253)

$$j_\mathcal{W} = -\nu^{-1}[\nabla \mathbb{Q} + (\mathcal{W}\,\mathrm{Id} + \Pi)\nabla V - \frac{\hbar^2}{8}n\,\nabla(\Delta V)], \tag{256}$$

which is again the formula to be proven and closes the formal proof that the collisional Wigner equation converges in the diffusive limit towards the Quantum Energy-Transport model.

We now look at the $\hbar$ expansion of the model up to order $O(\hbar^2)$. Again, the computations that lead to these formulae can be found in [23]. The expansion of Π is

$$\Pi_{rs} = \delta_{rs}\, n\, T$$

$$+ \frac{\hbar^2}{12d}\, n\, \delta_{rs}(\Delta_x \ln n + 2\Delta_x \ln T + 2\nabla_x \ln n \cdot \nabla_x \ln T - \tfrac{d+2}{2}\, |\nabla_x \ln T|^2)$$

$$+ \frac{\hbar^2}{12}\, n(-\partial^2_{rs} \ln n - 2\partial^2_{rs} \ln T - \partial_r \ln n\, \partial_s \ln T$$

$$- \partial_r \ln T\, \partial_s \ln n + \tfrac{d+2}{2}\, \partial_r \ln T\, \partial_s \ln T), \tag{257}$$

with $T = 2W/(dn)$. The expansion of $\mathbb{Q}$ gives:

$$\mathbb{Q}_{rs} = \tfrac{d+2}{2}\, \delta_{rs}\, n\, T^2$$

$$+ \frac{\hbar^2}{24d}\, n\, T\, \delta_{rs}(\Delta_x \ln n + (d+8)\Delta_x \ln T$$

$$+ 2(d+4)\nabla_x \ln n \cdot \nabla_x \ln T + \tfrac{d^2-4d-8}{2}\, |\nabla_x \ln T|^2) \tag{258}$$

$$+ \frac{\hbar^2}{24}\, (d+4)\, n\, T(-\partial^2_{rs} \ln n - 3\partial^2_{rs} \ln T$$

$$- \partial_r \ln n\, \partial_s \ln T - \partial_r \ln T\, \partial_s \ln n + \tfrac{d}{2}\, \partial_r \ln T\, \partial_s \ln T).$$

These expressions are quite complicated but if a small temperature variation assumption is made:

$$|\nabla \ln T|/|\nabla \ln n| \ll 1, \tag{259}$$

the model simplifies and gives the following formulae for the fluxes

$$J^n = -\nabla\left(n\, T + \frac{\hbar^2}{12d}\, n\, \Delta \ln n\right) - n\, \nabla(V + V_B[n]), \tag{260}$$

$$J^w = -\nabla\left(\frac{d+2}{2}\, n\, T^2 + \frac{\hbar^2}{24}\, \frac{d+4}{d}\, n\, T\, \Delta \ln n\right)$$

$$- \frac{d+4}{2}\, n\, T\, \nabla V_B[n] - \left(\frac{d+2}{2}\, n\, T + \frac{\hbar^2}{12d}\, n\, \Delta \ln n\right) \nabla V$$

$$+ \frac{\hbar^2}{12}\, n\, (\nabla\nabla \ln n)\, \nabla V + \frac{\hbar^2}{8}\, \nabla \Delta \ln n. \tag{261}$$

The derivation of these relations is beyond the scope of this review and we refer the reader to [23]. It is also unknown if this model possesses a strictly decaying entropy. Indeed, there is no reason why the expansion up to order $O(\hbar^2)$ of the quantum entropy of the Energy-Transport model would be time decaying. This property has not been proved right nor wrong.

We close this section about Quantum Energy-Transport models by a few remarks. The first one is that there is no rigorous proof neither for the existence of solutions nor for its derivation from the collisional Wigner equation. Numerical simulations have not been performed yet either. In the literature, quantum energy-transport models can be found but their derivation (and the model itself) are different. For instance, we refer to the energy-transport extension of the DG (Density-Gradient) model by Chen and Liu [19].

4.2 Quantum Drift-Diffusion Model

This section summarizes a series of works [23], [24], [32], [21].

In the classical setting, the Drift-Diffusion model is a simplification of the Energy-Transport model when the assumption of constant temperature is made. To derive a Quantum-Drift-Diffusion model, we start by a discussion of the appropriate BGK operator.

This operator will be defined as a relaxation to a quantum Maxwellian with a fixed temperature, and can be expressed by

$$Q(w)(v) = -\nu(w - \mathcal{E}xp(A - |p|^2/2)), \tag{262}$$

where the function $A(x)$ is such that the operator conserves mass. Here again, we take a constant temperature equal to unity for the sake of simplicity.

For a given density $n(x)$, the Quantum Maxwellian which has density n in Wigner form is given by

$$\mathcal{M}_n = \mathcal{E}xp(A - |p|^2/2), \tag{263}$$

$$\int \mathcal{E}xp(A - |p|^2/2) \, \widetilde{dp} = n. \tag{264}$$

In density operator form it is written

$$\rho_n = W^{-1}(\mathcal{M}_n) = \exp(W^{-1}(A - |p|^2/2)), \tag{265}$$

with, for all test function ϕ:

$$\text{Tr}\{\rho_n \, \phi\} = \int n\phi \, dx. \tag{266}$$

This Quantum Maxwellian satisfies the free energy minimization principle: $\rho_n = \exp(W^{-1}(A - |p|^2/2))$ is a solution of the problem: to find

$$\min \{G[\rho] = \text{Tr}\{\rho(\ln \rho - 1) + \mathcal{H}\rho\} \quad \text{subject to:}$$

$$\text{Tr}\{\rho_n \, \phi\} = \int n\phi \, dx, \quad \forall \text{ test fct } \phi\}, \tag{267}$$

where $\mathcal{H} = |p|^2/2 + V$ is the system Hamiltonian.

In Wigner form, the free energy is written

$$G[\rho] = \text{Tr}\{\rho(\ln \rho - 1) + \mathcal{H}\rho\} = \int [w(\mathcal{L}\text{n}\, w - 1) + \mathcal{H}w]\, dx\, \widetilde{dp}, \qquad (268)$$

with the quantum logarithm $\mathcal{L}\text{n}\, w = W[\ln(W^{-1}(w))]$.

For a given Wigner distribution w, we denote $\mathcal{M}_w := \mathcal{M}_n$ the Quantum Maxwellian which has the same density n as w:

$$\int \mathcal{M}_w\, dp = \int w\, dp. \qquad (269)$$

Then Quantum BGK operator is finally written

$$Q(w) = -\nu(w - \mathcal{M}_w). \qquad (270)$$

In density operator formulation, we denote by $\mathcal{M}_\rho$ the Quantum Maxwellian associated with ρ, and the BGK operator is written:

$$\mathcal{Q}(\rho) = -\nu(\rho - \mathcal{M}_\rho). \qquad (271)$$

The situation modeled by $Q(w)$ is that of a system where energy exchanges between the particles and the surrounding relax the temperature to the background temperature.

Again, two variables appear, the conservative variable n and the entropic variable A, with a functional change of variable between these two variables which can be expressed through the free energy and its Legendre dual.

We now list the properties of Q:

(a) Mass conservation:

$$\int Q(w)\, dp = 0, \qquad (272)$$

(b) Null set of Q (equilibria):

$$Q(w) = 0 \iff \exists A \text{ such that } w = \mathcal{E}\text{xp}(A - |p|^2/2), \qquad (273)$$

(c) Free energy decay:

$$\int Q(w)(\mathcal{L}\text{n}w + \mathcal{H})\, dx\, \widetilde{dp} = \text{Tr}\{\mathcal{Q}(\rho)(\ln \rho + \mathcal{H})\} \leq 0. \qquad (274)$$

The proof of (c) is similar to the energy-transport case and is omitted. We now look at the Wigner equation under diffusion scaling:

$$\eta^2 \frac{\partial w^\eta}{\partial t} + \eta(v \cdot \nabla_x w^\eta - \Theta(w^\eta)) = Q(w^\eta), \qquad (275)$$

The limit $\eta \to 0$ leads to the Quantum Drift-Diffusion model: more precisely, as $\eta \to 0$, $w^\eta \longrightarrow \mathcal{E}\text{xp}(A - |p|^2/2)$ where A satisfies the Energy-Transport model which consists of the mass conservation equation

$$\frac{\partial n}{\partial t} + \nabla_x \cdot j_n = 0. \tag{276}$$

with

$$\int \mathcal{E}\mathrm{xp}(A - |p|^2/2)\,\widetilde{dp} = n, \tag{277}$$

and the flux j_n given by

$$j_n = -\nu^{-1}[\nabla\Pi + n\nabla V], \tag{278}$$

with

$$\Pi(A) = \int \mathcal{E}\mathrm{xp}(A - |p|^2/2)\,p \otimes p\,\widetilde{dp}. \tag{279}$$

The derivation of this model from the collisional Wigner equation follows exactly the same lines as in the Energy-Transport case and is omitted.

Now, the fluid free energy is the kinetic free energy evaluated on the equilibrium $\mathcal{G}(n) = G(\mathcal{M}_n)$ and is given by

$$\begin{aligned}
\mathcal{G}(n) &= \int \mathcal{M}_{n,\mathcal{W}}\,(\mathcal{L}\mathrm{n}\mathcal{M}_{n,\mathcal{W}} - 1 + \mathcal{H})\,dx\,\widetilde{dp} \\
&= \int \mathcal{E}\mathrm{xp}(A - |p|^2/2)(A - |p|^2/2 - 1 + \mathcal{H})\,dx\,\widetilde{dp} \\
&= \int n(A + V - 1)\,dx. \tag{280}
\end{aligned}$$

Then if either V is independent of t or V is given by Poisson's equation

$$\Delta V = n, \tag{281}$$

then

$$\frac{d}{dt}\mathcal{G}(n) \leq 0, \tag{282}$$

(in the latter case, we have to multiply the term nV by a factor $1/2$).

We now give a more tractable expression of the pressure tensor Π than (279). We just remark that $\Pi(A) = \Pi(-A, 0)$ where $\Pi(A, B)$ is the pressure tensor of the Isentropic Quantum Euler model:

$$\Pi(A, B) = \int \mathcal{E}\mathrm{xp}(-H(A, B))\,p \otimes p\,\widetilde{dp}, \tag{283}$$

with $H(A, B) = |p|^2/2 - B \cdot p + A + |B|^2/2$. In that case, we proved (see Sect. 3.3) that

$$\nabla\Pi(A, B) = \nabla(nu \otimes B) + (\nabla B)nu - n\nabla(A + |B|^2/2). \tag{284}$$

In the present case here, $\nabla \Pi(A)$ is deduced through the identification $B = 0$ and $A \to -A$ which leads to

$$\nabla \Pi(A) = n \nabla A. \tag{285}$$

This leads to an equivalent formulation of the QDD model:

$$\frac{\partial n}{\partial t} + \nabla_x \cdot j_n = 0, \tag{286}$$

$$j_n = -\nu^{-1}(n \nabla(A + V)), \tag{287}$$

$$\int \mathcal{E} \mathrm{xp}(A - |p|^2/2)\, \widetilde{dp} = n. \tag{288}$$

The moment reconstruction problem (288) has also a simpler expression if we suppose that the Hamiltonian $H(A) = |p|^2/2 - A$ has a discrete spectrum with eigenvalues $\lambda_p(A)$ and eigenfunctions $\psi_p(A)$, $p = 1, \ldots, \infty$. Following (145), we have

$$n(A)(x) = \sum_{p=1}^{\infty} \exp(-\lambda_p(A)) |\psi_p(A)(x)|^2. \tag{289}$$

The "final" expression of the Quantum Drift-Diffusion model is therefore:

$$\frac{\partial n}{\partial t} + \nabla_x \cdot j_n = 0, \tag{290}$$

$$j_n = -\nu^{-1}(n \nabla(A + V)), \tag{291}$$

$$n(A)(x) = \sum_{p=1}^{\infty} \exp(-\lambda_p(A)) |\psi_p(A)(x)|^2. \tag{292}$$

with $\lambda_p(A)$ and $\psi_p(A)$ the eigenvalues and eigenvectors associated with the modified Hamiltonian $H(A) = |p|^2/2 - A$.

Now, we would like to consider the equilibrium states of the QDD model, defined by $j_n = 0$. This obviously implies $A = -V$ (up to a constant that we take equal to zero). Therefore, the moment reconstruction problem becomes

$$n(x) = \sum_{p=1}^{\infty} \exp(-\lambda_p) |\psi_p(x)|^2, \tag{293}$$

with λ_p, ψ_p the eigenvalue and eigenvector associated with the "true" system Hamiltonian $H(-V) = |p|^2/2 + V$. If additionally, n is related with V through Poisson's equation (281), this leads to the well-known Schrödinger–Poisson problem which characterizes equilibrium states.

Now, if we assume that we are close to equilibrium, we can make the approximation $A \approx -V$ and replace A by $-V$ in the moment reconstruction problem (288), which leads to the following system

$$\frac{\partial n}{\partial t} + \nabla_x \cdot j_n = 0, \tag{294}$$

$$j_n = \nu^{-1}(n\nabla(A + V)), \tag{295}$$

$$n(A)(x) = \sum_{p=1}^{\infty} \exp(A + V - \lambda_p(-V)) |\psi_p(-V)(x)|^2, \tag{296}$$

in which case, the spectral problem to be solved is associated with the "true" system Hamiltonian $H(-V) = |p|^2/2 + V$. This system is known as the Schrödinger–Poisson-Drift-Diffusion and has been investigated by Sacco and coauthors in [27, 51, 58].

We now investigate $\hbar$ expansions of the QDD model. Up to $O(\hbar^2)$ terms, the QDD model reads:

$$\partial_t n + \nabla \cdot j_n = 0, \tag{297}$$

$$j_n = -\nu^{-1}[\nabla n - n\nabla(V + V_B[n])), \tag{298}$$

$$V_B[n] = -\frac{\hbar^2}{6} \frac{1}{\sqrt{n}} \Delta(\sqrt{n}). \tag{299}$$

This model is called the Density-Gradient model and has first been proposed by Ancona and coauthors [1–3]. We note that this is just the classical drift-diffusion model with the addition of the Bohm potential (divided by a factor 3 as compared with the Bohm potential of the single-particle hydrodynamics). Usually, this factor is treated as a fitting parameter in the simulation codes.

It is a remarkable fact that the Density-Gradient model has an entropy, which is nothing but the free energy of the QDD model expanded up to $O(\hbar^2)$ terms:

$$\mathcal{G}_2(n) = \int_{\mathbb{R}^d} n(\ln n - 1 + V + V_B[n]) \, dx. \tag{300}$$

If V is independent of t it can be shown that

$$\frac{d}{dt}\mathcal{G}_2(n) = -\int_{\mathbb{R}^d} \frac{1}{\nu n} |\nabla n + n\nabla(V + V_B[n])|^2 \, dx \le 0. \tag{301}$$

A similar expression would hold if V is solved through Poisson's equation (281). The proof can be found in [23].

The Density-Gradient model has been widely investigated in the literature. The mathematical theory has been settled first by Ben Abdallah and Unterreiter in [11] and later by Pinnau [56]. Numerical methods have been developed by Pinnau and Unterreiter [57] and Jüngel and Pinnau [42]. The present approach provides a derivation of the DG model from first principles and proves (for the first time) that DG model is compatible with free energy decay.

About the full QDD model (i.e. with no $\hbar$ expansion), there is no rigorous proof, neither of existence nor of convergence.

Table 2 Parameters used for the modeling of an isolated RTD

Effect. mass (kg)	Mobility (m² V⁻¹ s⁻¹)	Permittivity (F m⁻¹)	Temperature (K)
$0.067 \times 9.11e - 31$	0.85	$11.44 \times 8.85e - 12$	300

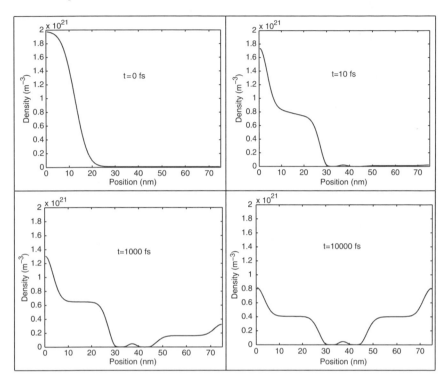

Fig. 6 Electron density at different times ($t = 0, 10, 1,000$ and $10,000\,\text{fs}$) for the QDD model

We now present some numerical simulations. We first show a Resonant Tunneling Diode with insulating boundary conditions. The parameters are all chosen independent of x in this case and are given in Table 2. The initial density is concentrated to the left of the double barrier and Fig. 6 shows the evolution of electrons for the QDD model coupled with Poisson equation under insulating boundary conditions. The next figure (Fig. 7) demonstrates that the quantum free energy is a decreasing function of time. Figure 8 displays the evolution of the electrochemical potential $\varphi(x) = A(x) - V(x)$.

Figure 9 permits to compare the QDD model, the Schrödinger–Poisson–Drift-Diffusion (SPDD) model and the stationary Schrödinger–Poisson (SP) model. The QDD and the SPDD model are closer than the QDD and the SP models.

Now, we look at open boundary conditions. We first analyze the influence of the effective mass on the shape of the current–voltage characteristic.

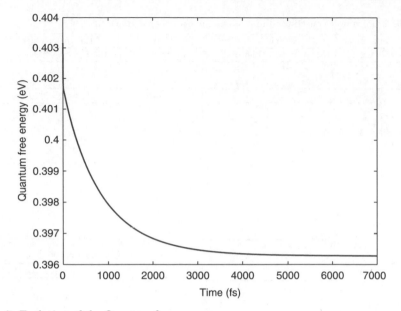

Fig. 7 Evolution of the Quantum free energy

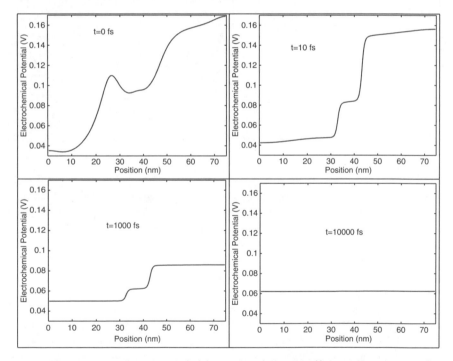

Fig. 8 Electrochemical potential ($\varphi(x) = A - (V_s + V_{ext})$) at different times ($t = 0, 10, 1,000$ and $10,000$ fs)

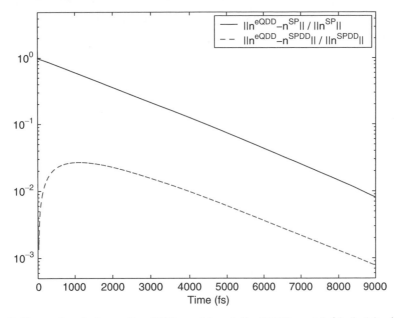

Fig. 9 Comparison between the QDD model and the SPDD model (*dashed line*), and between the QDD model and the SP model (*solid line*)

The temperature is chosen equal to $77\,\mathrm{K}$ and the mobility is supposed to be constant and equal to $0.85\,\mathrm{m^2\,V^{-1}\,s^{-1}}$. The permittivity is also supposed to be constant and equal to $11.44\,\epsilon_0$. Figure 10 shows four different IV curves with different values of the effective mass inside and outside the double barriers. These curves show a certain sensitivity of the model to the value of the effective mass inside the barrier.

Figure 11 shows the time evolution of the density from the peak to the valley when the effective mass is $m_2 = 1.5 \times 0.092 m_e$ inside the barriers and $m_1 = 1.5 \times 0.067 m_e$ outside it (corresponding to the IV curve at the bottom right of Fig. 10). To obtain this figure, we apply a voltage of $0.25\,\mathrm{V}$ and wait for the electrons to achieve the stationary state. Then we suddenly change the value of the applied bias to $0.29\,\mathrm{V}$ and we record the evolution of the density. As expected, the density inside the well grows significantly and the stationary state is achieved at about $1{,}500\,\mathrm{fs}$.

The next two figures (Figs. 12 and 13) display the details of the reconstruction of the density from the eigenstates ψ_p (for $p = 1 \cdots 6$) of the modified Hamiltonian $H[A]$. The density $e^{-\lambda_p}|\psi_p|^2$ corresponding to each eigenstate is plotted for two values of the applied bias, respectively corresponding to the current peak (Fig. 12) and to the valley (Fig. 13). Table 3 shows the values of the corresponding energies λ_p. Last, Fig. 14 shows the transient current at the left contact ($x = 0$). A detailed discussion of these results can be found in [21].

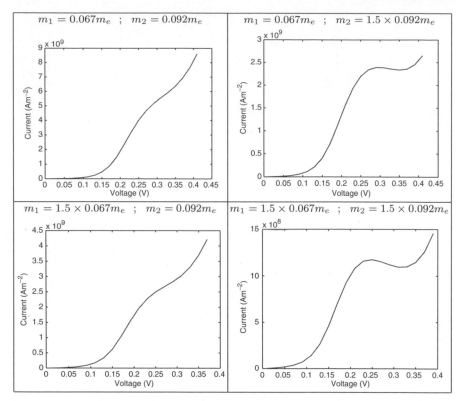

Fig. 10 Influence of the effective mass on the IV curve, m_1 being the mass outside the barriers, and m_2 being the mass inside

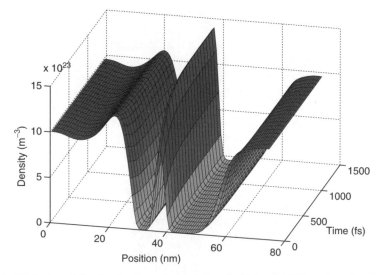

Fig. 11 Evolution of the density from the peak (applied bias: 0.25 V) to the valley (applied bias: 0.31 V)

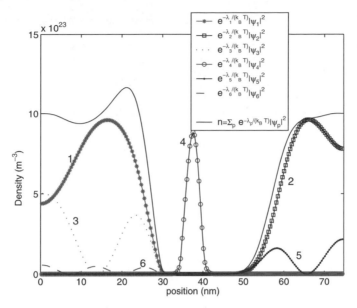

Fig. 12 Density at the peak (applied bias: 0.25 V)

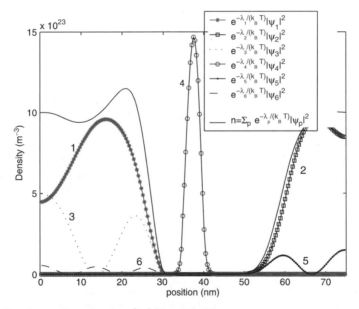

Fig. 13 Density at the valley (applied bias: 0.31 V)

Table 3 Eigenvalues (energies [eV]) of the modified Hamiltonian $H[A]$ at the peak and at the valley

	λ_1	λ_2	λ_3	λ_4	λ_5	λ_6	λ_7
Peak	0.87	1.05	1.56	2.03	2.28	3.03	4.47
Valley	0.87	1.11	1.57	1.70	2.54	3.05	5.03

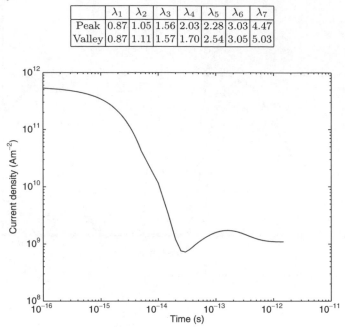

Fig. 14 Transient current density

In Fig. 15, we show the results obtained with the Density Gradient model using the same parameters as defined for the QDD model. As we can see, results are qualitatively similar but differ significantly. Even with a smoother external potential (replacing the two step functions by two gaussians), it appears that the current–voltage characteristics are still different for the two models as suggested by Fig. 16. To finish, Fig. 17 shows the role of the temperature on the current for an applied bias of 0.2 V and for the three models QDD, DG and CDD with a constant mass equal to $0.067m_e$.

5 Summary and Conclusion

In these lecture notes, after reviewing the basics of quantum statistical mechanics of nonequilibrium systems (density operator, quantum Liouville equation, Wigner transform and Wigner equation, mean-field limits, Hartree and Hartree–Fock systems), we have discussed the modeling of open systems interacting with a large and unperfectly known environment and have briefly summarized previous quantum hydrodynamic approaches.

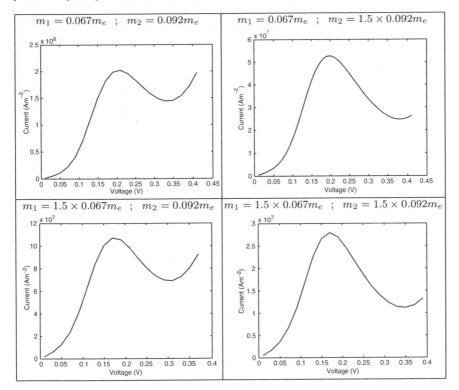

Fig. 15 IV curves obtained with the DG model (m_1 being the mass outside the barriers, and m_2 being the mass inside)

Then, we have developed our approach for deriving new quantum hydrodynamic systems based on the entropy minimization principle. This approach relies on a suitable extension of Levermore's moment method to the quantum case. It consists in taking local moments of the density operator equation and closing the resulting chain of equations by a minimizer of the entropy functional subject to moment constraints. It leads to a formulation of the entropy minimization problem as a global problem (whereas it is a local problem in classical mechanics) and results in a non-local closure of the so-obtained Quantum Hydrodynamic equations.

Then, we have considered the special case of the Isothermal Quantum Euler model, which is the quantum hydrodynamic model corresponding to the isothermal Euler system in classical mechanics. In this case, since the temperature is a constant, the role of the entropy is played by the free energy. Analytic computation of the expression of the pressure tensor were possible in terms of both the conservative variables (n, nu) and the entropic ones (A, B). The two systems are related one to each other by the free energy and its Legendre dual. A remarkable gauge invariance property for this system has been exhibited. As a by-product, a constraint between the velocity u

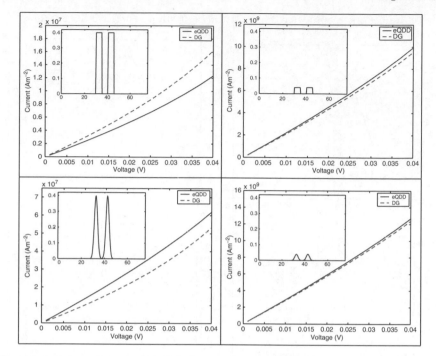

Fig. 16 Influence of the shape and the height of the double barrier on the Current–Voltage characteristics for the QDD and DG models

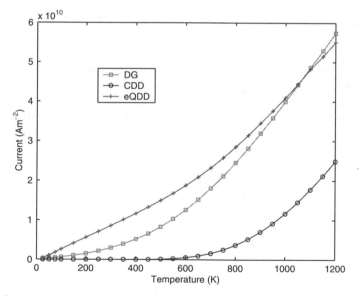

Fig. 17 Current–Temperature curve (applied bias: 0.2 V)

and its adjoint entropic variable B has been discovered. Several equivalent formulations of the model are possible. Irrotational flows present a special interest, since in this case, $u = B$ and the problem depends on A only, which reduces the size of the moment reconstruction problem (i.e. the inversion of the mapping $A \to n$). A particular interesting case of irrotational flows are one-dimensional flows. Preliminary numerical simulations seem to indicate that the model gives meaningful results in realistic situations.

Many open problems do remain. The most formidable one is obviously to show that the entropy minimization problem has a solution in a reasonable sense. An analytical computation of the closure relations in the case of the full QHD model (as done for the isothermal case) is certainly at reach. The investigation of gauge invariance properties for the full QHD problem is of course mandatory. Another interesting problem is the investigation of the small T asymptotics (which has formally been realized for the isothermal case in [22]). The $\hbar$ expansion up to order $\hbar^2$ is available in [22] or in the work of Jüngel et al.[41]). Finally, a normal mode analysis of the linearized model would at least give some indication about the well-posedness of the model.

In a last section, diffusion models have been derived by means of the same approach. We have first proposed a formulation of a quantum BGK operator (which models a relaxation of the Wigner distribution function towards a quantum equilibrium). Then, we have performed a diffusion approximation of the resulting Quantum Kinetic Equation and provided new Quantum Energy-Transport or Drift-Diffusion models. The Quantum Drift-Diffusion model has been analyzed in more detail. This model differs from classical models by the reconstruction of the density from the chemical potential (through an eigenvalue problem). We can recover the Density-Gradient (DG) model of Ancona and Iafrate [2] as an $O(\hbar^2)$ approximation, and the Schrödinger–Poisson Drift-Diffusion (SPDD) model of de Falco et al. [27] in situations close to equilibrium. A large set of numerical simulations have been realized and show that the qualitative behaviour of the model is fairly satisfactory, while a certain sensitivity to some physical parameters still needs to be understood.

The quantum Energy-Transport model needs to be analyzed in the same way. The first step would be to find a simplified expression of the model (having local values of the pressure tensors in terms of the conservative and entropic variables).

Multi-dimensional simulations will require more computing power but are within reach. A better account of the continuous spectrum of the operators would certainly improve the results, notably close to the boundaries.

Of the overall approach, some other extensions and applications will require further developments. One would wish to introduce many particle effects more accurately than through the use of the BGK collision operator. Using this approach, phonon–electron collision operators for electrons in crystals could be derived. Also, the introduction of confinement in one or more directions would lead to sub-band models which could be applied to systems such as quantum wires or quantum dots. Following the same lines, Born–Oppenheimer approximations in quantum chemistry could also be used in the framework in these models and would lead to hybrid quantum-classical models, in the spirit of [17]. Applications could span from reaction dynamics in chemistry to biology problems such as ionic channels in cell membrane physiology.

Acknowledgements The first author acknowledges support of the CIME for the preparation of these notes.

References

1. Ancona, M. G., Diffusion-Drift modeling of strong inversion layers, COMPEL *6*, 11–18 (1987)
2. Ancona, M. G., Iafrate, G. J., Quantum correction of the equation of state of an electron gas in a semiconductor, Phys. review B, *39*, 9536–9540 (1989)
3. Ancona, M. G., Tiersten, H. F., Macroscopic physics of the silicon inversion layer, Phys. review B, *35*, 7959–7965 (1987)
4. Argyres, P. N., Quantum kinetic equations for electrons in high electric and phonon fields, Physics Lett. A, *171*, 373–379 (1992)
5. Arnold, A., Lopez, J. L., Markowich, P., Soler, J., An analysis of quantum Fokker - Planck models: A Wigner function approach, Rev. Mat. Iberoamericana, *20* 771–814 (2004)
6. Bardos, C., Golse, F., Mauser, N. J., Weak coupling limit of the N-particle Schrodinger equation, Methods Appl. Anal., *7*, 275–293 (2000)
7. Bardos, C., Golse, F., Gottlieb, A. D., Mauser, N. J., Mean field dynamics of fermions and the time-dependent Hartree-Fock equation, J. Math. Pures Appl., *82*, 665–683 (2003)
8. Bardos, C., Golse, F., Gottlieb, A. D., Mauser, N. J., Accuracy of the time-dependent Hartree-Fock approximation for uncorrelated initial states, J. Statist. Phys., *115*, 1037–1055 (2004)
9. Ben Abdallah, N., Degond, P., On a hierarchy of macroscopic models for semiconductors, J. Math. Phys., *37*, 3306–3333 (1996)
10. Ben Abdallah, N., Degond, P., Génieys, S., An energy-transport model for semiconductors derived from the Boltzmann equation, J. Stat. Phys., *84*, 205-231 (1996)
11. Ben Abdallah, N., Unterreiter, A., On the stationary quantum drift-diffusion model, Z. Angew. Math. Phys., *49*, 251–275 (1998)
12. Brenier, Y., Grenier, E., Sticky particles and scalar conservation laws, SIAM J. Numer. Anal., *35*, 2317–2328 (1998)
13. Bouchut, F., On zero pressure gas dynamics, in Advances in kinetic theory and computing, B. Perthame (ed), World Scientific (1994)
14. Burghardt, I., Cederbaum, L. S., Hydrodynamic equations for mixed quantum states I. General formulation, Journal of Chemical Physics, *115*, 10303–10311 (2001)
15. Burghardt, I., Cederbaum, L. S., Hydrodynamic equations for mixed quantum states II. Coupled electronic states, Journal of Chemical Physics, *115*, 10312–10322 (2001)
16. Burghardt, I., Moller, K. B., Quantum dynamics for dissipative systems: a hydrodynamic perspective, Journal of Chemical Physics, *117*, 7409–7425 (2002)
17. Burghardt, I., Parlant, G., On the dynamics of coupled Bohmian and phase-space variables, a new hybrid quantum-classical approach, Journal of Chemical Physics, *120*, 3055–3058 (2004)

18. Car, R., Parrinello, M., Unified approach for molecular dynamics and density-functional theory, Phys. Rev. Lett., *55*, 2471–2474 (1985)
19. Chen, R-C., Liu, J-L., A quantum corrected energy-transport model for nanoscale semiconductor devices, J. Comput. Phys., *204*, 131–156 (2005)
20. Degond, P., Mathematical modelling of microelectronics semiconductor devices, AMS/IP Studies in Advanced Mathematics, AMS Society and International Press, 77–109, (2000)
21. Degond, P., Gallego, S., Méhats, F., An entropic quantum drift-diffusion model for electron transport in resonant tunneling diodes, J. Comp. Phys., to appear
22. Degond, P., Gallego, S., Méhats, F., Isothermal quantum hydrodynamics: derivation, asymptotic analysis and simulation, manuscript, submitted
23. Degond, P., Méhats, F., Ringhofer, C., Quantum energy-transport and drift-diffusion models, J. Stat. Phys., *118*, 625–667 (2005)
24. Degond, P., Méhats, F., Ringhofer, C., Quantum hydrodynamic models derived from the entropy principle, Contemp. Math., *371*, 107–131 (2005)
25. Degond, P., Ringhofer, C., Quantum moment hydrodynamics and the entropy principle, J. Stat. Phys. *112*, 587–628 (2003)
26. Degond, P., Ringhofer, C., A note on quantum moment hydrodynamics and the entropy principle, C. R. Acad. Sci. Paris Ser 1, *335*, 967–972 (2002)
27. de Falco, C., Gatti, E., Lacaita, A. L., Sacco, R., Quantum-Corrected Drift-Diffusion Models for Transport in Semiconductor Devices, J. Comput. Phys., *204* 533–561 (2005)
28. Dreizler, R. M., Gross, E. K. U., Density Functional Theory, Springer, Berlin (1990)
29. E, W., Rykov, Y. G., Sinai, Y. G., Generalized variational principles, global weak solutions and behavior with random initial data for systems of conservation laws arising in adhesion particle dynamics, Comm. Math. Phys. *177*, 349–380 (1996)
30. Fischetti, M. V., Theory of electron transport in small semiconductor devices using the Pauli Master equation, J. Appl. Phys., *83*, 270–291 (1998)
31. Fromlet, F., Markowich, P., Ringhofer, C., A Wignerfunction Approach to Phonon Scattering, VLSI Design, *9*, 339–350 (1999)
32. Gallego, S., Méhats, F., Entropic discretization of a quantum drift-diffusion model, SIAM J. Numer. Anal., *43*, 1828–1849 (2005)
33. Gallego, S., Méhats, F., Numerical approximation of a quantum drift-diffusion model, C. R. Math. Acad. Sci. Paris, *339*, 519–524 (2004)
34. Gardner, C., The quantum hydrodynamic model for semiconductor devices, SIAM J. Appl. Math. *54*, 409–427 (1994)
35. Gardner, C., Ringhofer, C., The smooth quantum potential for the hydrodynamic model, Phys. Rev. E *53*, 157–167 (1996)
36. Gardner, C., Ringhofer, C., The Chapman-Enskog Expansion and the Quantum Hydrodynamic Model for Semiconductor Devices, VLSI Design *10*, 415–435 (2000)
37. Gasser, I., Markowich, P. A., Quantum Hydrodynamics, Wigner Transforms and the Classical Limit, Asympt. Analysis, *14*, 97–116 (1997)
38. Gasser, I., Markowich, P. A., Ringhofer, C., Closure conditions for classical and quantum moment hierarchies in the small temperature limit, Transp. Th. Stat. Phys. *25* 409–423 (1996)
39. Hohenberg, P., Kohn, W. Inhomogeneous electron gas, Phys. Rev. B, *136*, 864–871 (1964)
40. Jüngel, A., Matthes, D., A derivation of the isothermal quantum hydrodynamic equations using entropy minimization, ZAMM Z. Angew. Math. Mech., *85*, 806–814 (2005)
41. Jüngel, A., Matthes, D., Milisic, P., Derivation of new quantum hydrodynamic equations using entropy minimization, submitted
42. Jüngel, A., Pinnau, R., A positivity preserving numerical scheme for a fourth order parabolic equation, SIAM J. Num. Anal., *39*, 385–406 (2001)
43. Kaiser, H-C., Rehberg, J., On stationary Schrödinger-Poisson equations modelling an electron gas with reduced dimension, Math. Methods Appl. Sci., *20*, 1283–1312 (1997)

44. C. Le Bris (ed), Handbook of numerical analysis. Vol. X. Special volume: computational chemistry, North-Holland, Amsterdam (2003)
45. Levermore, C. D., Moment closure hierarchies for kinetic theories, J. Stat. Phys., *83*, 1021–1065 (1996)
46. Luzzi, R., untitled, electronic preprint archive, reference arXiv:cond-mat/9909160 v2 11 Sep 1999
47. Lions, P-L., Paul, T., Sur les mesures de Wigner, Rev. Mat. Iberoamericana, *9*, 553–618 (1993)
48. Lopreore, C. L., Wyatt, R. E., Quantum Wave Packet Dynamics with Trajectories, Phys. Rev. Lett., *82*, 5190–5193 (1999)
49. Maddox, J. B., Bittner, E. R., Quantum dissipation in the hydrodynamic moment hierarchy, a semiclassical truncation stategy, J. Phys. Chem. B, *106*, 7981–7990 (2002)
50. Markowich, P. A., Mauser, N. J., The classical limit of a self-consistent quantum-Vlasov equation in 3D, Math. Models Methods Appl. Sci., *3*, 109–124 (1993)
51. Micheletti, S., Sacco, R., Simioni, P., Numerical Simulation of Resonant Tunnelling Diodes with a Quantum-Drift-Diffusion Model, Scientific Computing in Electrical Engineering, Lecture Notes in Computer Science, Springer-Verlag, pp. 313–321 (2004)
52. Morozov, V. G., Röpke, G., Zubarev's method of a nonequilibrium statistical operator and some challenges in the theory of irreversible processes, Condensed Matter Physics, *1*, 673–686 (1998)
53. Nier, F., A stationary Schrödinger-Poisson system arising from the modelling of electronic devices, Forum Math., *2*, 489–510 (1990)
54. Nier, F., A variational formulation of Schrödinger-Poisson systems in dimension $d \leq 3$, Comm. Partial Differential Equations, *18*, 1125–1147 (1993)
55. Nier, F., Schrödinger-Poisson systems in dimension $d \leq 3$: the whole-space case, Proc. Roy. Soc. Edinburgh Sect. A, *123*, 1179–1201 (1993)
56. Pinnau, R., The Linearized Transient Quantum Drift Diffusion Model - Stability of Stationary States, Z. Angew. Math. Mech., *80* 327–344 (2000)
57. Pinnau, R., Unterreiter, A., The Stationary Current-Voltage Characteristics of the Quantum Drift Diffusion Model, SIAM J. Numer. Anal., *37*, 211–245 (1999)
58. Pirovano, A., Lacaita, A., Spinelli, A., Two-Dimensional Quantum effects in Nanoscale MOSFETs, IEEE Trans. Electron Devices, *47*, 25–31 (2002)
59. Spohn, H., Large scale dynamics of interacting particles, Springer, Berlin (1991)
60. Wyatt, R. E., Bittner, E. R., Quantum wave packet dynamics with trajectories: Implementation with adaptive Lagrangian grids, The Journal of Chemical Physics, *113*, 8898–8907 (2000)
61. Zubarev, D. N., Morozov, V. G., Röpke, G., Statistical mechanics of nonequilibrium processes. Vol 1, basic concepts, kinetic theory, Akademie Verlag, Berlin (1996)

Multiscale Computations for Flow and Transport in Heterogeneous Media

Yalchin Efendiev and Thomas Y. Hou

Abstract Many problems of fundamental and practical importance have multiple scale solutions. The direct numerical solution of multiple scale problems is difficult to obtain even with modern supercomputers. The major difficulty of direct solutions is due to disparity of scales. From an engineering perspective, it is often sufficient to predict macroscopic properties of the multiple-scale systems, such as the effective conductivity, elastic moduli, permeability, and eddy diffusivity. Therefore, it is desirable to develop a method that captures the small scale effect on the large scales, but does not require resolving all the small scale features. The purpose of this lecture note is to review some recent advances in developing multiscale finite element (finite volume) methods for flow and transport in strongly heterogeneous porous media. Extra effort is made in developing a multiscale computational method that can be potentially used for practical multiscale for problems with a large range of nonseparable scales. Some recent theoretical and computational developments in designing global upscaling methods will be reviewed. The lectures can be roughly divided into four parts. In part 1, we review some homogenization theory for elliptic and hyperbolic equations. This homogenization theory provides a guideline for designing effective multiscale methods. In part 2, we review some recent developments of multiscale finite element (finite volume) methods. We also discuss the issue of upscaling one-phase, two-phase flows through heterogeneous porous media and the use of limited global information in multiscale finite element (volume) methods. In part 4, we will consider multiscale simulations of two-phase flow immiscible

Yalchin Efendiev
Department of Mathematics, Texas A&M University, College Station, TX 77843-3368, USA, e-mail: efendiev@math.tamu.edu
Thomas Yizhao Hou
Applied and Computational Math. 217-50, California Institute of Technology, 1200 California Blvd, Pasadena, CA 91125, USA, e-mail: hou@acm.caltech.edu

N. Ben Abdallah, G. Frosali (eds.), *Quantum Transport*.
Lecture Notes in Mathematics 1946.
© Springer-Verlag Berlin Heidelberg 2008

flows using a flow-based adaptive coordinate, and introduce a theoretical framework which enables us to perform global upscaling for heterogeneous media with long range connectivity.

1 Introduction

Many problems of fundamental and practical importance have multiple scale solutions. Composite materials, porous media, and turbulent transport in high Reynolds number flows are examples of this type. A complete analysis of these problems is extremely difficult. For example, the difficulty in analyzing groundwater transport is mainly caused by the heterogeneity of subsurface formations spanning over many scales. This heterogeneity is often represented by the multiscale fluctuations in the permeability of media. For composite materials, the dispersed phases (particles or fibers), which may be randomly distributed in the matrix, give rise to fluctuations in the thermal or electrical conductivity; moreover, the conductivity is usually discontinuous across the phase boundaries. In turbulent transport problems, the convective velocity field fluctuates randomly and contains many scales depending on the Reynolds number of the flow.

The direct numerical solution of multiple scale problems is difficult even with the advent of supercomputers. The major difficulty of direct solutions is the scale of computation. For groundwater simulations, it is common that millions of grid blocks are involved, with each block having a dimension of tens of meters, whereas the permeability measured from cores is at a scale of several centimeters. This gives more than 10^5 degrees of freedom per spatial dimension in the computation. Therefore, a tremendous amount of computer memory and CPU time are required, and this can easily exceed the limit of today's computing resources. The situation can be relieved to some degree by parallel computing; however, the size of the discrete problem is *not* reduced. The load is merely shared by more processors with more memory. Whenever one can afford to resolve all the small scale features of a physical problem, direct solutions provide quantitative information of the physical processes at all scales. On the other hand, from an engineering perspective, it is often sufficient to predict the macroscopic properties of the multiscale systems, such as the effective conductivity, elastic moduli, permeability, and eddy diffusivity. Therefore, it is desirable to develop a method that captures the small scale effect on the large scales, but does not require resolving all the small scale features. Upscaling procedures have been commonly applied for this purpose and are effective in many cases. More recently, a number of multiscale techniques have been developed and successfully applied to various areas, e.g., porous media flows. The main idea of upscaling techniques is to form coarse-scale equations with a prescribed analytical form that may differ from the underlying fine-scale equations. In multiscale methods, the fine-scale information is carried throughout the simulation and the coarse-scale equations are generally not expressed analytically, but rather formed and solved numerically.

The purpose of this lecture note is to review some recent advances in developing multiscale finite element (volume) methods for flow and transport in strongly heterogeneous porous media. Extra effort is made in developing a multiscale computational method that can be potentially used for practical multiscale problems with a large range of nonseparable scales. Substantial progress has been made in recent years by combining modern mathematical techniques such as multiscale analysis, adaptivity and multiresolution. The lectures can be roughly divided into four parts. In Sect. 2, we will review some homogenization theory for elliptic and hyperbolic equations as well as for incompressible flows. This homogenization theory provides a guideline for designing effective multiscale methods. In Sect. 3, we discuss some recent developments of multiscale finite element methods. We also discuss the issue of upscaling one-phase, two-phase flows through heterogeneous porous media and the use of limited global information in multiscale finite element methods. In Sect. 4, we discuss the generalization of the multiscale finite element methods to nonlinear partial differential equations. In Sect. 5, we will consider multiscale simulations of two-phase flow immiscible flows using a flow-based adaptive coordinate system. There are many other multiscale methods which we will not cover due to the limited scope of these lectures. The above methods are chosen because they are similar philosophically and the materials complement each other very well. This paper is not intended to be a detailed survey of all available multiscale methods. The discussion is limited by scope of the lectures and expertise of the authors.

2 Review of Homogenization Theory

In this section, we will review some classical homogenization theory for elliptic and hyperbolic PDEs. This homogenization theory will play a role in designing effective multiscale numerical methods for partial differential equations with multiscale solutions.

2.1 Homogenization Theory for Elliptic Problems

Consider the second order elliptic equation

$$\mathcal{L}(u_\epsilon) \equiv -\frac{\partial}{\partial x_i}\left(a_{ij}\left(x/\epsilon\right)\frac{\partial}{\partial x_j}\right)u_\epsilon + a_0(x/\epsilon)u_\epsilon = f, \ u_\epsilon|_{\partial\Omega} = 0, \quad (1)$$

where $a_{ij}(y)$ and $a_0(y)$ are 1-periodic in both variables of y, and satisfy $a_{ij}(y)\xi_i\xi_j \geq \alpha\xi_i\xi_i$, with $\alpha > 0$, and $a_0 > \alpha_0 > 0$. Here we have used the Einstein summation notation, i.e., repeated index means summation with respect to that index.

This model equation represents a common difficulty shared by several physical problems. For porous media, it is the pressure equation through Darcy's law, the coefficient a_ϵ representing the permeability tensor. For composite materials, it is the steady heat conduction equation and the coefficient a_ϵ represents the thermal conductivity. For steady transport problems, it is a symmetrized form of the governing equation. In this case, the coefficient a_ϵ is a combination of transport velocity and viscosity tensor.

Homogenization theory studies the limiting behavior $u_\epsilon \to u$ as $\epsilon \to 0$. The main task is to find the homogenized coefficients, a_{ij}^* and a_0^*, and the *homogenized equation* for the limiting solution u

$$-\frac{\partial}{\partial x_i}\left(a_{ij}^*\frac{\partial}{\partial x_j}\right)u + a_0^*u = f, \quad u|_{\partial\Omega} = 0. \tag{2}$$

Define the L^2 and H^1 norms over Ω as follows

$$\|v\|_0^2 = \int_\Omega |v|^2\,dx, \quad \|v\|_1^2 = \|v\|_0^2 + \|\nabla v\|_0^2. \tag{3}$$

Further, we define the bilinear form

$$a^\epsilon(u,v) = \int_\Omega a_{i,j}^\epsilon(x)\frac{\partial u}{\partial x_j}\frac{\partial v}{\partial x_i}\,dx + \int_\Omega a_0^\epsilon uv\,dx. \tag{4}$$

It is easy to show that

$$c_1\|u\|_1^2 \leq a^\epsilon(u,u) \leq c_2\|u\|_1^2, \tag{5}$$

with $c_1 = \min(\alpha, \alpha_0)$, $c_2 = \max(\|a_{ij}\|_\infty, \|a_0\|_\infty)$.

The elliptic problem can also be formulated as a variational problem: find $u_\epsilon \in H_0^1$

$$a^\epsilon(u_\epsilon, v) = (f, v), \quad \text{for all} \quad v \in H_0^1(\Omega), \tag{6}$$

where (f, v) is the usual L^2 inner product, $\int_\Omega fv\,dx$.

2.1.1 Special Case: One-Dimensional Problem

Let $\Omega = (x_0, x_1)$ and take $a_0 = 0$. We have

$$-\frac{d}{dx}\left(a(x/\epsilon)\frac{du_\epsilon}{dx}\right) = f, \quad \text{in} \quad \Omega, \tag{7}$$

where $u_\epsilon(x_0) = u_\epsilon(x_1) = 0$, and $a(y) > \alpha_0 > 0$ is y-periodic with period y_0.

By taking $v = u_\epsilon$ in the variational problem, we have

$$\|u_\epsilon\|_1 \leq c.$$

Therefore one can extract a subsequence, still denoted by u_ϵ, such that

$$u_\epsilon \rightharpoonup u \quad \text{in} \quad H_0^1(\Omega) \quad \text{weakly.} \tag{8}$$

Next, we introduce

$$\xi^\epsilon = a^\epsilon \frac{du^\epsilon}{dx}.$$

Since a^ϵ is bounded, and du^ϵ/dx is bounded in $L^2(\Omega)$, so ξ^ϵ is bounded in $L^2(\Omega)$. Moreover, since $-\frac{d\xi^\epsilon}{dx} = f$, we have $\xi^\epsilon \in H^1(\Omega)$. Thus we get

$$\xi^\epsilon \to \xi \quad \text{in} \quad L^2(\Omega) \quad \text{strongly,}$$

so that

$$\frac{1}{a^\epsilon}\xi^\epsilon \to m(1/a)\xi \quad \text{in} \quad L^2(\Omega) \quad \text{weakly.}$$

Further, we note that $\frac{1}{a^\epsilon}\xi^\epsilon = \frac{du^\epsilon}{dx}$. Therefore, we arrive at

$$\frac{du}{dx} = m(1/a)\xi.$$

On the other hand, $-\frac{d\xi^\epsilon}{dx} = f$ implies $-\frac{d\xi}{dx} = f$. This gives

$$-\frac{d}{dx}\left(\frac{1}{m(1/a)}\frac{du}{dx}\right) = f. \tag{9}$$

This is the correct homogenized equation for u. Note that $a^* = \frac{1}{m(1/a)}$ is the harmonic average of a^ϵ. It is in general not equal to the arithmetic average $\overline{a^\epsilon} = m(a)$.

2.1.2 Multiscale Asymptotic Expansions

The above analysis does not generalize to multi-dimensions. In this subsection, we introduce the multiscale expansion technique in deriving homogenized equations.

We shall look for $u_\epsilon(x)$ in the form of asymptotic expansion

$$u_\epsilon(x) = u_0(x, x/\epsilon) + \epsilon u_1(x, x/\epsilon) + \epsilon^2 u_2(x, x/\epsilon) + \cdots, \tag{10}$$

where the functions $u_j(x, y)$ are double periodic in y with period 1.

Denote by A^ϵ the second order elliptic operator

$$A^\epsilon = -\frac{\partial}{\partial x_i}\left(a_{ij}\left(x/\epsilon\right)\frac{\partial}{\partial x_j}\right). \tag{11}$$

When differentiating a function $\phi(x, x/\epsilon)$ with respect to x, we have

$$\frac{\partial}{\partial x_j} = \frac{\partial}{\partial x_j} + \frac{1}{\epsilon}\frac{\partial}{\partial y_j},$$

where y is evaluated at $y = x/\epsilon$. With this notation, we can expand A^ϵ as follows

$$A^\epsilon = \epsilon^{-2}A_1 + \epsilon^{-1}A_2 + \epsilon^0 A_3, \tag{12}$$

where

$$A_1 = -\frac{\partial}{\partial y_i}\left(a_{ij}(y)\frac{\partial}{\partial y_j}\right), \tag{13}$$

$$A_2 = -\frac{\partial}{\partial y_i}\left(a_{ij}(y)\frac{\partial}{\partial x_j}\right) - \frac{\partial}{\partial x_i}\left(a_{ij}(y)\frac{\partial}{\partial y_j}\right), \tag{14}$$

$$A_3 = -\frac{\partial}{\partial x_i}\left(a_{ij}(y)\frac{\partial}{\partial x_j}\right) + a_0. \tag{15}$$

Substituting the expansions for u_ϵ and A^ϵ into $A^\epsilon u_\epsilon = f$, and equating the terms of the same power, we get

$$A_1 u_0 = 0, \tag{16}$$
$$A_1 u_1 + A_2 u_0 = 0, \tag{17}$$
$$A_1 u_2 + A_2 u_1 + A_3 u_0 = f. \tag{18}$$

Equation (16) can be written as

$$-\frac{\partial}{\partial y_i}\left(a_{ij}(y)\frac{\partial}{\partial y_j}\right)u_0(x, y) = 0, \tag{19}$$

where u_0 is periodic in y. The theory of second order elliptic PDEs [44] implies that $u_0(x, y)$ is independent of y, i.e., $u_0(x, y) = u_0(x)$. This simplifies (17) for u_1,

$$-\frac{\partial}{\partial y_i}\left(a_{ij}(y)\frac{\partial}{\partial y_j}\right)u_1 = \left(\frac{\partial}{\partial y_i}a_{ij}(y)\right)\frac{\partial u}{\partial x_j}(x).$$

Define $\chi^j = \chi^j(y)$ as the solution to the following *cell problem*

$$\frac{\partial}{\partial y_i}\left(a_{ij}(y)\frac{\partial}{\partial y_j}\right)\chi^j = \frac{\partial}{\partial y_i}a_{ij}(y), \tag{20}$$

where χ^j is double periodic in y. The general solution of (17) for u_1 is then given by

$$u_1(x, y) = -\chi^j(y)\frac{\partial u}{\partial x_j}(x) + \tilde{u}_1(x) . \tag{21}$$

Finally, we note that the equation for u_2 is given by

$$\frac{\partial}{\partial y_i}\left(a_{ij}(y)\frac{\partial}{\partial y_j}\right)u_2 = A_2 u_1 + A_3 u_0 - f . \tag{22}$$

The solvability condition implies that the right-hand side of (22) must have mean zero in y over one periodic cell $Y = [0, 1] \times [0, 1]$, i.e.,

$$\int_Y (A_2 u_1 + A_3 u_0 - f)\, dy = 0.$$

This solvability condition for second order elliptic PDEs with periodic boundary condition [44] requires that the right-hand side of (22) have mean zero with respect to the fast variable y. This solvability condition gives rise to the homogenized equation for u:

$$-\frac{\partial}{\partial x_i}\left(a_{ij}^*\frac{\partial}{\partial x_j}\right)u + m(a_0)u = f , \tag{23}$$

where $m(a_0) = \frac{1}{|Y|}\int_Y a_0(y)\, dy$ and

$$a_{ij}^* = \frac{1}{|Y|}\left(\int_Y (a_{ij} - a_{ik}\frac{\partial \chi^j}{\partial y_k})\, dy\right) . \tag{24}$$

2.1.3 Justification of Formal Expansions

The above multiscale expansion is based on a formal asymptotic analysis. However, we can justify its convergence rigorously.

Let $z_\epsilon = u_\epsilon - (u + \epsilon u_1 + \epsilon^2 u_2)$. Applying A^ϵ to z_ϵ, we get

$$A^\epsilon z_\epsilon = -\epsilon r_\epsilon,$$

where $r_\epsilon = A_2 u_2 + A_3 u_1 + \epsilon A_3 u_2$. Thus we have $\|r_\epsilon\|_\infty \le c$.

On the other hand, we have

$$z_\epsilon|_{\partial\Omega} = -(\epsilon u_1 + \epsilon^2 u_2)|_{\partial\Omega}.$$

Thus, we obtain

$$\|z_\epsilon\|_{L^\infty(\partial\Omega)} \le c\epsilon.$$

It follows from the maximum principle [44] that

$$\|z_\epsilon\|_{L^\infty(\Omega)} \le c\epsilon$$

and therefore we conclude that

$$\|u_\epsilon - u\|_{L^\infty(\Omega)} \le c\epsilon.$$

2.1.4 Boundary Corrections

The above asymptotic expansion does not take into account the boundary condition of the original elliptic PDEs. If we add a boundary correction, we can obtain higher order approximations.

Let $\theta_\epsilon \in H^1(\Omega)$ denote the solution to

$$\nabla_x \cdot a^\epsilon \nabla_x \theta_\epsilon = 0 \text{ in } \Omega, \quad \theta_\epsilon = u_1(x, x/\epsilon) \text{ on } \partial\Omega.$$

Then we have

$$(u_\epsilon - (u + \epsilon u_1(x, x/\epsilon) - \epsilon\theta_\epsilon))|_{\partial\Omega} = 0.$$

Moskow and Vogelius [68] have shown that

$$\begin{aligned}
\|u_\epsilon - u - \epsilon u_1(x, x/\epsilon) + \epsilon\theta_\epsilon\|_0 &\le C_\omega \epsilon^{1+\omega}\|u\|_{2+\omega}, \\
\|u_\epsilon - u - \epsilon u_1(x, x/\epsilon) + \epsilon\theta_\epsilon\|_1 &\le C\epsilon\|u\|_2,
\end{aligned} \tag{25}$$

where we assume $u \in H^{2+\omega}(\Omega)$ with $0 \le \omega \le 1$, and Ω is assumed to be a bounded, convex curvilinear polygon of class C^∞. This improved estimate is used in the convergence analysis of the multiscale finite element method to be presented in Sect. 3.

2.2 Convection of Microstructure

It is most interesting to see if one can apply homogenization technique to obtain an averaged equation for the large scale quantity for incompressible Euler or Navier–Stokes equations. In 1985, McLaughlin et al. [67] attempted to obtain a homogenized equation for the 3-D incompressible Euler equations with highly oscillatory velocity field. More specifically, they considered the following initial value problem:

$$u_t + (u \cdot \nabla)u = -\nabla p,$$

with $\nabla \cdot u = 0$ and highly oscillatory initial data

$$u(x,0) = U(x) + W(x, x/\epsilon).$$

They then constructed multiscale expansions for both the velocity field and the pressure. In doing so, they made an important assumption that the microstructure is convected by the mean flow. Under this assumption, they constructed a multiscale expansion for the velocity field as follows:

$$u^\epsilon(x,t) = u(x,t) + w(\tfrac{\theta(x,t)}{\epsilon}, \tfrac{t}{\epsilon}, x, t) + \epsilon u_1(\tfrac{\theta(x,t)}{\epsilon}, \tfrac{t}{\epsilon}, x, t) + O(\epsilon^2).$$

The pressure field p^ϵ is expanded similarly. From this ansatz, one can show that θ is convected by the mean velocity:

$$\theta_t + u \cdot \nabla \theta = 0 , \qquad \theta(x,0) = x .$$

It is a very challenging problems to develop a systematic approach to study the large scale solution in three-dimensional Euler and Navier–Stokes equations. The work of McLaughlin, Papanicolaou and Pironneau provided some insightful understanding into how small scales interact with large scale and how to deal with the closure problem. However, the problem is still not completely resolved since the cell problem obtained this way does not have a unique solution. Additional constraints need to be enforced in order to derive a large scale averaged equation. With additional assumptions, they managed to derive a variant of the $k - \epsilon$ model in turbulence modeling.

Remark 2.1. One possible way to improve the work of [67] is take into account the oscillation in the Lagrangian characteristics, θ_ϵ. The oscillatory part of θ_ϵ in general could have order one contribution to the mean velocity of the incompressible Euler equation. In [51–53], Hou and Yang and co-workers have studied convection of microstructure of the 2-D and 3-D incompressible Euler equations using a new approach. They do not assume that the oscillation is propagated by the mean flow. In fact, they found that it is crucial to include the effect of oscillations in the characteristics on the mean flow. Using this new approach, they can derive a well-posed cell problem which can be used to obtain an effective large scale average equation.

More can be said for a passive scalar convection equation.

$$v_t + \frac{1}{\epsilon} \nabla \cdot (u(x/\epsilon)v) = \alpha \Delta v,$$

with $v(x,0) = v_0(x)$. Here $u(y)$ is a known incompressible periodic (or stationary random) velocity field with zero mean. Assume that the initial condition is smooth.

Expand the solution v^ϵ in powers of ϵ

$$v^\epsilon = v(t,x) + \epsilon v_1(t, x, x/\epsilon) + \epsilon^2 v_2(t, x, x/\epsilon) + \cdots .$$

The coefficients of ϵ^{-1} lead to

$$\alpha\Delta_y v_1 - u \cdot \nabla_y v_1 - u \cdot \nabla_x v = 0.$$

Let e_k, $k = 1, 2, 3$ be the unit vectors in the coordinate directions and let $\chi^k(y)$ satisfy the cell problem:

$$\alpha\Delta_y \chi^k - u \cdot \nabla_y \chi^k - u \cdot e_k = 0.$$

Then we have

$$v_1(t, x, y) = \sum_{k=1}^{3} \chi^k(y) \frac{v(t, x)}{\partial x_k}.$$

The coefficients of ϵ^0 give

$$\alpha\Delta_y v_2 - u \cdot \nabla_y v_2 = u \cdot \nabla_x v_1 - 2\alpha\nabla_x \cdot \nabla_y v_1 - \alpha\Delta_x v + v_t.$$

The solvability condition for v_2 requires that the right-hand side has zero mean with respect to y. This gives rise to the equation for homogenized solution v

$$v_t = \alpha\Delta_x v - \overline{u \cdot \nabla_x v_1}.$$

Using the cell problem, McLaughlin et al. obtained [67]

$$v_t = \sum_{i,j=1}^{3} (\alpha\delta_{ij} + \alpha_{T_{ij}}) \frac{\partial^2 v}{\partial x_i \partial x_j},$$

where $\alpha_{T_{ij}} = -\overline{u_i \chi^j}$.

2.2.1 Nonlocal Memory Effect of Homogenization

It is interesting to note that for certain degenerate problem, the homogenized equation may have a nonlocal memory effect.

Consider the simple 2-D linear convection equation:

$$\frac{\partial u_\epsilon(x, y, t)}{\partial t} + a_\epsilon(y) \frac{\partial u_\epsilon(x, y, t)}{\partial x} = 0,$$

with initial condition $u_\epsilon(x, y, 0) = u_0(x, y)$. Note that $y = x_2$ is not a fast variable here.

We assume that a_ϵ is bounded and u_0 has compact support. While it is easy to write down the solution explicitly,

$$u_\epsilon(x, y, t) = u_0(x - a_\epsilon(y)t, y),$$

it is not an easy task to derive the homogenized equation for the weak limit of u_ϵ.

Using Laplace Transform and measure theory, Tartar [78] showed that the weak limit u of u_ϵ satisfies

$$\frac{\partial}{\partial t}u(x,y,t) + A_1(y)\frac{\partial}{\partial x}u(x,y,t) = \int_0^t \int \frac{\partial^2}{\partial x^2}u(x - \lambda(t - s), y, s)d\mu_y(\lambda)\,ds,$$

with $u(x,y,0) = u_0(x,y)$, where $A_1(y)$ is the weak limit of $a_\epsilon(y)$, and μ_y is a probability measure of y and has support in $[\min(a_\epsilon), \max(a_\epsilon)]$.

As we can see, the convection induces a nonlocal history dependent diffusion term in the propagating direction (x). The homogenized equation is *not* amenable to computation in general since the measure μ_y cannot be expressed explicitly in terms of a_ϵ.

3 Numerical Upscaling Based on Multiscale Finite Element Methods

In this section, we review the multiscale finite element method (Ms-FEM) for solving partial differential equations with multiscale solutions, see [3, 19, 32, 35, 48–50, 79]. The central goal of this approach is to obtain the large scale solutions accurately and efficiently without resolving the small scale details. The main idea is to construct finite element basis functions which capture the small scale information within each element. The small scale information is then brought to the large scales through the coupling of the global stiffness matrix. Thus, the effect of small scales on the large scales is correctly captured. In our method, the basis functions are constructed from the leading order homogeneous elliptic equation in each element. As a consequence, the basis functions are adapted to the local microstructure of the differential operator. In the case of two-scale periodic structures, we have proved that the multiscale method indeed converges to the correct solution independent of the small scale in the homogenization limit [50].

In practical computations, a large amount of overhead time comes from constructing the basis functions. In general, these multiscale basis functions are constructed numerically, except for certain special cases. Since the basis functions are independent of each other, they can be constructed independently and can be done perfectly in parallel. This greatly reduces the overhead time in constructing these basis functions. In many applications, it is important to obtain a scale-up equation from the fine grid equation. For example, the high degree of variability and multiscale nature of formation properties in subsurface flows (such as permeability) pose significant challenges for subsurface flow modeling. Geological characterizations that capture these effects are typically developed at scales that are too fine for direct flow simulation, so techniques are required to enable the solution of flow problems in practice. Upscaling procedures have been commonly applied for this purpose and are effective in many cases (see, e.g., [58] for reviews and discussion). Our multiscale finite element method can be used for a similar purpose and successfully applied for problems of this type.

As discussed in [58], upscaling methods and multiscale numerical techniques (as applied within the context of subsurface flow modeling) have many similarities and some important differences. Upscaling techniques provide coefficients, which are typically computed in a pre-processing step, for coarse scale equations of prescribed analytical forms. In multiscale methods, the coarse scale equations are formed numerically and fine scale information may be carried throughout the simulation and used at various stages. For example, in multiscale procedures for subsurface flow applications, different grids are often used for flow and transport computations. The advantage of deriving a scale-up equation or performing multiscale computations is that one can perform many useful tests on the coarse model with different boundary conditions or source terms. This would be very expensive if we have to perform all these tests on a fine grid. For time dependent problems, the coarse-scale equation also allows for larger time steps. This results in additional computational saving.

It should be mentioned that many numerical methods have been developed with goals similar to ours. These include generalized finite element methods [8, 9, 11], wavelet based numerical homogenization methods [16, 22, 24, 62], methods based on the homogenization theory (cf. [14, 21, 28]), equation-free computations (e.g., [61]), variational multiscale methods [18, 56, 57, 74], heterogeneous multiscale methods [30], matrix-dependent multigrid based homogenization [22, 62], generalized p-FEM in homogenization [64, 65], and some upscaling methods based on simple physical and/or mathematical motivations (cf. [26, 66]). The methods based on the homogenization theory have been successfully applied to determine the effective conductivity and permeability of certain composite materials and porous media. However, their range of applications is usually limited by restrictive assumptions on the media, such as scale separation and periodicity [13, 60]. They are also expensive to use for solving problems with many separate scales since the cost of computation grows exponentially with the number of scales. But for the multiscale method, the number of scales does not increase the overall computational cost exponentially. The upscaling methods are more general and have been applied to problems with random coefficients with partial success (cf. [26, 66]). But the design principle is strongly motivated by the homogenization theory for periodic structures. Their application to nonperiodic structures is not always guaranteed to work.

Most multiscale methods presented to date have applied local calculations for the determination of basis functions. Though effective in many cases, global effects can be important for some problems. The importance of global information has been illustrated within the context of upscaling procedures as well as multiscale computations in recent investigations. These studies have shown that the use of limited global information in the calculation of the coarse-scale parameters (such as basis functions) can significantly improve the accuracy of the resulting coarse model. In this lecture notes, we describe the use of limited global information in multiscale simulations.

We remark that the idea of using basis functions governed by the differential equations has been applied to convection–diffusion equation with boundary layers (see, e.g., [12] and references therein). Babuska et al. applied a similar idea to 1-D problems [11] and to a special class of 2-D problems with the coefficient varying locally in one direction [9]. Most of these methods are based on the special property of one-dimensional properties of the coefficients. As indicated by our convergence analysis, there is a fundamental difference between one-dimensional problems and genuinely multi-dimensional problems. Special complications such as the resonance between the mesh scale and the physical scale never occur in the corresponding 1-D problems.

3.1 Multiscale Finite Element Methods for Elliptic PDEs

In this section we introduce the multiscale finite element method applied to the following problem

$$L_\epsilon p := -\nabla \cdot (k_\epsilon(x)\nabla p) = f \quad \text{in } \Omega, \quad p = 0 \quad \text{on } \Gamma = \partial\Omega, \tag{1}$$

where Ω is a domain in R^d and $k_\epsilon(x)$ is a heterogeneous field. We use ϵ to denote heterogeneous fields (assuming ϵ is the finest scale) and do not assume periodicity in general. The unknown is changed to p, since it will be used later in porous media flow simulations, where the solution represents the pressure field. $k_\epsilon(x) = (k_{ij}(x))$ is assumed to be symmetric and satisfies $\alpha|\xi|^2 \leq k_{ij}\xi_i\xi_j \leq \beta|\xi|^2$, for all $\xi \in \mathbb{R}^2$ and with $0 < \alpha < \beta$.

Next we introduce the multiscale finite element method. Let $\mathcal{T}_h$ be a regular partition of Ω into triangles. Let $\{x_j\}_{j=1}^J$ be the interior nodes of the mesh $\mathcal{T}_h$ and $\{\psi_j\}_{j=1}^J$ be the nodal basis of the standard linear finite element space $W_h \subset H_0^1(\Omega)$. Denote by $S_i = \text{supp}(\psi_i)$ and define ϕ^i with support in S_i as follows:

$$L_\epsilon \phi^i = 0 \quad \text{in } K, \quad \phi^i = \psi_i \quad \text{on } \partial K \quad \forall K \in \mathcal{T}_h, K \subset S_i. \tag{2}$$

It is obvious that $\phi^i \in H_0^1(S_i) \subset H_0^1(\Omega)$. Finally, let $V_h \subset H_0^1(\Omega)$ be the finite element space spanned by $\{\phi^i\}_{i=1}^J$.

Remark 3.1. We note that one can take K to be a representative volume element (RVE) that is smaller than K if the characteristic length scale is much smaller than the size of RVE.

With above notations, the multiscale finite element method is to find $p_h \in V_h$ such that

$$(k_\epsilon(x)\nabla p_h, \nabla v_h) = (f, v_h) \quad \forall \; v_h \in V_h, \tag{3}$$

where and hereafter we denote by $(\cdot, \cdot)$ the L^2 inner product in $L^2(\Omega)$.

Note that MsFEM consists of two parts, basis function construction and a choice of the global formulation that couples these basis functions. In the above discussion, we presented simplest basis function construction and global formulation. In general, the global formulation can be easily modified and various global formulations based on Petrov–Galerkin finite element method, finite volume, mixed finite element, discontinuous Galerkin finite element and other methods have been studied in the literature. Some of them will be discussed in these lecture notes.

As for basis functions, the choice of boundary conditions in defining the multiscale basis functions play a crucial role in approximating the multiscale solution. Intuitively, the boundary condition for the multiscale basis function should reflect the multiscale oscillation of the solution p across the boundary of the coarse grid element. By choosing a linear boundary condition for the basis function, we will create a mismatch between the exact solution p and the finite element approximation across the element boundary. In the next sections, we will discuss this issue further and introduce an oversampling technique to alleviate this difficulty. This technique enables us to remove the artificial numerical boundary layer across the coarse grid boundary element.

We would like also to note that MsFEM can be naturally extended to solve nonlinear partial differential equations. As in the case of linear problems, the main idea of MsFEM remains the same with the exception of basis function construction. Because of nonlinearities, the multiscale basis functions are replaced by multiscale maps, which are in general nonlinear maps from W_h to heterogeneous fields (see discussions in Sect. 4). Similar to linear problems, one can solve local problems in a representative volume element (RVE) that is smaller than K if the characteristic length scale is much smaller than the size of RVE.

To gain some insight into the multiscale finite element method, we next perform an error analysis for the multiscale finite element method in the simplest case, i.e., we use linear boundary conditions for the multiscale basis functions.

3.2 Convergence Analysis of MsFEM

For the analysis here, we restrict ourselves to a periodic case $k(x) = (k_{ij}(x/\epsilon))$. We assume $k_{ij}(y)$ are smooth periodic functions in y in a unit cube Y. We assume that $f \in L^2(\Omega)$. The assumptions on k_{ij} can be relaxed. Let p_0 be the solution of the homogenized equation

$$L_0 p_0 := -\nabla \cdot (k^* \nabla p_0) = f \quad \text{in } \Omega, \quad p_0 = 0 \quad \text{on } \Gamma, \tag{4}$$

where $\Gamma = \partial \Omega$ and

$$k_{ij}^* = \frac{1}{|Y|} \int_Y k_{ik}(y) (\delta_{kj} - \frac{\partial \chi^j}{\partial y_k}) \, dy,$$

and $\chi^j(y)$ is the periodic solution of the cell problem

$$\nabla_y \cdot (k(y) \nabla_y \chi^j) = \frac{\partial}{\partial y_i} k_{ij}(y) \quad \text{in } Y, \quad \int_Y \chi^j(y) \, dy = 0.$$

It is clear that $p_0 \in H^2(\Omega)$ since Ω is a convex polygon. Denote by $p_1(x, y) = -\chi^j(y) \frac{\partial p_0(x)}{\partial x_j}$ and let θ_ϵ be the solution of the problem

$$L_\epsilon \theta_\epsilon = 0 \quad \text{in } \Omega, \quad \theta_\epsilon(x) = p_1(x, \tfrac{x}{\epsilon}) \quad \text{on } \Gamma. \tag{5}$$

Our analysis of the multiscale finite element method relies on the following homogenization result obtained by Moskow and Vogelius [68].

Lemma 3.1. *Let $p_0 \in H^2(\Omega)$ be the solution of (4), $\theta_\epsilon \in H^1(\Omega)$ be the solution to (5) and $p_1(x) = -\chi^j(x/\epsilon)\partial p_0(x)/\partial x_j$. Then there exists a constant C independent of u_0, ϵ and Ω such that*

$$\| p - p_0 - \epsilon(u_1 - \theta_\epsilon) \|_{1,\Omega} \le C\epsilon(| p_0 |_{2,\Omega} + \| f \|_{0,\Omega}).$$

3.2.1 Error Estimates ($h < \epsilon$)

The starting point is the well-known Cea's lemma.

Lemma 3.2. *Let p be the solution of (1) and p_h be the solution of (3). Then we have*

$$\| p - p_h \|_{1,\Omega} \le C \inf_{v_h \in V_h} \| p - v_h \|_{1,\Omega}.$$

Let $\Pi_h : C(\bar{\Omega}) \to W_h \subset H_0^1(\Omega)$ be the usual Lagrange interpolation operator:

$$\Pi_h p(x) = \sum_{j=1}^J p(x_j) \psi_j(x) \quad \forall \, u \in C(\bar{\Omega})$$

and $I_h : C(\bar{\Omega}) \to V_h$ be the corresponding interpolation operator defined through the multiscale basis function ϕ

$$I_h p(x) = \sum_{j=1}^J p(x_j) \phi^j(x) \quad \forall \, u \in C(\bar{\Omega}).$$

From the definition of the basis function ϕ^i in (2) we have

$$L_\epsilon(I_h p) = 0 \quad \text{in } K, \quad I_h p = \Pi_h p \quad \text{on } \partial K, \tag{6}$$

for any $K \in \mathcal{T}_h$.

Lemma 3.3. *Let $p \in H^2(\Omega)$ be the solution of (1). Then there exists a constant C independent of h, ϵ such that*

$$\| p - I_h p \|_{0,\Omega} + h \| p - I_h p \|_{1,\Omega} \leq Ch^2(| p |_{2,\Omega} + \| f \|_{0,\Omega}). \tag{7}$$

Proof. At first it is known from the standard finite element interpolation theory that

$$\| p - \Pi_h p \|_{0,\Omega} + h \| p - \Pi_h p \|_{1,\Omega} \leq Ch^2(| p |_{2,\Omega} + \| f \|_{0,\Omega}). \tag{8}$$

On the other hand, since $\Pi_h p - I_h p = 0$ on ∂K, the standard scaling argument yields

$$\| \Pi_h p - I_h p \|_{0,K} \leq Ch |\Pi_h p - I_h p|_{1,K} \quad \forall \, K \in \mathcal{T}_h. \tag{9}$$

To estimate $|\Pi_h p - I_h p|_{1,K}$ we multiply the equation in (6) by $I_h p - \Pi_h p \in H_0^1(K)$ to get

$$(k(\tfrac{x}{\epsilon})\nabla I_h p, \nabla(I_h p - \Pi_h p))_K = 0,$$

where $(\cdot, \cdot)_K$ denotes the L^2 inner product of $L^2(K)$. Thus, upon using the equation in (1), we get

$$(k(\tfrac{x}{\epsilon})\nabla(I_h p - \Pi_h p), \nabla(I_h p - \Pi_h p))_K$$

$$= (k(\tfrac{x}{\epsilon})\nabla(p - \Pi_h p), \nabla(I_h p - \Pi_h p))_K - (k(\tfrac{x}{\epsilon})\nabla p, \nabla(I_h p - \Pi_h p))_K$$

$$= (k(\tfrac{x}{\epsilon})\nabla(p - \Pi_h p), \nabla(I_h p - \Pi_h p))_K - (f, I_h p - \Pi_h p)_K.$$

This implies that

$$|I_h p - \Pi_h p|_{1,K} \leq Ch| p |_{2,K} + \| I_h p - \Pi_h p \|_{0,K} \| f \|_{0,K}.$$

Hence

$$|I_h p - \Pi_h p|_{1,K} \leq Ch(| p |_{2,K} + \| f \|_{0,K}), \tag{10}$$

where we have used (9). Now the lemma follows from (8)–(10). $\qquad\square$

In conclusion, we have the following estimate by using Lemmas 3.2–3.3.

Theorem 3.1. *Let $p \in H^2(\Omega)$ be the solution of (1) and $p_h \in V_h$ be the solution of (3). Then we have*

$$\| p - p_h \|_{1,\Omega} \leq Ch(| p |_{2,\Omega} + \| f \|_{0,\Omega}). \tag{11}$$

Note that the estimate (11) blows up like h/ϵ as $\epsilon \to 0$ since $|p|_{2,\Omega} = O(1/\epsilon)$. This is insufficient for practical applications. In next subsection we derive an error estimate which is uniform as $\epsilon \to 0$.

3.2.2 Error Estimates ($h > \epsilon$)

In this section, we will show that the multiscale finite element method gives a convergence result uniform in ϵ as ϵ tends to zero. This is the main feature of this multiscale finite element method over the traditional finite element method. The main result in this subsection is the following theorem.

Theorem 3.2. *Let $p \in H^2(\Omega)$ be the solution of (1) and $p_h \in V_h$ be the solution of (3). Then we have*

$$\| p - p_h \|_{1,\Omega} \leq C(h + \epsilon)\| f \|_{0,\Omega} + C\left(\frac{\epsilon}{h}\right)^{1/2}\| p_0 \|_{1,\infty,\Omega}, \qquad (12)$$

where $p_0 \in H^2(\Omega) \cap W^{1,\infty}(\Omega)$ is the solution of the homogenized equation (4).

To prove the theorem, we first denote by

$$p_I(x) = I_h p_0(x) = \sum_{j=1}^{J} p_0(x_j)\phi^j(x) \in V_h.$$

From (6) we know that $L_\epsilon p_I = 0$ in K and $p_I = \Pi_h p_0$ on ∂K for any $K \in T_h$. The homogenization theory (see (25)) implies that

$$\| p_I - p_{I0} - \epsilon(p_{I1} - \theta_{I\epsilon}) \|_{1,K} \leq C\epsilon(\| f \|_{0,K} + | p_{I0} |_{2,K}), \qquad (13)$$

where p_{I0} is the solution of the homogenized equation on K:

$$L_0 p_{I0} = 0 \quad \text{in } K, \quad p_{I0} = \Pi_h p_0 \quad \text{on } \partial K, \qquad (14)$$

p_{I1} is given by the relation

$$p_{I1}(x, y) = -\chi^j(y)\frac{\partial p_{I0}}{\partial x_j} \quad \text{in } K, \qquad (15)$$

and $\theta_{I\epsilon} \in H^1(K)$ is the solution of the problem:

$$L_\epsilon \theta_{I\epsilon} = 0 \quad \text{in } K, \quad \theta_{I\epsilon}(x) = p_{I1}(x, \tfrac{x}{\epsilon}) \quad \text{on } \partial K. \qquad (16)$$

It is obvious from (14) that

$$p_{I0} = \Pi_h p_0 \quad \text{in } K, \qquad (17)$$

since $\Pi_h p_0$ is linear on K. From (13) we obtain that

$$\| p - p_{\mathrm{I}} \|_{1,\Omega} \le \| p_0 - p_{\mathrm{I}0} \|_{1,\Omega} + \| \epsilon(p_1 - p_{\mathrm{I}1}) \|_{1,\Omega}$$
$$+ \| \epsilon(\theta_\epsilon - \theta_{\mathrm{I}\epsilon}) \|_{1,\Omega} + C\epsilon \| f \|_{0,\Omega}, \tag{18}$$

where we have used the regularity estimate $\| p_0 \|_{2,\Omega} \le C \| f \|_{0,\Omega}$. Now it remains to estimate the terms at the right-hand side of (18).

Lemma 3.4. *We have*

$$\| p_0 - p_{\mathrm{I}0} \|_{1,\Omega} \le Ch \| f \|_{0,\Omega}, \tag{19}$$
$$\| \epsilon(p_1 - p_{\mathrm{I}1}) \|_{1,\Omega} \le C(h + \epsilon) \| f \|_{0,\Omega}. \tag{20}$$

Proof. The estimate (19) is a direct consequence of the standard finite element interpolation theory since $p_{\mathrm{I}0} = \Pi_h p_0$ by (17). Next we note that $\chi^j(x/\epsilon)$ satisfies

$$\| \chi^j \|_{0,\infty,\Omega} + \epsilon \| \nabla \chi^j \|_{0,\infty,\Omega} \le C \tag{21}$$

for some constant C independent of h and ϵ. Thus we have, for any $K \in \mathcal{T}_h$,

$$\| \epsilon(p_1 - p_{\mathrm{I}1}) \|_{0,K} \le C\epsilon \| \chi^j \frac{\partial}{\partial x_j}(p_0 - \Pi_h p_0) \|_{0,K} \le Ch\epsilon |p_0|_{2,K},$$

$$\| \epsilon \nabla(p_1 - p_{\mathrm{I}1}) \|_{0,K} = \epsilon \| \nabla(\chi^j \frac{\partial(p_0 - \Pi_h p_0)}{\partial x_j}) \|_{0,K}$$
$$\le C \| \nabla(p_0 - \Pi_h p_0) \|_{0,K} + C\epsilon |p_0|_{2,K}$$
$$\le C(h + \epsilon) |p_0|_{2,K}.$$

This completes the proof.   □

Lemma 3.5. *We have*

$$\| \epsilon \theta_\epsilon \|_{1,\Omega} \le C \sqrt{\epsilon} \| p_0 \|_{1,\infty,\Omega} + C\epsilon |p_0|_{2,\Omega}. \tag{22}$$

Proof. Let $\zeta \in C_0^\infty(\mathbb{R}^2)$ be the cut-off function which satisfies $\zeta \equiv 1$ in $\Omega \setminus \Omega_{\delta/2}$, $\zeta \equiv 0$ in Ω_δ, $0 \le \zeta \le 1$ in $\mathbb{R}^2$, and $|\nabla \zeta| \le C/\delta$ in Ω, where for any $\delta > 0$ sufficiently small, we denote by Ω_δ as

$$\Omega_\delta = \{x \in \Omega : \mathrm{dist}(x, \partial\Omega) \ge \delta\}.$$

With this definition, it is clear that $\theta_\epsilon - \zeta p_1 = \theta_\epsilon + \zeta(\chi^j \partial p_0/\partial x_j) \in H_0^1(\Omega)$. Multiplying the equation in (5) by $\theta_\epsilon - \zeta p_1$, we get

$$(k(\tfrac{x}{\epsilon})\nabla\theta_\epsilon, \nabla(\theta_\epsilon + \zeta\chi^j \frac{\partial p_0}{\partial x_j})) = 0,$$

which yields, by using (21),

$$
\begin{aligned}
\| \nabla \theta_\epsilon \|_{0,\Omega} &\leq C \| \nabla (\zeta \chi^j \partial p_0/\partial x_j) \|_{0,\Omega} \\
&\leq C \| \nabla \zeta \cdot \chi^j \partial p_0/\partial x_j \|_{0,\Omega} + C \| \zeta \nabla \chi^j \partial p_0/\partial x_j \|_{0,\Omega} \\
&\quad + C \| \zeta \chi^j \partial^2 p_0/\partial^2 x_j \|_{0,\Omega} \\
&\leq C \sqrt{|\partial \Omega| \cdot \delta} \frac{D}{\delta} + C \sqrt{|\partial \Omega| \cdot \delta} \frac{D}{\epsilon} + C |p_0|_{2,\Omega},
\end{aligned}
\tag{23}
$$

where $D = \| p_0 \|_{1,\infty,\Omega}$ and the constant C is independent of the domain Ω. From (23) we have

$$
\begin{aligned}
\| \epsilon \theta_\epsilon \|_{0,\Omega} &\leq C \Big(\frac{\epsilon}{\sqrt{\delta}} + \sqrt{\delta} \Big) \| p_0 \|_{1,\infty,\Omega} + C \epsilon |p_0|_{2,\Omega} \\
&\leq C \sqrt{\epsilon} \| p_0 \|_{1,\infty,\Omega} + C \epsilon |p_0|_{2,\Omega}.
\end{aligned}
\tag{24}
$$

Moreover, by applying the maximum principle to (5), we get

$$
\| \theta_\epsilon \|_{0,\infty,\Omega} \leq \| \chi^j \partial p_0/\partial x_j \|_{0,\infty,\partial\Omega} \leq C \| p_0 \|_{1,\infty,\Omega}.
\tag{25}
$$

Combining (24) and (25) completes the proof.   □

Lemma 3.6. *We have*

$$
\| \epsilon \theta_{I\epsilon} \|_{1,\Omega} \leq C \Big(\frac{\epsilon}{h} \Big)^{1/2} \| p_0 \|_{1,\infty,\Omega}.
\tag{26}
$$

Proof. First we remember that for any $K \in \mathcal{T}_h$, $\theta_{I\epsilon} \in H^1(K)$ satisfies

$$
L_\epsilon \theta_{I\epsilon} = 0 \ \text{ in } K, \quad \theta_{I\epsilon} = -\chi^j \Big(\frac{x}{\epsilon} \Big) \frac{\partial (\Pi_h p_0)}{\partial x_j} \ \text{ on } \partial K.
\tag{27}
$$

By applying maximum principle and (21) we get

$$
\| \theta_{I\epsilon} \|_{0,\infty,K} \leq \| \chi^j \partial (\Pi_h p_0)/\partial x_j \|_{0,\infty,\partial K} \leq C \| p_0 \|_{1,\infty,K}.
$$

Thus we have

$$
\| \epsilon \theta_{I\epsilon} \|_{0,\Omega} \leq C \epsilon \| p_0 \|_{1,\infty,\Omega}.
\tag{28}
$$

On the other hand, since the constant C in (23) is independent of Ω, we can apply the same argument leading to (23) to obtain

$$
\begin{aligned}
\| \epsilon \nabla \theta_{I\epsilon} \|_{0,K} &\leq C \epsilon \| \Pi_h p_0 \|_{1,\infty,K} (\sqrt{|\partial K|}/\sqrt{\delta} + \sqrt{|\partial K| \cdot \delta}/\epsilon) + C \epsilon |\Pi_h p_0|_{2,K} \\
&\leq C \sqrt{h} \| p_0 \|_{1,\infty,K} \Big(\frac{\epsilon}{\sqrt{\delta}} + \sqrt{\delta} \Big) \\
&\leq C \sqrt{h\epsilon} \| p_0 \|_{1,\infty,K},
\end{aligned}
$$

which implies that

$$\| \epsilon \nabla \theta_{I\epsilon} \|_{0,\Omega} \le C \Big(\frac{\epsilon}{h} \Big)^{1/2} \| p_0 \|_{1,\infty,\Omega}.$$

This completes the proof.   □

Proof (of Theorem 3.2.). The theorem is now a direct consequence of (18) and the Lemmas 3.4–3.6 and the regularity estimate $\| p_0 \|_{2,\Omega} \le C \| f \|_{0,\Omega}$.
□

Remark 3.2. As we pointed out earlier, the multiscale FEM indeed gives correct homogenized result as ϵ tends to zero. This is in contrast with the traditional FEM which does not give the correct homogenized result as $\epsilon \to 0$. The error would grow like $O(h^2/\epsilon^2)$. On the other hand, we also observe that when $h \sim \epsilon$, the multiscale method attains large error in both H^1 and L^2 norms. This is what we call the *resonance* effect between the grid scale (h) and the small scale (ϵ) of the problem. This estimate reflects the intrinsic scale interaction between the two scales in the *discrete* problem. Our extensive numerical experiments confirm that this estimate is indeed generic and sharp. From the viewpoint of practical applications, it is important to reduce or completely remove the resonance error for problems with many scales since the chance of hitting a resonance sampling is high. In the next subsection, we propose an oversampling method to overcome this difficulty.

3.3 The Oversampling Technique

As illustrated by our error analysis, large errors result from the "resonance" between the grid scale and the scales of the continuous problem. For the two-scale problem, the error due to the resonance manifests as a ratio between the wavelength of the small scale oscillation and the grid size; the error becomes large when the two scales are close. A deeper analysis shows that the boundary layer in the first order corrector seems to be the main source of the resonance effect. By a judicious choice of boundary conditions for the basis function, we can eliminate the boundary layer in the first order corrector. This would give a nice conservative difference structure in the discretization, which in turn leads to *cancellation of resonance errors* and gives an improved rate of convergence.

Motivated by our convergence analysis, we propose an *over-sampling* method to overcome the difficulty due to scale resonance [48]. The idea is quite simple and easy to implement. Since the boundary layer in the first order corrector is thin, $O(\epsilon)$, we can sample in a domain with size larger than $h + \epsilon$ and use only the interior sampled information to construct the basis functions; here, h is the mesh size and ϵ is the small scale in the solution. By doing this, we can reduce the influence of the boundary layer in the larger sample domain on the basis functions significantly. As a consequence, we obtain an improved rate of convergence.

Specifically, let ψ^j be the basis functions satisfying the homogeneous elliptic equation in the larger domain $S \supset K$. We then form the actual basis ϕ^i by linear combination of ψ^j,

$$\phi^i = \sum_{j=1}^{d} c_{ij} \psi^j \ .$$

The coefficients c_{ij} are determined by condition $\phi^i(\mathbf{x}_j) = \delta_{ij}$. The corresponding θ_ϵ^i for ϕ^i are now free of boundary layers. Our extensive numerical experiments have demonstrated that the oversampling technique does improve the numerical error substantially in many applications. On the other hand, the oversampling technique results in a *non-conforming* MsFEM method. In [35], we perform a careful estimate of the nonconforming errors in both H^1 norm and the L^2 norm. The analysis shows that the non-conforming error is indeed small, consistent with our numerical results [48, 49]. Our analysis also reveals another source of resonance, which is the mismatch between the mesh size and the "perfect" sample size. In case of a periodic structure, the "perfect" sample size is the length of an integer multiple of the period. We call the new resonance the "cell resonance". In the error expansion, this resonance effect appears as a *higher* order correction. In numerical computations, we found that the cell resonance error is generically small, and is rarely observed in practice. Nonetheless, it is possible to completely eliminate this cell resonance error by using the oversampling technique to construct the basis functions but using piecewise linear functions as test functions. This reduces the nonconforming error and eliminates the resonance error completely (see [46]).

3.4 Performance and Implementation Issues

The multiscale method given in the previous section is fairly straightforward to implement. Here, we outline the implementation and define some notations that are used in the discussion below. We consider solving problems in a unit square domain. Let N be the number of elements in the x and y directions. The mesh size is thus $h = 1/N$. To compute the basis functions, each element is discretized into $M \times M$ subcell elements with mesh size $h_s = h/M$. To implement the oversampling method, we partition the domain into sampling domains and each of them contains many elements. From the analysis and numerical tests, the size of the sampling domains can be chosen freely as long as the boundary layer is avoided. In practice, though, one wants to maximize the efficiency of oversampling by choosing the largest possible sample size which reduces the redundant computation of overlapping domains to a minimum.

In general, the multiscale (sampling) basis functions are constructed numerically, except for certain special cases. They are solved in each K or S using standard FEM. The linear systems are solved using a robust multigrid method with matrix dependent prolongation and ILLU smoothing (MG-ILLU, see [80]). The global linear system on Ω is solved using the same method. Numerical tests show that the accuracy of the final solution is insensitive to the accuracy of basis functions.

Since the basis functions are independent of each other, their construction can be carried out in parallel perfectly. In our parallel implementation of oversampling, the sample domains are chosen such that they can be handled within each processor without communication. The multigrid solver is also modified to better suit the parallelization. In particular, the ILLU smoothing is replaced by Gauss–Seidel iterations. More implementation details can be found in [48].

3.4.1 Cost and Performance

In practical computations, a large amount of overhead time comes from constructing the basis functions. On a sequential machine, the operation count of our method is about twice that of a conventional FEM for a 2-D problem. However, due to good parallel efficiency, this difference is reduced significantly on a massively parallel computer. For example, using 256 processors on an Intel Paragon, our method with $N = 32$ and $M = 32$ only spends 9% more CPU time than the conventional linear FEM method using $1,024 \times 1,024$ elements [48]. Note that this comparison is made for a single solve of the problem. In practice, multiple solves are often required, then the overhead of basis construction is negligible. A detailed study of MsFEM's parallel efficiency has been conducted in [48]. It was also found that MsFEM is helpful for improving multigrid convergence when the coefficient k_ϵ has very large contrast (i.e., the ratio between the maximum and minimum of k_ϵ).

Significant computational savings can be obtained for time dependent problems (such as two-phase flows) by constructing the multiscale basis functions adaptively. Multiscale basis functions are updated only for those coarse grid elements where the saturation changes significantly. In practice, the number of such coarse grid elements are small. They are concentrated near sharp interfaces. Also, the cost of solving a basis function in a small cell is more efficient than solving the fine grid problem globally because the condition number for solving the local basis function in each coarse grid element is much smaller than that of the corresponding global fine grid pressure system. Thus, updating a small number of multiscale basis functions dynamically is much cheaper than updating the fine grid pressure field globally. Another advantage of the multiscale finite element method is its ability to scale down the size of a large scale problem.

3.4.2 MsFEM for Problems with Scale Separation

If there is a scale separation in representative volumes smaller than the coarse block, then multiscale finite element basis functions can be computed based on the smaller regions. To demonstrate this, we first consider a periodic case. In this case, the basis functions can be approximated by

$$\phi^j(x) = \phi_0^j(x) + \epsilon \chi^i \nabla_i \phi_0^j.$$

Consequently, the approximation of the basis functions can be carried out in a domain of size ϵ via the computation of χ^i. This reduces the computational cost. Moreover, the assembly of stiffness matrix can be also performed in a period, because $k(x/\epsilon)\nabla\phi^i \cdot \nabla\phi^j$ is a periodic function. The results obtained by this approximation give the classical numerical homogenization procedure that is based on the computation of effective coefficients based on periodic problems. We would like to note that this approximation procedure is not limited to periodic problems and can be applied to random homogeneous problems with the strong scale separation, i.e., the size of representative volume is much smaller than the coarse mesh size. In general, this holds for problems where homogenization by periodization (see [60]) is true. Random homogeneous case with ergodicity is one of them. We note that a number of methods used in practice employs this strategy (e.g., [30, 42, 61, 73]). Also, we would like to note that in the case of scale separation, one can take K to be representative volume element.

3.4.3 Convergence and Accuracy

Since we need to use an additional grid to compute the basis function numerically, it makes sense to compare our MsFEM with a traditional FEM at the subcell grid, $h_s = h/M$. Note that MsFEM only captures the solution at the coarse grid h, while FEM tries to resolve the solution at the fine grid h_s. Our extensive numerical experiments demonstrate that the accuracy of MsFEM on the coarse grid h is comparable to that of FEM on the fine grid. In some cases, MsFEM is even more accurate than the FEM (see below and the next section).

As an example, in Table 1 we present the result for

$$k(\mathbf{x}/\epsilon) = \frac{2 + P\sin(2\pi x/\epsilon)}{2 + P\cos(2\pi y/\epsilon)} + \frac{2 + \sin(2\pi y/\epsilon)}{2 + P\sin(2\pi x/\epsilon)} \quad (P = 1.8), \qquad (29)$$

$$f(\mathbf{x}) = -1 \quad \text{and} \quad u|_{\partial\Omega} = 0. \qquad (30)$$

The convergence of three different methods are compared for fixed $\epsilon/h = 0.64$, where "-L" indicates that linear boundary condition is imposed on the multiscale basis functions, "os" indicates the use of oversampling, and LFEM

Table 1 Convergence for periodic case

N	ϵ	MsFEM-L		MsFEM-os-L		LFEM	
		$\|E\|_{l^2}$	Rate	$\|E\|_{l^2}$	Rate	MN	$\|E\|_{l^2}$
16	0.04	3.54e-4		7.78e-5		256	1.34e-4
32	0.02	3.90e-4	-0.14	3.83e-5	1.02	512	1.34e-4
64	0.01	4.04e-4	-0.05	1.97e-5	0.96	1,024	1.34e-4
128	0.005	4.10e-4	-0.02	1.03e-5	0.94	2,048	1.34e-4

stands for standard FEM with linear basis functions. We see clearly the scale
resonance in the results of MsFEM-L and the (almost) first order convergence
(i.e., no resonance) in MsFEM-os-L. Evident also is the error of MsFEM-os-L
being smaller than those of LFEM obtained on the fine grid. In [48, 50], more
extensive convergence tests have been presented.

3.5 Brief Overview of Mixed Finite Element and Finite Volume Element Methods

3.5.1 Control Volume Multiscale Finite Element Method

In this section, we discuss multiscale finite volume element method. Finite
volume method is chosen because, by its construction, it satisfies the numerical local conservation which is important in groundwater and reservoir
simulations. Let $\mathcal{T}^h$ denote the collection of coarse elements/rectangles K.
Consider a coarse element K, and let ξ_K be its center. The element K is divided into four rectangles of equal area by connecting ξ_K to the midpoints of
the element's edges. We denote these quadrilaterals by K_ξ, where $\xi \in Z_h(K)$,
are the vertices of K. Also, we denote $Z_h = \bigcup_K Z_h(K)$ and $Z_h^0 \subset Z_h$ the vertices which do not lie on the Dirichlet boundary of Ω. The control volume
V_ξ is defined as the union of the quadrilaterals K_ξ sharing the vertex ξ (see
Fig. 1).

The key idea of the method is the construction of basis functions on
the coarse grids, such that these basis functions capture the small-scale
information on each of these coarse grids. As before, the basis functions
are constructed from the solution of the leading order homogeneous elliptic
equation on each coarse element with some specified boundary conditions.
We consider a coarse element K that has d vertices, the local basis functions
$\phi_i, i = 1, \ldots, d$ are set to satisfy the following elliptic problem:

$$-\nabla \cdot (k \cdot \nabla \phi_i) = 0 \quad \text{in } K$$
$$\phi_i = g_i \quad \text{on } \partial K, \tag{31}$$

for some function g_i defined on the boundary of the coarse element K. As we
discussed earlier, Hou and Wu [48] have demonstrated that a careful choice of

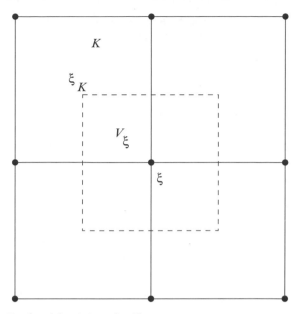

Fig. 1 Schematic of nodal points and grid

boundary conditions would improve the accuracy of the method. In previous findings, the function g_i for each i is chosen to vary linearly along ∂K or to be the solution of the local one-dimensional problems [59] or the solution of the problem in a slightly larger domain is chosen to define the boundary conditions. For simplicity, we consider linear boundary conditions and also discuss the boundary conditions obtained from a global solution. We will require $\phi_i(x_j) = \delta_{ij}$. Finally, a nodal basis function associated with the vertex x_i in the domain Ω is constructed from the combination of the local basis functions that share this x_i and zero elsewhere. We would like to note that one can use an approximate solution of (31) as discussed before. For example, in the case of periodic or random homogeneous cases, the basis functions can be approximated using homogenization expansion $\phi_i = \phi_i^0 + \epsilon \chi_k \nabla_k \phi_i^0$, where χ_k is the solution of the cell problem and ϕ_i^0 is standard finite element basis on the coarse mesh (see [33]).

Next, we denote by V^h the space of our approximate pressure solution, which is spanned by the basis functions $\{\phi_j\}_{x_j \in Z_h^0}$. Then we formulate the finite dimensional problem corresponding to finite volume element formulation of pressure equation. A statement of mass conservation on a coarse-control volume V_x is formed from pressure equation, where the approximate solution is written as a linear combination of the basis functions. Assembly of this conservation statement for all control volumes would give the corresponding linear system of equations that can be solved accordingly. The resulting linear system has incorporated the fine-scale information through the involvement of the nodal basis functions on the approximate solution. To be specific, the problem now is to seek $p^h \in V^h$ with $p^h = \sum_{x_j \in Z_h^0} p_j \phi_j$ such that

$$\int_{\partial V_\xi} k \cdot \nabla p^h \cdot n \, dl = 0, \tag{32}$$

for every control volume $V_\xi \subset \Omega$. Here n defines the normal vector on the boundary of the control volume, ∂V_ξ and S is the fine-scale saturation field at this point. The resulting multiscale method differs from the multiscale finite element method, since it employs the finite volume element method as a global solver, and it is called multiscale finite volume element method (MsFVEM). We would like to note that the coarse-scale velocity field obtained using MsFVEM is conservative in control volume elements V_ξ (not in $\mathcal{T}^h$).

3.5.2 Mixed Multiscale Finite Element Methods

For simplicity, we assume Neumann boundary conditions. First, we review the mixed multiscale finite element formulation following [19] (see also [5], [1], and [6]). We can rewrite two-phase flow equation as

$$\begin{aligned} k^{-1}u - \nabla p &= 0 \quad \text{in} \quad \Omega \\ div(u) &= 0 \quad \text{in} \quad \Omega \\ k(x)\nabla p \cdot n &= g(x) \quad \text{on} \quad \partial\Omega. \end{aligned} \tag{33}$$

The variational problem associated with (33) is to seek $(u, p) \in H(div, \Omega) \times L^2(\Omega)/R$ such that $u \cdot n = g$ on $\partial\Omega$ and

$$\begin{aligned} (k^{-1}u, v) + (div\,v, p) &= 0 \quad \forall v \in H_0(div, \Omega) \\ (div\,u, q) &= 0 \quad \forall q \in L^2(\Omega)/R. \end{aligned} \tag{34}$$

where $H_0(div, \Omega)$ is $H(div, \Omega)$ with homogeneous Neumann boundary conditions. By defining

$$a(u, v) = (k^{-1}u, v), \quad b(v, q) = (div\,v, q), \tag{35}$$

we can rewrite the weak formulation as

$$\begin{aligned} a(u, v) + b(v, p) &= 0 \quad \forall v \in H_0(div, \Omega), \\ b(u, q) &= 0 \quad \forall q \in L^2(\Omega)/R. \end{aligned}$$

Let $V_h \subset H(div, \Omega)$ and $Q_h \subset L^2(\Omega)/R$ be finite dimensional spaces and $V_h^0 = V_h \cap H_0(div, \Omega)$. The numerical approximation problem associated with (34) is to find $(u_h, p_h) \in V_h \times Q_h$ such that $u_h \cdot n = g_h$ on $\partial\Omega$, where $g_h = g_{0,h} n$ on $\partial\Omega$ and $g_{0,h} = \sum_{e \in \{\partial K \cap \partial\Omega, K \in \mathcal{T}_h\}} (\int_e g\,ds) N_e$, $N_e \in V_h$, is corresponding basis function to edge e,

$$(k^{-1}u_h, v_h) + (div v_h, p_h) = \quad 0 \quad \forall v_h \in V_h^0$$
$$(div u_h, q_h) = \quad 0 \quad \forall q_h \in Q_h. \tag{36}$$

One can define a linear operator $B_h : V_h^0 \rightarrow Q_h'$ by $b(u_h, q_h) = (B_h u_h, q_h)$. Suppose that the following conditions are satisfied

$$a(u_h, u_h) \quad \text{is} \quad \ker B_h - \text{coercive} \tag{37}$$

$$\inf_{q_h \in Q_h} \sup_{v_h \in V_h} \frac{b(v_h, q_h)}{\|v_h\|_{H(div,\Omega)} \|q_h\|_{L^2(\Omega)}} \geq C. \tag{38}$$

Then the following approximation property follows (see, e.g., [17]).

Lemma 3.1 *If (u, p) and (u_h, p_h) respectively solve the problem (34) and (36) and the conditions (37) and (38) hold, then*

$$\|u - u_h\|_{H(div,\Omega)} + \|p - p_h\|_{0,\Omega} \leq \inf_{\substack{v_h \in V_h \\ v_h - g_{0,h} \in V_h^0}} \|u - v_h\|_{H(div,\Omega)} + \inf_{q_h \in Q_h} \|p - q_h\|_{0,\Omega}.$$
$$\tag{39}$$

Following Chen and Hou [19] (see also [5]), one can construct multiscale basis functions for velocity in each coarse block K

$$div(k(x)\nabla w_i^K) = \quad \frac{1}{|K|} \quad \text{in} \quad K$$
$$k(x)\nabla w_i^K n^K = \quad \begin{cases} g_i^K & \text{on } e_i^K \\ 0 & \text{else,} \end{cases} \tag{40}$$

where $g_i^K = \frac{1}{|e_i^K|}$ and e_i^K are the edges of K. Then, we can define the finite dimensional space for velocity by

$$V_h = \bigoplus_K \{\Psi_i^K\},$$
$$V_h^0 = V_h \cap H_0(div, \Omega),$$

where $\Psi_i^K = k(x)\nabla w_i^K$.

3.6 Applications

3.6.1 Flow in Porous Media

One of the main application of our multiscale method is the flow and transport through porous media. This is a fundamental problem in subsurface applications. Here, we apply MsFEM to solve the single phase flow, which is a good test problem in practice.

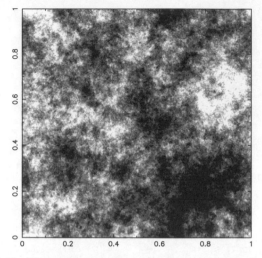

Fig. 2 Porosity field with fractal dimension of 2.8 generated using the spectral method

We model the porous media by random distributions of k_ϵ generated using a spectral method. In fact, $k_\epsilon = \alpha 10^{\beta p}$, where p is a random field represents porosity, and α and β are scaling constants to give the desired contrast of k_ϵ. In particular, we have tested the method for a porous medium with a statistically fractal porosity field (see Fig. 2). The fractal dimension is 2.8. This is a model of flow in an oil reservoir or aquifer with uniform injection in the domain and outflow at the boundaries. We note that the problem has a continuous scale because of the fractal distribution.

The pressure field due to uniform injection is solved and the error is shown in Fig. 3. The horizontal dash line indicates the error of the LFEM solution with $N = 2,048$. The coarse-grid solutions are obtained with different number of elements, N, but fixed $NM = 2,048$. We note that error of MsFEM-os-L almost coincide with that of the well-resolved solution obtained using LFEM. However, MsFEM without oversampling is less accurate. MsFEM-O indicates that oscillatory boundary conditions, obtained from solving reduced 1-D elliptic equations along ∂K (see [48]), are imposed on the basis functions. The decay of error in MsFEM is because of the decay of small scales in k_ϵ. Figure 4 shows the results for a log-normally distributed k_ϵ. In this case, the effect of scale resonance shows clearly for MsFEM-L, i.e., the error increases as h approaches ϵ. Here $\epsilon \sim 0.004$ roughly equals the correlation length. Using the oscillatory boundary conditions (MsFEM-O) gives better results, but it does not completely eliminate resonance. On the other hand, the multiscale method with oversampling agrees extremely well with the well-resolved calculation. One may wonder why the errors do not decrease as the number of coarse grid elements increase. This is because we use the same subgrid mesh size, which is the same as the well-resolved grid size, to construct the basis

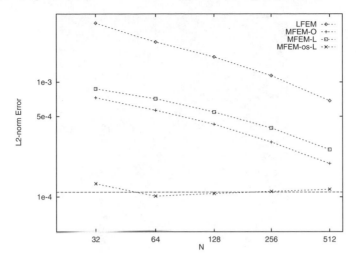

Fig. 3 The l^2-norm error of the solutions using various schemes for a fractal distributed permeability field

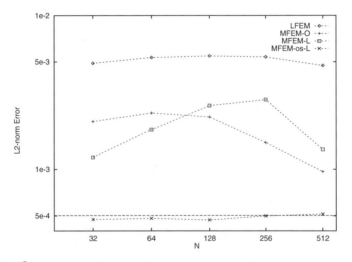

Fig. 4 The l^2-norm error of the solutions using various schemes for a log-normally distributed permeability field

functions for various coarse grid sizes ($N = 32, 64, 128$, etc.). In some special cases, one can construct multiscale basis functions analytically. In this case, the errors for the coarse grid computations will indeed decrease as the number of coarse grid elements increase.

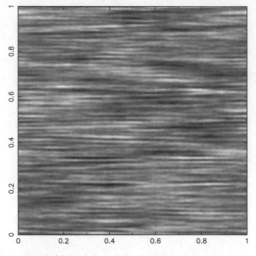

Fig. 5 A random porosity field with layered structure

3.6.2 Fine Scale Recovery

To solve transport problems in the subsurface formations, as in oil reservoir simulations, one needs to compute the velocity field from the elliptic equation for pressure, i.e., $v = -k_\epsilon \nabla p$, where p is pressure. In some applications involving isotropic media, the cell-averaged velocity is sufficient, as shown by some computations using the local upscaling methods (cf. [26]). However, for anisotropic media, especially layered ones (Fig. 5), the velocity in some thin channels can be much higher than the cell average, and these channels often have dominant effects on the transport solutions. In this case, the information about fine scale velocity becomes important. Therefore, an important question for all upscaling methods is how to take those fast-flow channels into account. For MsFEM, the fine scale velocity can be easily recovered from the multiscale basis functions, noting that they provide interpolations from the coarse h-grid to the fine h_s-grid.

To demonstrate the accuracy of the recovered velocity and effect of small-scale velocity on the transport problem, we show the fractional flow result of a "tracer" test using the layered medium in Fig. 5: a fluid with red color originally saturating the medium is displaced by the same fluid with blue color injected by flow in the medium at the left boundary, where the flow is created by a unit horizontal pressure drop. The linear convection equation is solved to compute the saturation of the red fluid (for details, see [27]). To demonstrate that we can recover the fine grid velocity field from the coarse grid pressure calculation, we plot the horizontal velocity fields obtained by two methods. In Fig. 6a, we plot the horizontal velocity field obtained by using a fine grid ($N = 1,024$) calculation. In Fig. 6b, we plot the same horizontal velocity field obtained by using the coarse grid pressure calculation with $N = 64$ and

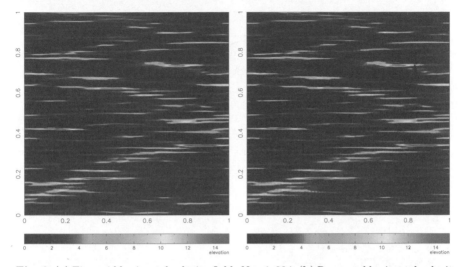

Fig. 6 (a) Fine grid horizontal velocity field, $N = 1,024$. (b) Recovered horizontal velocity field from the coarse grid $N = 64$ calculation using multiscale basis functions

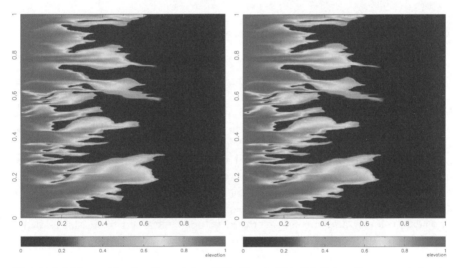

Fig. 7 (a) Fine grid saturation at $t = 0.06$, $N = 1,024$. (b) Saturation computed using the recovered velocity field from the coarse grid calculation

using the multiscale finite element basis functions to interpolate the fine grid velocity field. We can see that the recovered velocity field captures very well the layer structure in the fine grid velocity field. Further, we use the recovered fine grid velocity field to compute the saturation in time. In Fig. 7a, we plot the saturation at $t = 0.06$ obtained by the fine grid calculation. Figure 7b shows the corresponding saturation obtained using the recovered velocity field from the coarse grid calculation. The agreement is striking.

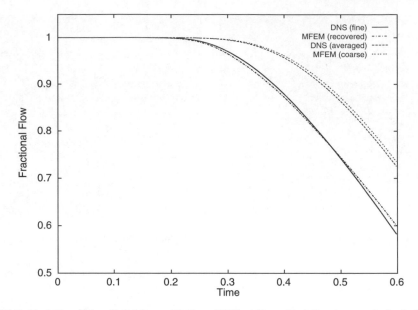

Fig. 8 Variation of fractional flow with time. DNS: well-resolved direct numerical solution using LFEM ($N = 512$). MsFEM: oversampling is used ($N = 64$, $M = 8$)

We also check the fractional flow curves obtained by the two calculations. The fractional flow of the red fluid, defined as $F = \int S_{red} v_x \, dy / \int v_x \, dy$ (S being the saturation), at the right boundary is shown in Fig. 8. The top pair of curves are the solutions of the transport problem using the cell-averaged velocity obtained from a well-resolved solution and from MsFEM; the bottom pair are solutions using well-resolved fine scale velocity and the recovered fine scale velocity from the MsFEM calculation. Two conclusions can be made from the comparisons. First, the cell-averaged velocity may lead to a large error in the solution of the transport equation. Second, both recovered fine scale velocity and the cell-averaged velocity obtained from MsFEM give faithful reproductions of respective direct numerical solutions.

3.6.3 Scale-Up of One-Phase Flows

The multiscale finite element method has been used in conjunction with some moment closure models to obtain an upscaled method for one-phase flows, see, e.g., [19, 31, 39]. Note that the multiscale finite element method presented above does not conserve mass. For long time integration, it may lead to significant loss of mass. This is an undesirable feature of the method. In a recent work [19], we have designed and analyzed a mixed multiscale finite element method, and we have applied this mixed method to study the scale up of one-phase flows and found that mass is conserved very well even for long time integration. Below we describe our results in some detail.

In its simplest form, neglecting the effect of gravity, compressibility, capillary pressure, and considering constant porosity and unit mobility, the governing equations for the flow transport in highly heterogeneous porous media can be described by the following partial differential equations [63], [81], and [31]

$$\text{div}(k(x)\nabla p) = 0, \tag{41}$$

$$\frac{\partial S}{\partial t} + v \cdot \nabla S = 0, \tag{42}$$

where p is the pressure, S is the water saturation, $k(x) = (k_{ij}(x))$ is the permeability tensor, and $v = -k(x)\nabla p$ is the Darcy velocity. The highly heterogeneous properties of the medium are built into the permeability tensor $k(x)$ which is generated through the use of geological and geostatistical modeling tools. The detailed structure of the permeability coefficients makes the direct simulation of the above model infeasible. For example, it is common in real simulations to use millions of grid blocks, with each block having a dimension of tens of meters, whereas the permeability measured from cores is at a scale of centimeters [66]. This gives more than 10^5 degrees of freedom per spatial dimension in the computation. This makes a direct simulation to resolve all small scales prohibitive even with today's most powerful supercomputers. On the other hand, from an engineering perspective, it is often sufficient to predict the macroscopic properties of the solutions. Thus it is highly desirable to derive effective coarse grid models to capture the correct large scale solution without resolving the small scale features. Numerical upscaling is one of the commonly used approaches in practice.

Now we describe how the (mixed) multiscale finite element can be combined with the existing upscaling technique for the saturation equation (42) to get a complete coarse grid algorithm for the problem (41)–(42). The numerical upscaling of the saturation equation has been under intensive study in the literature [27, 39, 45, 63, 82, 84]. Here, we use the upscaling method proposed in [39] and [31] to design an overall coarse grid model for the problem (41)–(42). The work of [39] for upscaling the saturation equation involves a moment closure argument. The velocity and the saturation are separated into a local mean quantity and a small scale perturbation with zero mean. For example, the Darcy velocity is expressed as $v = v_0 + v'$ in (42), where v_0 is the average of velocity v over each coarse element, $v' = (v'_1, v'_2)$ is the deviation of the fine scale velocity from its coarse scale average. After some manipulations, an average equation for the saturation S can be derived as follows [39]:

$$\frac{\partial S}{\partial t} + v_0 \cdot \nabla S = \frac{\partial}{\partial x_i}\left(D_{ij}(x,t)\frac{\partial S}{\partial x_j}\right), \tag{43}$$

where the diffusion coefficients $D_{ij}(x,t)$ are defined by

$$D_{ii}(x,t) = \langle |v_i'(x)| \rangle L_i^0(x,t), \quad D_{ij}(x,t) = 0, \quad \text{for } i \neq j,$$

$\langle |v_i'(x)| \rangle$ stands for the average of $|v_i'(x)|$ over each coarse element. $L_i^0(x,t)$ is the length of the coarse grid streamline in the x_i direction which starts at time t at point x, i.e.,

$$L_i^0(x,t) = \int_0^t y_i(s) \, ds,$$

where $y(s)$ is the solution of the following system of ODEs

$$\frac{dy(s)}{ds} = v_0(y(s)), \quad y(t) = x.$$

Note that the hyperbolic equation (42) is now replaced by a convection–diffusion equation. The convection-dominant parabolic equation (43) is solved by the characteristic linear finite element method [25], [72] in our simulation. The flow transport model (41)–(42) is solved in the coarse grid as follows:

1. Solve the pressure equation (41) by the oversampling mixed multiscale finite element method and obtain the fine scale velocity field using the multiscale basis functions
2. Compute the coarse grid average v_0 and the fine scale deviation $\langle |v_i'(x)| \rangle$ on the coarse grid
3. At each time step, solve the convection–diffusion equation (43) by the characteristic linear finite element method on the coarse grid in which the lengths $L_i^0(x,t)$ of the streamline are computed for the center of each coarse grid element

The mixed multiscale finite element method can be readily combined with the above upscaling model for the saturation equation. The local fine grid velocity v' will be constructed from the multiscale finite element basis functions. The main cost in the above algorithm lies in the computation of multiscale basis functions which can be done a priori and completely in parallel. This algorithm is particularly attractive when multiple simulations must be carried out due to the change of boundary and source distribution as it is often the case in engineering applications. In such a situation, the cost of computing the multiscale basis functions is just an over-head. Moreover, once these basis functions are computed, they can be used for subsequent time integration of the saturation. Because the evolution equation is now solved on a coarse grid, a larger time step can be used. This also offers additional computational saving. For many oil recovery problems, due to the excessively large fine grid data, upscaling is a necessary step before performing many simulations and realizations on the upscaled coarse grid model. If one can coarsen the fine grid by a factor of 10 in each dimension, the computational saving of the coarse grid model over the original fine model could be as large as a factor 10,000 (three space dimensions plus time).

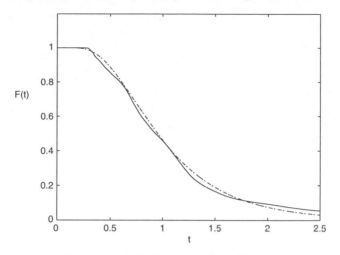

Fig. 9 The accuracy of the coarse grid algorithm. *Solid line* is the "exact" fractional flow curve using mixed finite element method solving the pressure equation. The *slash-dotted line* is the fractional flow curve using above coarse grid algorithm

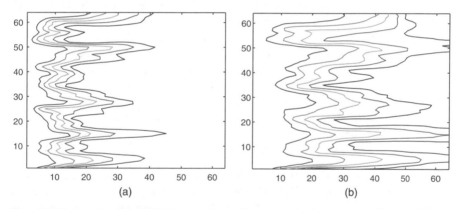

Fig. 10 The contour plots of the saturation S computed using the upscaled model on a 64×64 mesh at time $t = 0.25$ (*left*) and $t = 0.5$ (*right*)

We perform a coarse grid computation of the above algorithm on the coarse 64×64 mesh. The fractional flow curve using the above algorithm is depicted in Fig. 9. It gives excellent agreement with the "exact" fractional flow curve. The contour plots of the saturation S on the fine $1,024 \times 1,024$ mesh at time $t = 0.25$ and $t = 0.5$ computed by using the "exact" velocity field are displayed in Fig. 11. In Fig. 10, we show the contour plots of the saturation obtained using the recovered velocity field from the coarse grid pressure calculation $N = 64$. We can see that the contour plots in Fig. 10 approximate the "exact" ones in Fig. 11 in certain accuracy but the sharp oil/water interfaces in Fig. 11 are smeared out. This is due to the parabolic

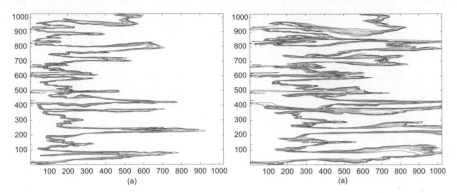

Fig. 11 The contour plots of the saturation S computed on the fine $1,024 \times 1,024$ mesh using the "exact" velocity field at time $t = 0.25$ (*left*) and $t = 0.5$ (*right*)

nature of the upscaled equation (43). We have also performed many other numerical experiments to test the robustness of this combined coarse grid model. We found that for permeability fields with strong layered structure, the above coarse grid model is very robust. The agreement with the fine grid calculations is very good. We are currently working toward some qualitative and quantitative understanding of this upscaling model.

Finally, we remark that the upscaling equation (43) uses small scale information v' of the velocity field to define the diffusion coefficients. This information can be constructed *locally* through the mixed multiscale basis functions. This is an important property of our multiscale finite element method. It is clear that solving directly the homogenized pressure equation

$$\operatorname{div}(k^*(x)\nabla p^*) = 0$$

will not provide such small scale information. On the other hand, whenever one can afford to resolve all the small scale feature by a fine grid, one can use fast linear solvers, such as multigrid methods, to solve the pressure equation (41) on the fine mesh. From the fine grid computation, one can easily construct the average velocity v_0 and its deviation v'. However, when multiple simulations must be carried out due to the change of boundary conditions, the pressure equation (41) will then have to be solved again on the fine mesh. The multiscale finite element method only solves the pressure equation once on a coarse mesh, and the fine grid velocity can be constructed locally through the finite element basis functions. This is the main advantage of the mixed multiscale finite element method. This process can be used for the nonlinear two-phase flow due to the dynamic coupling between the pressure and the saturation.

It should be noted that some adaptive scale-up strategies have also been developed [27, 84]. The idea is to refine the mesh around the fast-flow channels in order to capture their effect directly. The approach seems to work well when the channels are isolated.

3.7 MsFEM Using Limited Global Information

3.7.1 Motivation

Multiscale finite element methods and their modifications are used in two-phase flow simulations through heterogeneous porous media. First, we briefly describe the underlying fine-scale equations. We present two-phase flow equations neglecting the effects of gravity, compressibility, capillary pressure and dispersion on the fine scale. Porosity, defined as the volume fraction of the void space, will be taken to be constant and therefore serves only to rescale time. The two phases will be referred to as water and oil and designated by the subscripts w and o, respectively. We can then write Darcy's law, with all quantities dimensionless, for each phase j as follows:

$$v_j = -\lambda_j(S)k\nabla p, \tag{44}$$

where $\mathbf{v}_j$ is phase velocity, S is water saturation (volume fraction), p is pressure, $\lambda_j = k_{rj}(S)/\mu_j$ is phase mobility, where k_{rj} and μ_j are the relative permeability and viscosity of phase j respectively, and k is the permeability tensor.

Combining Darcy's law with conservation of mass, $div(\mathbf{v}_w + \mathbf{v}_o)=0$, allows us to write the flow equation in the following form

$$div(\lambda(S)k\nabla p) = f, \tag{45}$$

where the total mobility $\lambda(S)$ is given by $\lambda(S) = \lambda_w(S) + \lambda_o(S)$ and f is a source term. The saturation dynamics affects the flow equations. One can derive the equation describing the dynamics of the saturation

$$\frac{\partial S}{\partial t} + div(F) = 0, \tag{46}$$

where $F = vf_w(S)$, with $f_w(S)$, the fractional flow of water, given by $f_w = \lambda_w/(\lambda_w + \lambda_o)$, and the total velocity v by:

$$v = v_w + v_o = -\lambda(S)k\nabla p. \tag{47}$$

In the presence of capillary effects, an additional diffusion term is present in (46).

If $k_{rw} = S$, $k_{ro} = 1 - S$ and $\mu_w = \mu_o$, then the flow equation reduces to

$$div(k\nabla p^{sp}) = f.$$

This equation, the linear advection pollutant transport equation, will be referred to as the single-phase flow equation associated with (45), and p^{sp} will be referred to as the single-phase flow solution.

For two-phase flow simulations, we first solve the coarse scale pressure equation using multiscale finite element volume method. The fine scale velocity is then reconstructed by solving a local fine scale problem over each dual cell V_ξ with flux boundary conditions, as determined from the pressure solution and the Ψ_i, prescribed. This velocity is then used in the explicit solution of the saturation equation. The overall procedure is thus an IMPES (implicit pressure, explicit saturation) approach. We will also use coarse-scale velocity to update the saturation field.

As we see from (45) and (46), the pressure equation is solved many times for different saturation profiles. Thus, computing the basis functions once at time zero is very beneficial and the basis functions are only updated near sharp interfaces. In fact, our numerical results show that only slight improvement can be achieved by updating the basis functions near sharp fronts. However, we have found that for heterogeneous permeability fields with very strong non-local effects, the use of some type of global information can improve multiscale finite element results significantly, which will be discussed next.

We present a representative numerical example for a permeability field generated using two-point geostatistics. To generate this permeability field, we have used GSLIB algorithm [23]. The permeability is log-normally distributed with prescribed variance $\sigma^2 = 1.5$ (σ^2 here refers to the variance of $\log k$) and some correlation structure. The correlation structure is specified in terms of dimensionless correlation lengths in the x and z-directions, $l_x = 0.4$ and $l_z = 0.04$, nondimensionalized by the system length. Linear boundary conditions are used for constructing multiscale basis function in (31). Spherical variogram is used [23]. In this numerical example, the fine-scale field is 120×120, while the coarse-scale field is 12×12 defined in the rectangle with the length 5 and the width 1. For the two-phase flow simulations, the system is considered to initially contain only oil ($S = 0$) and water is injected at inflow boundaries ($S = 1$ is prescribed), i.e., we specify $p = 1$, $S = 1$ along the $x = 0$ edge and $p = 0$ along the $x = 5$ edge, and no flow boundary conditions on the lateral boundaries. Relative permeability functions are specified as $k_{rw} = S^2$, $k_{ro} = (1 - S)^2$; water and oil viscosities are set to $\mu_w = 1$ and $\mu_o = 5$. Porosity is constant and serves only to nondimensionalize time. Results are presented in terms of the fraction of oil in the produced fluid (i.e., oil cut, designated F) against pore volume injected (PVI). PVI represents dimensionless time and is computed via $\int Q dt / V_p$ where V_p is the total pore volume of the system and Q is the total flow rate.

In our first numerical test, Fig. 12, we compare the fractional flows. The dashed line corresponds to the calculations performed using a simple saturation upscaling (no subgrid treatment), while dotted line corresponds to the calculations performed by solving the saturation equation on the fine grid using the reconstructed fine-scale velocity field. We observe from this figure that the second approach is very accurate, while the first approach over-predicts

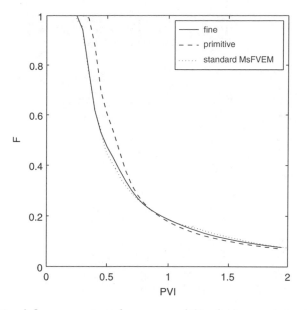

Fig. 12 Fractional flow comparison for a permeability field generated using two-point geostatistics

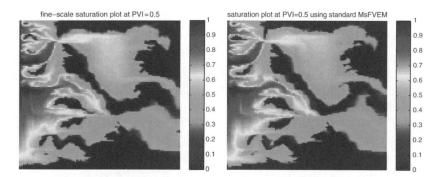

Fig. 13 Saturation maps at PVI = 0.5 for fine-scale solution (*left figure*) and standard MsFVEM (*right figure*)

the breakthrough time. The saturation snapshots are compared in Fig. 13. One can observe that there is a very good agreement.

In the next set of numerical results, we consider strongly channelized permeability fields. These permeability fields are proposed in some recent benchmark tests, such as the SPE comparative solution project [20]. In Fig. 14, one of the layers of this 3-D permeability field is depicted. All the layers have 220×60 fine-scale resolution, and we take the coarse grid to be 22×6. As it can be observed, the permeability field contains a high permeability channel, where most flow will occur in our simulation. In Fig. 15, the fractional flows

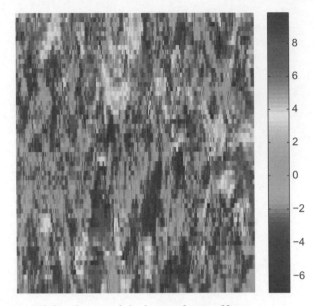

Fig. 14 Log-permeability for one of the layers of upper Ness

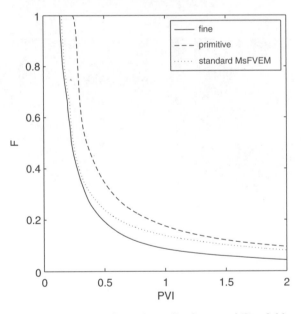

Fig. 15 Fractional flow comparison for a channelized permeability field

are compared. The boundary conditions are taken to be $p = 1$, $S = 1$ along the $x = 0$ edge and $p = 0$ along the $x = 5$ edge, and no flow boundary conditions on the lateral boundaries. Again, the dashed line corresponds to

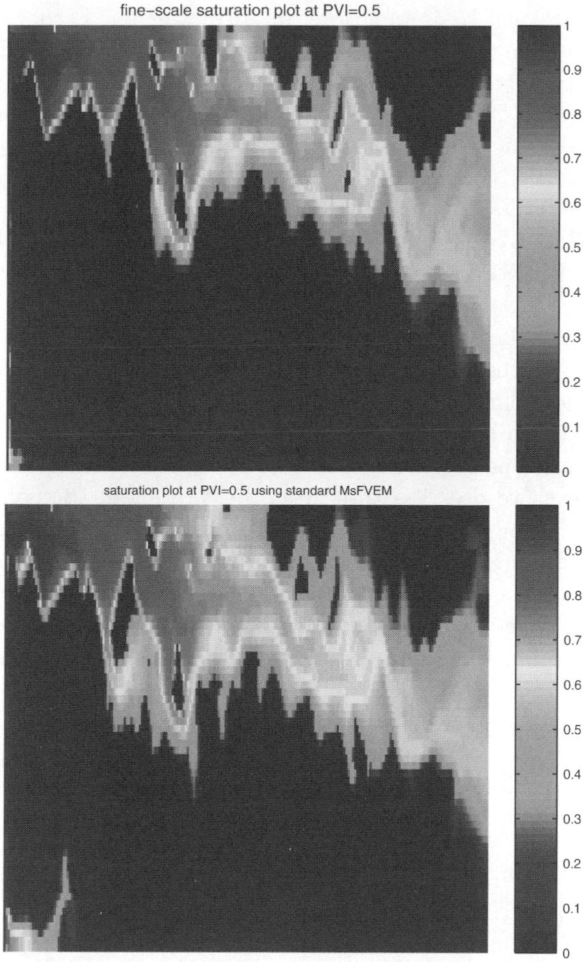

Fig. 16 Saturation maps at PVI = 0.5 for fine-scale solution (*left figure*) and standard MsFVEM (*right figure*)

the calculations performed using a simple saturation upscaling (no subgrid treatment), while dotted line corresponds to the calculations performed by solving the saturation equation on the fine grid using the reconstructed fine-scale velocity field. We observe from this figure that the second approach is not very accurate in contrast to the permeability field generated using two-point geostatistics [23]. This is because the local basis functions can not account accurately the global connectivity of the media. Indeed, in the next figure, Fig. 16, the saturation fields at time PVI = 0.5 are compared. We see that multiscale finite element methods with local basis functions introduce some errors. In the bottom left corner, there is a saturation pocket which is not in the reference solution computed using a fine grid. The reason for this is that the local basis functions in the lower left corner contains high permeability region. However, this high permeability region does not have global connectivity, and the local basis functions can not take this effect into account. Next, we discuss how global information can be incorporated into multiscale basis functions to improve the accuracy of the computations.

3.7.2 Modified Multiscale Finite Volume Element Method Using Limited Global Information

The main idea of the modified multiscale finite volume element method (Ms-FVEM) is to use the solution of the fine-scale problem at time zero to determine the boundary conditions for the basis functions. This approach is proposed in [32] to handle the permeability fields which are strongly channelized. For this type of permeability fields, some type of global information is needed. Next, we describe the method. We denote the solution of pressure equation at time zero by $p^{sp}(x)$. In defining $p^{sp}(x)$, we use the actual boundary conditions of the global problem. $p^{sp}(x)$ depends on global boundary conditions, and, generally, is updated each time when global boundary conditions are changed. The boundary conditions in (31) for modified basis functions are defined in the following way. For each rectangular element K with vertices x_i ($i = 1, 2, 3, 4$) denote by $\phi_i(x)$ a restriction of the nodal basis on K, such that $\phi_i(x_j) = \delta_{ij}$. At the edges where $\phi_i(x) = 0$ at both vertices, we take boundary condition for $\phi_i(x)$ to be zero. Consequently, the basis functions are localized. We only need to determine the boundary condition at two edges which have the common vertex x_i ($\phi_i(x_i) = 1$). Denote these two edges by $[x_{i-1}, x_i]$ and $[x_i, x_{i+1}]$ (see Fig. 17). We only need to describe the boundary condition, $g_i(x)$, for the basis function $\phi_i(x)$, along the edges $[x_i, x_{i+1}]$ and $[x_i, x_{i-1}]$. If $p^{sp}(x_i) \neq p^{sp}(x_{i+1})$, then

$$g_i(x)|_{[x_i, x_{i+1}]} = \frac{p^{sp}(x) - p^{sp}(x_{i+1})}{p^{sp}(x_i) - p^{sp}(x_{i+1})}, \quad g_i(x)|_{[x_i, x_{i-1}]} = \frac{p^{sp}(x) - p^{sp}(x_{i-1})}{p^{sp}(x_i) - p^{sp}(x_{i-1})}.$$

If $p^{sp}(x_i) = p^{sp}(x_{i+1}) \neq 0$, then

$$g_i(x)|_{[x_i, x_{i+1}]} = \phi_i^0(x) + \frac{1}{2p^{sp}(x_i)}(p^{sp}(x) - p^{sp}(x_{i+1})),$$

where $\phi_i^0(x)$ is a linear function on $[x_i, x_{i+1}]$ such that $\phi_i^0(x_i) = 1$ and $\phi_i^0(x_{i+1}) = 0$. Similarly,

$$g_{i+1}(x)|_{[x_i, x_{i+1}]} = \phi_{i+1}^0(x) + \frac{1}{2p^{sp}(x_{i+1})}(p^{sp}(x) - p^{sp}(x_{i+1})), \qquad (48)$$

where $\phi_{i+1}^0(x)$ is a linear function on $[x_i, x_{i+1}]$ such that $\phi_{i+1}^0(x_{i+1}) = 1$ and $\phi_{i+1}^0(x_i) = 0$. If $p^{sp}(x_i) = p^{sp}(x_{i+1}) \neq 0$, then one can also use simply

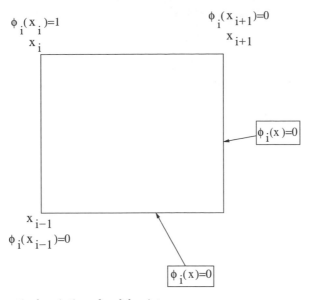

Fig. 17 Schematic description of nodal points

linear boundary conditions. If $p^{sp}(x_i) = p^{sp}(x_{i+1}) = 0$ then linear boundary conditions are used. In the applications considered in this paper, the initial pressure is always positive. Finally, the basis function $\phi_i(x)$ is constructed by solving (31). The choice of the boundary conditions for the basis functions is motivated by the analysis. In particular, we would like to recover the exact fine-scale solution along each edge if the nodal values of the pressure are equal to the values of exact fine-scale pressure. This is the underlying idea for the choice of boundary conditions. Using this property and Cea's lemma one can show that the pressure obtained from the numerical solution is equal to the underlying fine-scale pressure.

3.7.3 Mixed Multiscale Finite Element Methods

Next, following [1], we present a mixed multiscale finite element method that employs single-phase flow information. Suppose that p^{sp} solves the single-phase flow equation. We set $b_i^K = (k\nabla p^{sp}|_{e_i^K}) \cdot n^K$ and assume that b_i^K is uniformly bounded. Then the new basis functions for velocity is constructed by solving the following local problems (40) with $g_i^K = b_i^K / \beta_i^K$, where $\beta_i^K = \int_{e_i^K} k\nabla p^{sp} \cdot n^K ds$. For further analysis, we assume that $\beta_i^K \neq 0$. In general, if $\beta_i^K = 0$ one can use standard mixed multiscale finite element basis functions. Let $N_i^K = k(x)\nabla w_i^K$ and the multiscale finite dimensional space V_h^0 for velocity be defined by

$$V_h := \bigoplus_K \{N_i^K\} \subset H(div, \Omega),$$

$$V_h^0 := V_h \cap H_0(div, \Omega).$$

It can be shown that the resulting multiscale finite element solution for velocity is exact for single-phase flow (i.e., $\lambda(x) = 1$). Let $v_h|_K = \beta_i^K N_i^K$, then β_i^K is the interpolation value of the fine scale solution. Furthermore, a direct calculation yields $(v_h|_{e_i^K}) \cdot n^K = k\nabla p^{sp} \cdot n^K$. Because

$$divv_h = \beta_i^K div N_i^K = \frac{1}{|K|} \int_{\partial K} k\nabla p^{sp} \cdot n^K ds = \frac{1}{|K|} \int_K div(k\nabla p^{sp})dx = 0,$$

the following equation is obtained immediately

$$divv_h = 0 \quad \text{in} \quad K \tag{49}$$

$$v_h \cdot n^K = k\nabla p^{sp} \cdot n^K \quad \text{on} \quad \partial K \tag{50}$$

Because $div(k\nabla p^{sp}) = 0$, we get $v_h = k\nabla p^{sp}$ and the following proposition.

Proposition Let $\beta_i^K = \int_{e_i^K} k\nabla p^{sp} \cdot n^K ds$, then on each coarse block K

$$k\nabla p^{sp} = \beta_i^K N_i^K. \tag{51}$$

Lemma 3.2 *If* $|\beta_i^K| \geq Ch$ *with* C *is independent of* h, *then*
(1) $a(u_h, u_h)$ *is* $\ker B_h$-*coercive;*
(2) $\inf_{q_h \in Q_h} \sup_{v_h \in V_h^0} \frac{b(v_h, q_h)}{\|v_h\|_{H(div, \Omega)} \|q_h\|_{L^2(\Omega)}} \geq C.$

3.7.4 Numerical Results

Next, we show the numerical results obtained using modified multiscale finite element type methods for the permeability layer depicted in Fig. 14 and two-phase flow parameters presented earlier. We consider two types of boundary conditions in a rectangular region $[0, 5] \times [0, 1]$. For the first type of boundary conditions, we specify $p = 1$, $S = 1$ along the $x = 0$ edge and $p = 0$ along the $x = 5$ edge. On the rest of the boundaries, we assume no flow boundary condition. We call this type of the boundary condition as side-to-side. The other type of boundary conditions is obtained by specifying $p = 1$, $S = 1$ along the $x = 0$ edge for $0.5 \leq z \leq 1$ and $p = 0$ along the $x = 5$ edge for $0 \leq z \leq 0.5$. On the rest of the boundaries, we assume no flow boundary condition.

In Fig. 18, the fractional flows are plotted for standard and modified Ms-FVEM. We observe from this figure that modified MsFVEM is more accurate and provides nearly the same fractional flow response as the direct fine-scale

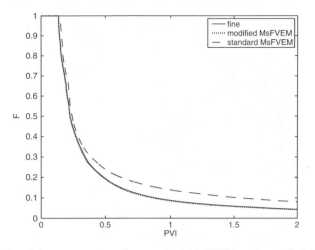

Fig. 18 Fractional flow comparison for standard MsFVEM and modified MsFVEM for side-to-side flow

calculations. In Fig. 19, we compare the saturation fields at PVI = 0.5. As we see, the saturation field obtained using modified MsFVEM is very accurate and there is no longer the saturation pocket at the left bottom corner. Thus, the modified MsFVEM captures the connectivity of the media accurately.

In the next set of numerical results, we test the modified multiscale finite element methods for a different layer (layer 40) of SPE comparative solution project. In Figs. 20 and 21, the fractional flows and total flow rates (Q) are compared for two different boundary conditions. One can see clearly that the modified MsFVEM method gives nearly exact results for these integrated responses. The standard MsFVEM tends to over-predict the total flow rate at time zero. This initial error persists at later times. This phenomena is often observed in upscaling of two-phase flows. More numerical results and discussions can be found in [32]. These numerical results demonstrate that modified multiscale finite element methods which use a limited global information are more accurate. Moreover, modified multiscale finite element methods are capable of capturing the long-range flow features accurately for channelized permeability fields.

For the next set of results, we consider another layer of the upper Ness (layer 59). In Fig. 22, both fractional flow (left figure) and total flow (right figure) are plotted. We observe that the modified MsFVEM gives almost the exact results for these quantities, while the standard MsFVEM overpredicts the total flow rate, and there are deviations in the fractional flow curve around $PVI \approx 0.6$. Note that unlike the previous case, fractional flow for standard MsFVEM is nearly exact at later times $(PVI \approx 2)$. In Fig. 23, the saturation maps are plotted at PVI = 0.5. The left figure represents the fine-scale, the

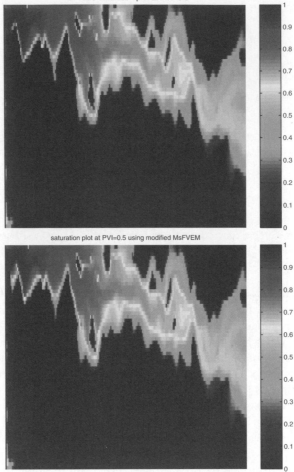

Fig. 19 Saturation maps at PVI = 0.5 for fine-scale solution (*left figure*) and modified MsFVEM (*right figure*). Side-to-side boundary condition is used

middle figure represents the results obtained using standard MsFVEM, and the right figure represents the results obtained using the modified MsFVEM. We observe from this figure that the saturation map obtained using standard MsFVEM has some errors. These errors are more evident near the lower left corner. The results of the saturation map obtained using the modified Ms-FVEM is nearly the same as the fine-scale saturation field. It is evident from these figures that the modified MsFVEM performs better than the standard MsFVEM.

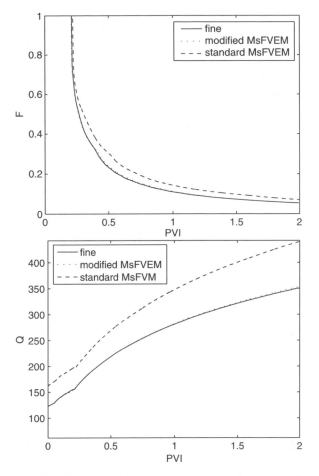

Fig. 20 Fractional flow (*left figure*) and total production (*right figure*) comparison for standard MsFVEM and modified MsFVEM for side-to-side flow (layer 40)

3.7.5 Galerkin Finite Element Methods with Limited Global Information

We have proposed some analysis for modified multiscale finite element method in [32] and [2]. The main idea is to show that the pressure evolution in two-phase flow simulations is strongly influenced by the initial pressure. To demonstrate this, we consider a channelized permeability field, where the value of the permeability in the channel is large. We assume the permeability has the form kI, where I is an identity matrix. In a channelized medium, the dominant flow is within the channels. Our analysis assumes a single channel and restricted to 2-D. Here, we briefly mention the main findings. Denote the

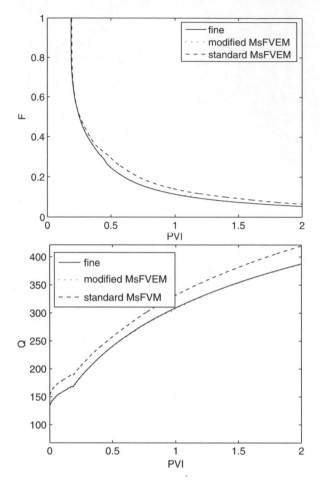

Fig. 21 Fractional flow (*left figure*) and total production (*right figure*) comparison for standard MsFVEM and modified MsFVEM for corner-to-corner (layer 40)

initial stream function and pressure by $\eta = \psi(x, t = 0)$ and $\zeta = p(x, t = 0)$ (ζ is also denoted by p^{sp} previously). The stream function is defined

$$\partial\psi/\partial x_1 = -v_2, \quad \partial\psi/\partial x_2 = v_1. \tag{52}$$

Then the equation for the pressure can be written as

$$\frac{\partial}{\partial\eta}\left(|k|^2\lambda(S)\frac{\partial p}{\partial\eta}\right) + \frac{\partial}{\partial\zeta}\left(\lambda(S)\frac{\partial p}{\partial\zeta}\right) = 0. \tag{53}$$

For simplicity, $S = 0$ at time zero is assumed. We consider a typical boundary condition that gives high flow within the channel, such that the high flow

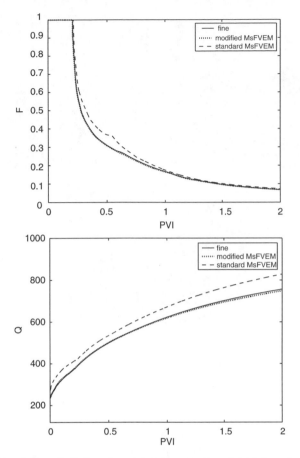

Fig. 22 Fractional flow (*left figure*) and total production (*right figure*) comparison for standard MsFVEM and modified MsFVEM for corner-to-corner flow

channel will be mapped into a large slab in (η, ζ) coordinate system. If the heterogeneities within the channel in η direction is not strong (e.g., narrow channel in Cartesian coordinates), the saturation within the channel will depend on ζ. In this case, the leading order pressure will depend only on ζ, and it can be shown that

$$p(\eta, \zeta, t) = p_0(\zeta, t) + \text{high order terms}, \tag{54}$$

where $p_0(\zeta, t)$ is the dominant pressure. This asymptotic expansion shows that the time-varying pressure strongly depends on the initial pressure (i.e., the leading order term in the asymptotic expansion is a function of initial pressure and time only). In our analysis, we will assume that $|p(x,t) - \hat{p}(p^{sp}, t)|_{H^1}$ is small.

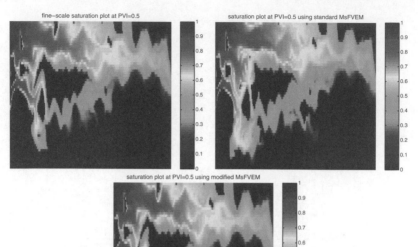

Fig. 23 Saturation maps at PVI = 0.5 for fine-scale solution (*left figure*), standard Ms-FVEM (*middle figure*), and modified MsFVEM (*right figure*). Corner-to-corner boundary condition is used

Since the analysis of the multiscale finite element methods is carried out only for the pressure equation, we will assume t (time) is fixed. Then, assuming the function $\hat{p}$ is sufficiently smooth, one can state the following. There exists A_K in each K, such that $\|\nabla p(x) - A_K \nabla p^{sp}(x)\|_{L^2(\Omega)}$ is small. Note that this assumption indicates that the fine-scale features of pressure solutions of two-phase equations does not change significantly during a simulation (e.g., streamlines do not vary significantly in each coarse block). This phenomena can be observed in numerical simulations of two-phase flows when $\mu_o/\mu_w > 1$.

The assumption for the case with scale separation indicates that the coarse-scale features of two-phase flow and single-phase flow are similar (e.g., coarse-scale streamlines do not vary significantly). We will use the following assumption.

Assumption G. There exists a sufficiently smooth scalar valued function $G(\eta)$ *(*$G \in C^3$*), such that*

$$|p - G(p^{sp})|_{1,\Omega} \leq C\delta, \tag{55}$$

where p^{sp} *is single-phase flow pressure and* δ *is sufficiently small.*

We note G is $p_0(\zeta, t)$ at fixed t in (54). Moreover, one does not need to know the function G for computing the multiscale approximation of the solution.

It is only necessary that G has certain smoothness properties, however, it is important that the basis functions span p^{sp} in each coarse block.

Theorem 3.3 *Under Assumption G and $p^{sp} \in W^{1,s}(\Omega)$ (s > 2), multiscale finite element method converges with the rate given by*

$$|p - p_h|_{1,\Omega} \leq C\delta + Ch^{1-2/s}|p^{sp}|_{W^{1,s}(\Omega)} + Ch^{1-2/s}|p^{sp}|_{1,\Omega} + Ch\|f\|_{0,\Omega}$$
$$\leq C\delta + Ch^{1-2/s}. \quad (56)$$

The proof of this theorem is given in [2]. Note that Theorem 3.3 shows that MsFEM converges for problems without any scale separation and the proof of this theorem does not use homogenization techniques. Next, we present the proof.

Proof. Following standard practice of finite element estimation, we seek $q_h = c_i \phi_i^K$, where ϕ_i^K are single-phase flow based multiscale finite element basis functions. Then from Cea's lemma, we have

$$|p - p_h|_{1,\Omega} \leq |p - G(p^{sp})|_{1,\Omega} + |G(p^{sp}) - c_i\phi_i^K|_{1,\Omega}. \quad (57)$$

Next, we present an estimate for the second term. We choose $c_i = G(p^{sp}(x_i))$, where x_i are vertices of K. Furthermore, using Taylor expansion of G around $\bar{p}_K$, which is the average of p^{sp} over K,

$$G(p^{sp}(x_i)) = G(\bar{p}_K) + G'(\bar{p}_K)(p^{sp}(x_i) - \bar{p}_K) + \frac{1}{2}G''(\xi_{x_i})(p^{sp}(x_i) - \bar{p}_K)^2,$$

where $\xi_{x_i} = \bar{p}_K + \theta(p^{sp}(x_i) - \bar{p}_K)$, $0 < \theta < 1$, we have in each K

$$c_i\phi_i^K = G(\bar{p}_K)\phi_i^K + G'(\bar{p}_K)(p^{sp}(x_i) - \bar{p}_K)\phi_i^K + \frac{1}{2}G''(\xi_{x_i})(p^{sp}(x_i) - \bar{p}_K)^2\phi_i^K =$$
$$G(\bar{p}_K) + G'(\bar{p}_K)(p^{sp}(x_i)\phi_i^K - \bar{p}_K) + \frac{1}{2}G''(\xi_{x_i})(p^{sp}(x_i) - \bar{p}_K)^2\phi_i^K. \quad (58)$$

In the last step, we have used $\sum_i \phi_i^K = 1$. Similarly, in each K,

$$G(p^{sp}(x)) = G(\bar{p}_K) + G'(\bar{p}_K)(p^{sp}(x) - \bar{p}_K) + \frac{1}{2}G''(\xi_x)(p^{sp}(x) - \bar{p}_K)^2, \quad (59)$$

where $\xi_x = \bar{p}_K + \theta(p^{sp}(x) - \bar{p}_K)$, $0 < \theta < 1$. Using (58) and (59), we get

$$|G(p^{sp}) - c_i\phi_i^K|_{1,K} \leq |G'(\bar{p}_K)(p^{sp}(x) - p^{sp}(x_i)\phi_i^K)|_{1,K} +$$
$$|\frac{1}{2}G''(\xi_{x_i})(p^{sp}(x_i) - \bar{p}_K)^2\phi_i^K|_{1,K} + |\frac{1}{2}G''(\xi_x)(p^{sp}(x) - \bar{p}_K)^2|_{1,K}. \quad (60)$$

Because of $|p^{sp}(x) - p^{sp}(x_i)|_{1,K} \leq Ch\|f\|_{0,K}$, the estimate of the first term is the following

$$|G'(\overline{p}_K)(p^{sp}(x) - p^{sp}(x_i)\phi_i^K)|_{1,K} \leq Ch\|f\|_{0,K}.$$

For the second term on the right-hand side of (60), assuming $p^{sp}(x) \in W^{1,s}(\Omega)$, we have

$$|G''(\xi_{x_i})(p^{sp}(x_i) - \overline{p}_K)^2\phi_i^K|_{1,K} \leq Ch^{2-4/s}|p^{sp}|^2_{W^{1,s}(K)} \leq Ch^{1-2/s}|p^{sp}|_{W^{1,s}(K)}.$$

where $s > 2$. Here, we have used the inequality (e.g., [4])

$$|u(x) - u(y)| \leq C|x-y|^{1-2/s}|u|_{W^{1,s}},$$

for $s > 2$, where C depends only on s.

For the third term, since G'' and G''' are bounded, we have the following estimate:

$$
\begin{aligned}
|G''(\xi_x)(p^{sp}(x) - \overline{p}_K)^2|_{1,K} &\leq C\|(p^{sp}(x) - \overline{p}_K)^2\nabla p^{sp}(x)\|_{0,K} + \\
C\|(p^{sp}(x) - \overline{p}_K)\nabla p^{sp}(x)\|_{0,K} &\leq Ch^{2-4/s}|p^{sp}|^2_{W^{1,s}(K)}|p^{sp}|_{1,K} + \\
Ch^{1-2/s}|p^{sp}|_{1,K} &\leq Ch^{2-4/s}|p^{sp}|^2_{W^{1,s}(\Omega)}|p^{sp}|_{1,K} + \\
Ch^{1-2/s}|p^{sp}|_{1,K} &\leq Ch^{1-2/s}|p^{sp}|_{1,K}.
\end{aligned}
\tag{61}
$$

Combining the above estimates, we have for (60),

$$|G(p^{sp}) - c_i\phi_i^K|_{1,K} \leq Ch^{1-2/s}|p^{sp}|_{W^{1,s}(K)} + Ch^{1-2/s}|p^{sp}|_{1,K} + Ch\|f\|_{0,K}.
\tag{62}$$

Summing (62) over all K and taking into account Assumption G, we have

$$|p - p_h|_{1,\Omega} \leq C\delta + Ch^{1-2/s}|p^{sp}|_{W^{1,s}(\Omega)} + Ch^{1-2/s}|p^{sp}|_{1,\Omega} + Ch\|f\|_{0,\Omega} \leq \\
C\delta + Ch^{1-2/s}.
\tag{63}$$

Consequently, if $s > 2$ (see, e.g., [7]) single-phase flow based multiscale finite element method converges. □

Remark 3.3. We can relax the assumption on G. In particular, it is sufficient to assume $G \in W^{2,m}$ ($m \geq 1$). In this case, the proof can be carried out using Taylor polynomials in Sobolev spaces. Also, if we assume $\nabla p^{sp} \in L^\infty(\Omega)$, then the convergence rate in (56) is $C\delta + Ch$.

3.7.6 Extensions of Galerkin Finite Element Methods with Limited Global Information

The multiscale finite element methods considered above employ information from only one single-phase flow solution. In general, depending on the source term, boundary data, and mobility $\lambda(S)$ (if it contains sharp variations), it might be necessary to use information from multiple global solutions for the computation of accurate two-phase flow solution. The previous multiscale finite element methods can be extended ([2]) to take into account additional global information. Next, we present an extension of the Galerkin multiscale finite element method that uses the partition of unity method [10] (also see, e.g., [43, 55, 77]).

Assume that u_1, $u_2,\ldots$, u_N are the global functions such that $|p - G(u_1, u_2, \ldots, u_N)|_{1,\Omega}$ is sufficiently small. Here, $u_1, \ldots, u_N$ can be possible pressure snapshots for different mobility $\lambda(S)$ or pressure fields corresponding to different source terms and/or boundary conditions. We would like to note that in a very interesting paper [69], the authors prove under certain conditions on f (source term) and $\lambda = 1$ that p is a smooth function of single-phase flow solutions (elliptic pressure equations) with boundary conditions x_1 and x_2 (it is also extended to multi-dimensional case). In this case, u_1 and u_2 are the solutions of single-phase flow equations with boundary conditions $u_i = x_i$ $(i = 1, 2)$, and it was shown that $p(u_1, u_2) \in H^2$. Next, we will formulate the method.

Let ω_i be a patch (see Fig. 24), and define ϕ_i^0 to be piecewise linear basis function in patch ω_i, such that $\phi_i^0(x_j) = \delta_{ij}$. For simplicity of notation, denote $u_1 = 1$. Then, the multiscale finite element method for each patch ω_i is constructed by

$$\psi_{ij} = \phi_i^0 u_j \tag{64}$$

where $j = 1, \ldots, N$ and i is the index of nodes (see Fig. 24). First, we note that in each K, $\sum_{i=1}^n \psi_{ij} = u_j$ is the desired single-phase flow solution.

We will use the following assumption. *There exists a sufficiently smooth scalar valued function* $G(\eta)$, $\eta \in R^N$ $(G \in C^3)$, *such that*

$$|p - G(u_1, \ldots, u_N)|_{1,\Omega} \leq C\delta, \tag{65}$$

where δ is sufficiently small.

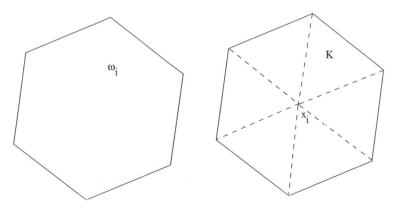

Fig. 24 Schematic description of patch

As before the form of the function G is not important for the computations, however, it is crucial that the basis functions span $u_1, \ldots, u_N$ in each coarse block. The next theorem shows that MsFEM converges for problems without scale separation in this case.

Theorem 3.4 *Assume (65) and $u_i \in W^{1,s}(\Omega)$, $s > 2$, $i = 1, \ldots, N$. Then*

$$|p - p_h|_{1,\Omega} \le C\delta + Ch^{1-2/s}.$$

The proof of this theorem is given in [2].

3.7.7 Mixed Finite Element Methods with Limited Global Information

One can carry out the analysis of mixed multiscale finite element method with limited global information. First, we re-formulate our assumption for the analysis of mixed multiscale finite element methods. From (55), it follows that

$$\|\nabla p - G'(p^{sp})\nabla p^{sp}\|_{0,\Omega} \le C\delta.$$

Using the fact that k and $\lambda(x)$ are bounded, we have

$$\|\lambda(x)k\nabla p - G'(p^{sp})\lambda(x)k\nabla p^{sp}\|_{0,\Omega} \le C\delta.$$

Noting that $u = \lambda(x)k\nabla p$ and $u^{sp} = k\nabla p^{sp}$, it follows that there exists a coarse-scale function scalar $A(x)$ such that

$$\|u - A(x)u^{sp}\|_{0,\Omega} \le \delta. \tag{66}$$

Since $A(x)u^{sp}$ approximates u, we assume that it has small divergence,

$$\left| \int_K div(A(x)u^{sp})dx \right| \le C\delta_1 h^2 \tag{67}$$

in each K, where δ_1 is a small number. For our analysis, we note that (67) gives

$$\left| \int_{\partial K} A(x)u^{sp}n^K ds \right| \le C\delta_1 h^2. \tag{68}$$

We will assume that $A(x) \in C^\gamma$ ($0 < \gamma \le 1$). (68) can be written as

$$\left| \sum_i A_i \int_{\partial e_i^K} u^{sp}n^K ds \right| \le C\delta_1 h^2. \tag{69}$$

Here A_i's are defined as $A_i = \int_{\partial e_i^K} A(x)u^{sp}n^K ds / \int_{\partial e_i^K} u^{sp}n^K ds$, since $\int_{\partial e_i^K} u^{sp}n^K ds = \beta_i^K \ne 0$. Note that *not* for any $A(x)$, A_i is necessarily a value of $A(x)$ along the edge e_i^K because $u^{sp}n^K$ can change sign. However, we only need to define $A(x)$ for each edge by its value A_i (e.g., the value of $A(x)$ at the center of edge). Then, for any such $A(x)$, (66) is satisfied provided $\delta < h^\gamma$. This can be directly verified. Thus, our main assumption will be (66) and (69), where $A(x)$ is defined, for example, at the center of each edge e_i^K. We would like to note that from the fact that $div(A(x)u^{sp})$ is small in each K, it follows that $A(x)$, for example, can be taken as an approximation of stream function corresponding to u^{sp}. As before, the form of $A(x)$ is not important for the computations of multiscale solutions.

The following theorem about the convergence of mixed multiscale finite element methods for problems without scale separation is proven in [2].

Theorem 3.5 *Assume (66) and (69) and $A(x) \in C^\gamma$, $0 < \gamma \le 1$. Let (u, p) and (u_h, p_h) respectively solve the problem (34) and (36) with single-phase flow based mixed multiscale finite element, then*

$$\|u - u_h\|_{H(div, \Omega)} + \|p - p_h\|_{0,\Omega} \le C\delta + C\delta_1 + Ch^\gamma. \qquad (70)$$

4 Multiscale Finite Element Methods for Nonlinear Partial Differential Equations

Next, we show that MsFEM can be naturally generalized to nonlinear partial differential equations. The goal of MsFEM is to find a numerical approximation of a homogenized solution *without* solving auxiliary problems (e.g., periodic cell problems) that arise in homogenization. The homogenized solutions are sought on a coarse grid space S^h, where $h \gg \epsilon$. Let K^h be a partition of Ω. We denote by S^h standard family of finite dimensional space, which possesses approximation properties, e.g., piecewise linear functions over triangular elements,

$$S^h = \{v_h \in C^0(\overline{\Omega}) : \text{the restriction } v_h \text{ is linear for each element } K \text{ and } v_h = 0 \text{ on } \partial\Omega\} \qquad (1)$$

In further presentation, K is a triangular element that belongs to K^h. To formulate MsFEM for general nonlinear problems, we will need (1) a *multiscale mapping* that gives us the desired approximation containing the small scale information and (2) a *multiscale numerical formulation* of the equation.

We consider the formulation and analysis of MsFEM for general nonlinear elliptic equations, $u_\epsilon \in W_0^{1,p}(\Omega)$

$$-div a_\epsilon(x, u_\epsilon, \nabla u_\epsilon) + a_{0,\epsilon}(x, u_\epsilon, \nabla u_\epsilon) = f, \qquad (2)$$

where $a_\epsilon(x, \eta, \xi)$ and $a_{0,\epsilon}(x, \eta, \xi)$, $\eta \in \mathbb{R}$, $\xi \in \mathbb{R}^d$ satisfy the following assumptions:

$$|a_\epsilon(x, \eta, \xi)| + |a_{0,\epsilon}(x, \eta, \xi)| \le C(1 + |\eta|^{p-1} + |\xi|^{p-1}), \qquad (3)$$

$$(a_\epsilon(x, \eta, \xi_1) - a_\epsilon(x, \eta, \xi_2))(\xi_1 - \xi_2) \geq C |\xi_1 - \xi_2|^p, \tag{4}$$

$$a_\epsilon(x, \eta, \xi)\xi + a_{0,\epsilon}(x, \eta, \xi)\eta \geq C|\xi|^p. \tag{5}$$

Denote

$$H(\eta_1, \xi_1, \eta_2, \xi_2, r) = (1 + |\eta_1|^r + |\eta_2|^r + |\xi_1|^r + |\xi_2|^r), \tag{6}$$

for arbitrary η_1, $\eta_2 \in \mathbb{R}$, ξ_1, $\xi_2 \in \mathbb{R}^d$, and $r > 0$. We further assume that

$$\begin{aligned}
&|a_\epsilon(x, \eta_1, \xi_1) - a_\epsilon(x, \eta_2, \xi_2)| + |a_{0,\epsilon}(x, \eta_1, \xi_1) - a_{0,\epsilon}(x, \eta_2, \xi_2)| \\
&\leq C\, H(\eta_1, \xi_1, \eta_2, \xi_2, p-1)\, \nu(|\eta_1 - \eta_2|) \\
&+ C\, H(\eta_1, \xi_1, \eta_2, \xi_2, p-1-s) |\xi_1 - \xi_2|^s,
\end{aligned} \tag{7}$$

where $\eta \in \mathbb{R}$ and $\xi \in \mathbb{R}^d$, $s > 0$, $p > 1$, $s \in (0, \min(p-1, 1))$ and ν is the modulus of continuity, a bounded, concave, and continuous function in $\mathbb{R}_+$ such that $\nu(0) = 0$, $\nu(t) = 1$ for $t \geq 1$ and $\nu(t) > 0$ for $t > 0$. These assumptions guarantee the well-posedness of the nonlinear elliptic problem (2). Here $\Omega \subset \mathbb{R}^d$ is a Lipschitz domain and ϵ denotes the small scale of the problem. The homogenization of nonlinear partial differential equations has been studied previously (see, e.g., [70]). It can be shown that a solution u_ϵ converges (up to a sub-sequence) to u in an appropriate norm, where $u \in W_0^{1,p}(\Omega)$ is a solution of a homogenized equation

$$-diva^*(x, u, Du) + a_0^*(x, u, Du) = f. \tag{8}$$

Multiscale mapping. Introduce the mapping $E^{MsFEM} : S^h \to V_\epsilon^h$ in the following way. For each element $v_h \in S^h$, $v_{\epsilon,h} = E^{MsFEM} v_h$ is defined as the solution of

$$-diva_\epsilon(x, \eta^{v_h}, \nabla v_{\epsilon,h}) = 0 \text{ in } K, \tag{9}$$

$v_{\epsilon,h} = v_h$ on ∂K and $\eta^{v_h} = \frac{1}{|K|} \int_K v_h dx$ for each K. We would like to point out that different boundary conditions can be chosen to obtain more accurate solutions and this will be discussed later. Note that for linear problems, E^{MsFEM} is a linear operator, where for each $v_h \in S^h$, $v_{\epsilon,h}$ is the solution of the linear problem. Consequently, V_ϵ^h is a linear space that can be obtained by mapping a basis of S^h. This is precisely the construction presented in [48] for linear elliptic equations.

Multiscale numerical formulation. Multiscale finite element formulation of the problem is the following.
Find $u_h \in S^h$ (consequently, $u_{\epsilon,h} = E^{MsFEM} u_h) \in V_\epsilon^h$) such that

$$\langle A_{\epsilon,h} u_h, v_h \rangle = \int_\Omega f v_h dx \quad \forall v_h \in S^h, \tag{10}$$

where

$$\langle A_{\epsilon,h} u_h, v_h \rangle = \sum_{K \in K^h} \int_K (a_\epsilon(x, \eta^{u_h}, \nabla u_{\epsilon,h}) \nabla v_h + a_{0,\epsilon}(x, \eta^{u_h}, \nabla u_{\epsilon,h}) v_h) dx. \tag{11}$$

Note that the above formulation of MsFEM is a generalization of the Petrov–Galerkin MsFEM introduced in [46] for linear problems. MsFEM, introduced above, can be generalized to different kinds of nonlinear problems and this will be discussed later.

4.1 Multiscale Finite Volume Element Method (MsFVEM)

The formulation of multiscale finite element (MsFEM) can be extended to a finite volume method. By its construction, the finite volume method has local conservative properties [41] and it is derived from a local relation, namely the balance equation / conservation expression on a number of subdomains which are called control volumes. Finite volume element method can be considered as a Petrov–Galerkin finite element method, where the test functions are constants defined in a dual grid. Consider a triangle K, and let z_K be its barycenter. The triangle K is divided into three quadrilaterals of equal area by connecting z_K to the midpoints of its three edges. We denote these quadrilaterals by K_z, where $z \in Z_h(K)$ are the vertices of K. Also we denote $Z_h = \bigcup_K Z_h(K)$, and Z_h^0 are all vertices that do not lie on Γ_D, where Γ_D is Dirichlet boundaries. The control volume V_z is defined as the union of the quadrilaterals K_z sharing the vertex z (see Fig. 25). The multiscale finite volume element method (MsFVEM) is to find $u_h \in S^h$ (consequently, $u_{\epsilon,h} = E^{MsFVEM} u_h$ such that

$$-\int_{\partial V_z} a_\epsilon\left(x, \eta^{u_h}, \nabla u_{\epsilon,h}\right) \cdot n \, dS + \int_{V_z} a_{0,\epsilon}\left(x, \eta^{u_h}, \nabla u_{\epsilon,h}\right) \, dx = \int_{V_z} f \, dx \quad \forall z \in Z_h^0,$$

$$(12)$$

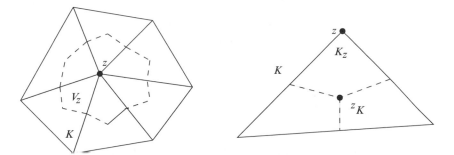

Fig. 25 *Left*: portion of triangulation sharing a common vertex z and its control volume. *Right*: partition of a triangle K into three quadrilaterals

where n is the unit normal vector pointing outward on ∂V_z. Note that the number of control volumes that satisfies (12) is the same as the dimension of S^h. We will present numerical results for both multiscale finite element and multiscale finite volume element methods.

4.2 Examples of V_ϵ^h

Linear case. For linear operators, V_ϵ^h can be obtained by mapping a basis of S^h. Define a basis of S^h, $S^h = \text{span}(\phi_0^i)$, where ϕ_0^i are standard linear basis functions. In each element $K \in K^h$, we define a set of nodal basis $\{\phi_\epsilon^i\}$, $i = 1, \ldots, n_d$ with $n_d(= 3)$ being the number of nodes of the element, satisfying

$$-\text{div} a_\epsilon(x) \nabla \phi_\epsilon^i = 0 \quad \text{in} \ \ K \in K^h \tag{13}$$

and $\phi_\epsilon^i = \phi_0^i$ on ∂K. Thus, we have

$$V_\epsilon^h = \text{span}\{\phi_\epsilon^i; \ \ i = 1, \ldots, n_d, \ \ K \subset K^h\} \subset H_0^1(\Omega).$$

Oversampling technique can be used to improve the method [48].

Special nonlinear case. For the special case, $a_\epsilon(x, u_\epsilon, \nabla u_\epsilon) = a_\epsilon(x) b(u_\epsilon)$ ∇u_ϵ, V_ϵ^h can be related to the linear case. Indeed, for this case, the local problems associated with the multiscale mapping E^{MsFEM} (see (9)) have the form

$$-\text{div} a_\epsilon(x) b(\eta^{v_h}) \nabla v_{\epsilon,h} = 0 \ \text{in} \ K.$$

Because η^{v_h} are constants over K, the local problems satisfy the linear equations,

$$-\text{div} a_\epsilon(x) \nabla \phi_\epsilon^i = 0 \quad \text{in} \ \ K,$$

and V_ϵ^h can be obtained by mapping a basis of S^h as it is done for the first example. Thus, for this case one can construct the basis functions in the beginning of the computations.

V_ϵ^h *using subdomain problems.* One can use the solutions of smaller (than $K \in \mathbf{K}^h$) subdomain problems to approximate the solutions of the local problems (9). This can be done in various ways based on a homogenization expansion. For example, instead of solving (9) we can solve (9) in a subdomain S with boundary conditions v_h restricted onto the subdomain boundaries, ∂S. Then the gradient of the solution in a subdomain can be extended periodically to K to approximate $\nabla v_{\epsilon,h}$ in (11). $v_{\epsilon,h}$ can be easily reconstructed based on $\nabla v_{\epsilon,h}$. When the multiscale coefficient has a periodic structure, the multiscale mapping can be constructed over one periodic cell with a specified average.

4.3 Convergence of MsFEM for Nonlinear Partial Differential Equations

In [37] it was shown using G-convergence theory that

$$\lim_{h\to 0}\lim_{\epsilon\to 0}\|u_h - u\|_{W_0^{1,p}(\Omega)} = 0, \tag{14}$$

(up to a subsequence) where u is a solution of (8) and u_h is a MsFEM solution given by (10). This result can be obtained without any assumption on the nature of the heterogeneities and can not be improved because there could be infinitely many scales, $\alpha(\epsilon)$, present such that $\alpha(\epsilon) \to 0$ as $\epsilon \to 0$.

For the periodic case, it can be shown that the convergence of MsFEM in the limit as $\epsilon/h \to 0$. To show the convergence for $\epsilon/h \to 0$, we consider $h = h(\epsilon)$, such that $h(\epsilon) \gg \epsilon$ and $h(\epsilon) \to 0$ as $\epsilon \to 0$. We would like to note that this limit as well as the proof of the periodic case is different from (14), where the double-limit is taken. In contrast to the proof of (14), the proof of the periodic case requires the correctors for the solutions of the local problems.

Next we will present the convergence results for MsFEM solutions. For general nonlinear elliptic equations under the assumptions (3)–(7) the strong convergence of MsFEM solutions can be shown. In the proof of this theorem we show the form of the truncation error (in a weak sense) in terms of the resonance errors between the mesh size and small scale ϵ. The resonance errors are derived explicitly. To obtain the convergence rate from the truncation error, one needs some lower bounds. Under the general conditions, such as (3)–(7), one can prove strong convergence of MsFEM solutions without an explicit convergence rate (cf. [75]). To convert the obtained convergence rates for the truncation errors into the convergence rate of MsFEM solutions, additional assumptions, such as monotonicity, are needed.

Next, we formulate convergence theorems. The proofs can be found in [33].

Theorem 4.1 *Assume $a_\epsilon(x,\eta,\xi)$ and $a_{0,\epsilon}(x,\eta,\xi)$ are periodic functions with respect to x and let u be a solution of (8) and u_h is a MsFEM solution given by (10). Moreover, we assume that ∇u_h is uniformly bounded in $L^{p+\alpha}(\Omega)$ for some $\alpha > 0$. Then*

$$\lim_{\epsilon\to 0}\|u_h - u\|_{W_0^{1,p}(\Omega)} = 0 \tag{15}$$

where $h = h(\epsilon) \gg \epsilon$ and $h \to 0$ as $\epsilon \to 0$ (up to a subsequence).

Theorem 4.2 *Let u and u_h be the solutions of the homogenized problem (8) and MsFEM (10), respectively, with the coefficient $a_\epsilon(x,\eta,\xi) = a(x/\epsilon,\xi)$ and $a_{0,\epsilon} = 0$. Then*

$$\|u_h - u\|_{W_0^{1,p}(\Omega)}^p \le c\left(\frac{\epsilon}{h}\right)^{\frac{s}{(p-1)(p-s)}} + c\left(\frac{\epsilon}{h}\right)^{\frac{p}{p-1}} + ch^{\frac{p}{p-1}}. \tag{16}$$

4.4 Multiscale Finite Element Methods for Nonlinear Parabolic Equations

We consider

$$\frac{\partial}{\partial t}u_\epsilon - div(a_\epsilon(x,t,u_\epsilon,\nabla u_\epsilon)) + a_{0,\epsilon}(x,t,u_\epsilon,\nabla u_\epsilon) = f, \qquad (17)$$

where ϵ is a small scale. Our motivation in considering (17) mostly stems from the applications of flow in porous media (multi-phase flow in saturated porous media, flow in unsaturated porous media) though many applications of nonlinear parabolic equations of these kinds occur in transport problems. Many problems in subsurface modeling have multiscale nature where the heterogeneities associated with the media is no longer periodic. It was shown that a solution u_ϵ converges to u (up to a subsequence) in an appropriate sense where u is a solution of

$$\frac{\partial}{\partial t}u - div(a^*(x,t,u,\nabla u)) + a_0^*(x,t,u,\nabla u) = f. \qquad (18)$$

In [36] the homogenized fluxes a^* and a_0^* are computed under the assumption that the heterogeneities are strictly stationary random fields with respect to both space and time.

The numerical homogenization procedure presented in the previous section can be extended to parabolic equations. To do this we will first formulate MsFEM in a slightly different manner from that presented in [48] for *the linear problem.* Consider a standard finite dimensional S^h space over a coarse triangulation of Ω, (1) and define $E^{MsFEM} : S^h \to V_\epsilon^h$ in the following way. For each $u_h \in S^h$ there is a corresponding element $u_{h,\epsilon}$ in V_ϵ^h that is defined by

$$\frac{\partial}{\partial t}u_{h,\epsilon} - div(a_\epsilon(x,t)\nabla u_{h,\epsilon}) = 0 \text{ in } K \times [t_n, t_{n+1}], \qquad (19)$$

with boundary condition $u_{h,\epsilon} = u_h$ on ∂K, and $u_{h,\epsilon}(t = t_n) = u_h$. For the linear equations E^{MsFEM} is a linear operator and the obtained multiscale space, V_ϵ^h is a linear space on $\Omega \times [t_n, t_{n+1}]$. Moreover, the basis in the space V_ϵ^h can be obtained by mapping the basis functions of S^h. For*the nonlinear parabolic equations* considered in this paper the operator E^{MsFEM} is constructed similar to (19) using the local problems, i.e., for each $u_h \in S^h$ there is a corresponding element $u_{h,\epsilon}$ in V_ϵ^h that is defined by

$$\frac{\partial}{\partial t}u_{h,\epsilon} - div(a_\epsilon(x,t,\eta,\nabla u_{h,\epsilon})) = 0 \text{ in } K \times [t_n, t_{n+1}], \qquad (20)$$

with boundary condition $u_{h,\epsilon} = u_h$ on ∂K, and $u_{h,\epsilon}(t = t_n) = u_h$. Here $\eta = \frac{1}{|K|}\int_K u_h dx$. Note E^{MsFEM} is a nonlinear operator and V_ϵ^h is no longer a linear space.

The following method that can be derived from general multiscale finite element framework. Find $u_h \in V_\epsilon^h$ such that

$$\int_{t_n}^{t_{n+1}} \int_\Omega \frac{\partial}{\partial t} u_h v_h \, dx \, dt + A(u_h, v_h) = \int_{t_n}^{t_{n+1}} \int_\Omega f v_h \, dx \, dt, \ \forall v_h \in S^h,$$

where

$$A(u_h, w_h) = \sum_K \int_{t_n}^{t_{n+1}} \int_K (a_\epsilon(x, t, \eta^{u_h}, \nabla v_\epsilon) \nabla w_h + a_{0,\epsilon}(x, t, \eta^{u_h}, \nabla v_\epsilon) w_h) \, dx \, dt,$$

where v_ϵ is the solution of the local problem (20), $u_h = l^{u_h}$ in each K, $\eta^{u_h} = \frac{1}{|K|} \int_K l^{u_h} dx$, and u_h is known at $t = t_n$.

We would like to note that the operator E^{MsFEM} can be constructed using larger domains as it is done in MsFEM with oversampling [48]. This way one reduces the effects of the boundary conditions and initial conditions. In particular, for the temporal oversampling it is only sufficient to start the computations before t_n and end them at t_{n+1}. Consequently, the oversampling domain for $K \times [t_n, t_{n+1}]$ consists of $[\tilde{t}_n, t_{n+1}] \times S$, where $\tilde{t}_n < t_n$ and $K \subset S$. More precise formulation and detail numerical studies of oversampling technique for nonlinear equations are currently under investigation. Further we would like to note that oscillatory initial conditions can be imposed (without using oversampling techniques) based on the solution of the elliptic part of the local problems (20). These initial conditions at $t = t_n$ are the solutions of

$$-div(a_\epsilon(x, t, \eta, \nabla u_{h,\epsilon})) = 0 \text{ in } K, \tag{21}$$

or

$$-div(\bar{a}_\epsilon(x, \eta, \nabla u_{h,\epsilon})) = 0 \text{ in } K, \tag{22}$$

where $\bar{a}_\epsilon(x, \eta, \xi) = \frac{1}{t_{n+1} - t_n} \int_{t_n}^{t_{n+1}} a(T(x/\epsilon^\beta, \tau/\epsilon^\alpha) \omega, \eta, \xi) d\tau$ and $u_{h,\epsilon} = u_h$ on ∂K. The latter can become efficient depending on the inter-play between the temporal and spatial scales. This issue is discussed below.

Note that in the case of periodic media the local problems can be solved in a single period in order to construct $A(u_h, v_h)$. In general, one can solve the local problems in a domain different from K (an element) to calculate $A(u_h, v_h)$, and our analysis is applicable to these cases. Note that the numerical advantages of our approach over the fine scale simulation is similar to that of MsFEM. In particular, for each Newton's iteration a linear system of equations on *a coarse grid* is solved.

For some special cases the operator E^{MsFEM} introduced in the previous section can be simplified (see [37]). In general one can avoid solving the local parabolic problems if the ratio between temporal and spatial scales is known, and solve instead a simplified equation. For example, assuming that $a_\epsilon(x, t, \eta, \xi) = a(x/\epsilon^\beta, t/\epsilon^\alpha, \eta, \xi)$, we have the following. If $\alpha < 2\beta$ one can solve instead of (20) the local problem $-div(a_\epsilon(x, t, \eta^{u_h}, \nabla v_\epsilon)) = 0$, if $\alpha > 2\beta$

one can solve instead of (20) the local problem $-div(\overline{a}_\epsilon(x, \eta^{u_h}, \nabla v_\epsilon)) = 0$, where $\overline{a}_\epsilon(x, \eta, \xi)$ is an average over time of $a_\epsilon(x, t, \eta, \xi)$, while if $\alpha = 2\beta$ we need to solve the parabolic equation in $K \times [t_n, t_{n+1}]$, (20).

We would like to note that, in general, one can use (21) or (22) as oscillatory initial conditions and these initial conditions can be efficient for some cases. For example, for $\alpha > 2\beta$ with initial conditions given by (22) the solutions of the local problems (20) can be computed easily since they are approximated by (22). Moreover, one can expect better accuracy with (22) for the case $\alpha > 2\beta$ because this initial condition is more compatible with the local heterogeneities compare to the artificial linear initial conditions (cf. (20)). The comparison of various oscillatory initial conditions including the ones obtained by oversampling method is a subject of future studies.

Finally, we would like to mention that one can prove the following theorem.

Theorem 4.3 $u_h = \sum_i \theta_i(t)\phi_i^0(x)$ *converges to* u, *a solution of the homogenized equation in* $V_0 = L^p(0, T, W_0^{1,p}(\Omega))$ *as* $\lim_{h \to 0} \lim_{\epsilon \to 0}$ *under additional not restrictive assumptions (see [37]).*

Remark 4.1. The proof of the theorem uses the convergence of the solutions and the fluxes, and consequently it is applicable for the case of general heterogeneities that uses G-convergence theory. Since the G-convergence of the operators occurs up to a subsequence the numerical solution converges to a solution of a homogenized equation (up to a subsequence of ϵ).

4.5 Numerical Results

In this section we present several ingredients pertaining to the implementation of multiscale finite element method for nonlinear elliptic equations. More numerical examples relevant to subsurface applications can be found in [33]. We will present numerical results for both MsFEM and multiscale finite volume element method (MsFVEM). We use an Inexact-Newton algorithm as an iterative technique to tackle the nonlinearity. For the numerical examples below, we use $a_\epsilon(x, u_\epsilon, \nabla u_\epsilon) = a_\epsilon(x, u_\epsilon)\nabla u_\epsilon$. Let $\{\phi_0^i\}_{i=1}^{N_{dof}}$ be the standard piecewise linear basis functions of S^h. Then MsFEM solution may be written as

$$u_h = \sum_{i=1}^{N_{dof}} \alpha_i \, \phi_0^i \tag{23}$$

for some $\alpha = (\alpha_1, \alpha_2, \ldots, \alpha_{N_{dof}})^T$, where α_i depends on ϵ. Hence, we need to find α such that

$$F(\alpha) = 0, \tag{24}$$

where $F : \mathbb{R}^{N_{dof}} \to \mathbb{R}^{N_{dof}}$ is a nonlinear operator such that

$$F_i(\alpha) = \sum_{K \in K^h} \int_K a_\epsilon(x, \eta^{u_h}) \nabla u_{\epsilon,h} \nabla \phi_0^i \, dx - \int_\Omega f \, \phi_0^i \, dx. \qquad (25)$$

We note that in (25) α is implicitly buried in η^{u_h} and $u_{\epsilon,h}$. An inexact-Newton algorithm is a variation of Newton's iteration for nonlinear system of equations, where the Jacobian system is only approximately solved. To be specific, given an initial iterate α^0, for $k = 0, 1, 2, \ldots$ until convergence do the following:

- Solve $F'(\alpha^k)\delta^k = -F(\alpha^k)$ by some iterative technique until $\|F(\alpha^k) + F'(\alpha^k)\delta^k\| \leq \beta_k \|F(\alpha^k)\|$.
- Update $\alpha^{k+1} = \alpha^k + \delta^k$.

In this algorithm $F'(\alpha^k)$ is the Jacobian matrix evaluated at iteration k. We note that when $\beta_k = 0$ then we have recovered the classical Newton iteration. Here we have used

$$\beta_k = 0.001 \left(\frac{\|F(\alpha^k)\|}{\|F(\alpha^{k-1})\|} \right)^2, \qquad (26)$$

with $\beta_0 = 0.001$. Choosing β_k this way, we avoid over-solving the Jacobian system when α^k is still considerably far from the exact solution.

Next we present the entries of the Jacobian matrix. For this purpose, we use the following notations. Let $K_i^h = \{K \in K^h : z_i \text{ is a vertex of } K\}$, $I^i = \{j : z_j \text{ is a vertex of } K \in K_i^h\}$, and $K_{ij}^h = \{K \in K_i^h : K \text{ shares } \overline{z_i z_j}\}$. We note that we may write $F_i(\alpha)$ as follows:

$$F_i(\alpha) = \sum_{K \in K_i^h} \left(\int_K a_\epsilon(x, \eta^{u_h}) \nabla u_{\epsilon,h} \nabla \phi_0^i \, dx - \int_K f \, \phi_0^i \, dx \right), \qquad (27)$$

with

$$-\mathrm{div} a_\epsilon(x, \eta^{u_h}) \nabla u_{\epsilon,h} = 0 \text{ in } K \quad \text{and} \quad u_{\epsilon,h} = \sum_{z_m \in Z_K} \alpha_m \phi_0^m \text{ on } \partial K, \quad (28)$$

where Z_K is all the vertices of element K. It is apparent that $F_i(\alpha)$ is not fully dependent on all $\alpha_1, \alpha_2, \ldots, \alpha_d$. Consequently, $\frac{\partial F_i(\alpha)}{\partial \alpha_j} = 0$ for $j \notin I^i$. To this end, we denote $\psi_\epsilon^j = \frac{\partial u_{\epsilon,h}}{\partial \alpha_j}$. By applying chain rule of differentiation to (28) we have the following local problem for ψ_ϵ^j:

$$-\mathrm{div} a_\epsilon(x, \eta^{u_h}) \nabla \psi_\epsilon^j = \frac{1}{3} \mathrm{div} \frac{\partial a_\epsilon(x, \eta^{u_h})}{\partial u} \nabla u_{\epsilon,h} \text{ in } K \quad \text{and} \quad \psi_\epsilon^j = \phi_\epsilon^j \text{ on } \partial K. \qquad (29)$$

The fraction $1/3$ comes from taking the derivative in the chain rule of differentiation. In the formulation of the local problem, we have replaced the nonlinearity in the coefficient by η^{v_h}, where for each triangle K $\eta^{v_h} = 1/3 \sum_{i=1}^{3} \alpha_i^K$, which gives $\partial \eta^{v_h} / \partial \alpha_i = 1/3$. Moreover, for a rectangular element the fraction $1/3$ should be replaced by $1/4$.

Thus, provided that $v_{\epsilon,h}$ has been computed, then we may compute ψ_ϵ^j using (29). Using the above descriptions we have the expressions for the entries of the Jacobian matrix:

$$\frac{\partial F_i}{\partial \alpha_i} = \sum_{K \in K_i^h} \left(\frac{1}{3} \int_K \frac{\partial a_\epsilon(x, \eta^{u_h})}{\partial u} \nabla u_{\epsilon,h} \nabla \phi_0^i \, dx + \int_K a_\epsilon(x, \eta^{u_h}) \nabla \psi_i \nabla \phi_0^i \, dx \right)$$
(30)

$$\frac{\partial F_i}{\partial \alpha_j} = \sum_{K \in K_{ij}^h} \left(\frac{1}{3} \int_K \frac{\partial a_\epsilon(x, \eta^{u_h})}{\partial u} \nabla u_{\epsilon,h} \nabla \phi_\epsilon^i \, dx + \int_K a_\epsilon(x, \eta^{u_h}) \nabla \psi_\epsilon^j \nabla \phi_0^i \, dx \right)$$
(31)

for $j \neq i$, $j \in I^i$.

The implementation of the oversampling technique is similar to the procedure presented earlier, except the local problems in larger domains are used. As in the non-oversampling case, we denote $\psi_\epsilon^j = \frac{\partial v_{\epsilon,h}}{\partial \alpha_j}$, such that after applying chain rule of differentiation to the local problem we have:

$$-div a_\epsilon(x, \eta^{u_h}) \nabla \psi_\epsilon^j = \frac{1}{3} \, div \frac{\partial a_\epsilon(x, \eta^{u_h})}{\partial u} \nabla v_{\epsilon,h} \text{ in } S \quad \text{and} \quad \psi_\epsilon^j = \phi_0^j \text{ on } \partial S,$$
(32)

where η^{u_h} is computed over the corresponding element K and ϕ_0^j is understood as the nodal basis functions on oversampled domain S. Then all the rest of the inexact-Newton algorithms are the same as in the non-oversampling case. Specifically, we also use (30) and (31) to construct the Jacobian matrix of the system. We note that we will only use ψ_ϵ^j from (32) pertaining to the element K.

From the derivation (both for oversampling and non-oversampling) it is obvious that the Jacobian matrix is not symmetric but sparse. Computation of this Jacobian matrix is similar to computing the stiffness matrix resulting from standard finite element, where each entry is formed by accumulation of element by element contribution. Once we have the matrix stored in memory, then its action to a vector is straightforward. Because it is a sparse matrix, devoting some amount of memory for entries storage is inexpensive. The resulting linear system is solved using preconditioned bi-conjugate gradient stabilized method.

We want to solve the following problem:

$$-div a(x/\epsilon, u_\epsilon) \nabla u_\epsilon = -1 \quad \text{in } \Omega \subset \mathbb{R}^2,$$
$$u_\epsilon = 0 \quad \text{on } \partial \Omega,$$
(33)

where $\Omega = [0,1] \times [0,1]$, $a(x/\epsilon, u_\epsilon) = k(x/\epsilon)/(1 + u_\epsilon)^{l(x/\epsilon)}$, with

$$k(x/\epsilon) = \frac{2 + 1.8 \sin(2\pi x_1/\epsilon)}{2 + 1.8 \cos(2\pi x_2/\epsilon)} + \frac{2 + \sin(2\pi x_2/\epsilon)}{2 + 1.8 \cos(2\pi x_1/\epsilon)}$$
(34)

and $l(x/\epsilon)$ is generated from $k(x/\epsilon)$ such that the average of $l(x/\epsilon)$ over Ω is 2. Here we use $\epsilon = 0.01$. Because the exact solution for this problem is not available, we use a well resolved numerical solution using standard finite element method as a reference solution. The resulting nonlinear system is solved using inexact-Newton algorithm. The reference solution is solved on 512×512 mesh. Tables 2 and 4 present the relative errors of the solution with and without oversampling, respectively. In Tables 3 and 5, the relative errors for multiscale finite volume element method are presented. The relative errors are computed as the corresponding error divided by the norm of the solution. In each table, the second, third, and fourth columns list the relative error in L^2, H^1, and L^∞ norm, respectively. As we can see from these two tables, the oversampling significantly improves the accuracy of the multiscale method.

For our next example, we consider the problem with non-periodic coefficients, where $a_\epsilon(x, \eta) = k_\epsilon(x)/(1+\eta)^{\alpha_\epsilon(x)}$. $k_\epsilon(x) = \exp(\beta_\epsilon(x))$ is chosen such that $\beta_\epsilon(x)$ is a realization of a random field with the spherical variogram [23]

Table 2 Relative MsFEM errors without oversampling

N	L^2-norm		H^1-norm		L^∞-norm	
	Error	Rate	Error	Rate	Error	Rate
32	0.029		0.115		0.03	
64	0.053	−0.85	0.156	−0.44	0.0534	−0.94
128	0.10	−0.94	0.234	−0.59	0.10	−0.94

Table 3 Relative MsFVEM errors without oversampling

N	L^2-norm		H^1-norm		L^∞-norm	
	Error	Rate	Error	Rate	Error	Rate
32	0.03		0.13		0.04	
64	0.05	−0.65	0.19	−0.60	0.05	−0.24
128	0.058	−0.19	0.25	−0.35	0.057	−0.19

Table 4 Relative MsFEM errors with oversampling

N	L^2-norm		H^1-norm		L^∞-norm	
	Error	Rate	Error	Rate	Error	Rate
32	0.0016		0.036		0.0029	
64	0.0012	0.38	0.019	0.93	0.0016	0.92
128	0.0024	−0.96	0.0087	1.14	0.0026	−0.71

Table 5 Relative MsFVEM errors with oversampling

N	L^2-norm		H^1-norm		L^∞-norm	
	Error	Rate	Error	Rate	Error	Rate
32	0.002		0.038		0.005	
64	0.003	−0.43	0.021	0.87	0.003	0.72
128	0.001	1.10	0.009	1.09	0.001	1.08

Table 6 Relative MsFEM errors for random heterogeneities, spherical variogram, $l_x = 0.20$, $l_z = 0.02$, $\sigma = 1.0$

N	L^2-norm		H^1-norm		L^∞-norm		Hor. flux	
	Error	Rate	Error	Rate	Error	Rate	Error	Rate
32	0.0006		0.0505		0.0025		0.025	
64	0.0002	1.58	0.029	0.8	0.001	1.32	0.017	0.57
128	0.0001	1	0.016	0.85	0.0005	1	0.011	0.62

Table 7 Relative MsFVEM errors for random heterogeneities, spherical variogram, $l_x = 0.20$, $l_z = 0.02$, $\sigma = 1.0$

N	L^2-norm		H^1-norm		L^∞-norm		Hor. flux	
	Error	Rate	Error	Rate	Error	Rate	Error	Rate
32	0.0006		0.0515		0.0025		0.027	
64	0.0002	1.58	0.029	0.81	0.0013	0.94	0.018	0.58
128	0.0001	1	0.016	0.85	0.0005	1.38	0.012	0.58

and with the correlation lengths $l_x = 0.2$, $l_y = 0.02$ and with the variance $\sigma = 1$. $\alpha_\epsilon(x)$ is chosen such that $\alpha_\epsilon(x) = k_\epsilon(x) + const$ with the spatial average of 2. As for the boundary conditions we use "left-to-right flow" in $\Omega = [0,5] \times [0,1]$ domain, $u_\epsilon = 1$ at the inlet ($x_1 = 0$), $u_\epsilon = 0$ at the outlet ($x_1 = 5$), and no flow boundary conditions on the lateral sides $x_2 = 0$ and $x_2 = 1$. In Table 6 we present the relative error for multiscale method with oversampling. Similarly, in Table 7 we present the relative error for multiscale finite volume method with oversampling. Clearly, the oversampling method captures the effects induced by the large correlation features. Both H^1 and horizontal flux errors are under 5%. Similar results have been observed for various kinds of non-periodic heterogeneities.

In the next set of numerical examples, we test MsFEM for problems with fluxes $a_\epsilon(x, \eta)$ that are discontinuous in space. The discontinuity in the fluxes is introduced by multiplying the underlying permeability function, $k_\epsilon(x)$, by a constant in certain regions, while leaving it unchanged in the rest of the domain. As an underlying permeability field, $k_\epsilon(x)$, we choose the random field used for the results in Table 6. In the numerical example, the discontinuities are introduced along the boundaries of the coarse elements. In particular, $k_\epsilon(x)$ on the left half of the domain is multiplied by a constant J, where $J = \exp(4)$. The results in Table 8 show that MsFEM converges and the error falls below 5% for relatively large coarsening. For the second numerical example (Table 9), the discontinuities are not aligned with the boundaries of the coarse elements. In particular, the discontinuity boundary is given by $y = x\sqrt{2} + 0.5$, i.e., the discontinuity line intersects the coarse grid blocks. Similar to the aligned case, $\exp(4)$ jump magnitude is considered. These results demonstrate the robustness of our approach for anisotropic fields where h and ϵ are nearly the same, and the fluxes that are discontinuous spatial functions.

As for CPU comparisons, we have observed more than 92% CPU savings when using MsFEM without oversampling. With the oversampling approach, the CPU savings depend on the size of the oversampled domain. For

Table 8 Relative MsFEM errors for random heterogeneities, spherical variogram, $l_x = 0.20$, $l_z = 0.02$, $\sigma = 1.0$, aligned discontinuity, jump = $\exp(4)$

N	L^2-norm		H^1-norm		L^∞-norm		Hor. flux	
	Error	Rate	Error	Rate	Error	Rate	Error	Rate
32	0.0011		0.1010		0.0068		0.195	
64	0.0006	0.87	0.0638	0.66	0.0045	0.59	0.109	0.84
128	0.0003	1.00	0.0349	0.87	0.0024	0.91	0.063	0.79

Table 9 Relative MsFEM errors for random heterogeneities, spherical variogram, $l_x = 0.20$, $l_z = 0.02$, $\sigma = 1.0$, nonaligned discontinuity, jump = $\exp(4)$

N	L^2-norm		H^1-norm		L^∞-norm		Hor. flux	
	Error	Rate	Error	Rate	Error	Rate	Error	Rate
32	0.0067		0.1775		0.1000		0.164	
64	0.0016	2.07	0.0758	1.23	0.0288	1.80	0.077	1.09
128	0.0009	0.83	0.0687	0.14	0.0423	-0.55	0.039	0.98

example, if the oversampled domain size is two times larger than the target coarse block (half coarse block extension on each side) we have observed 70% CPU savings for 64×64 and 80% CPU savings for 128×128 coarse grid. In general, the computational cost will decrease if the oversampled domain size is close to the target coarse block size, and this cost will be close to the cost of MsFEM without oversampling. Conversely, the error decreases if the size of the oversampled domains increases. In the numerical examples studied in our paper, we have observed the same errors for the oversampling methods using either one coarse block extension or half coarse block extensions. The latter indicates that the leading resonance error is eliminated by using a smaller oversampled domain. Oversampled domains with one coarse block extension are previously used in simulations of flow through heterogeneous porous media. As it is indicated in [48], one can use large oversampled domains for simultaneous computations of the several local solutions. Moreover, parallel computations will improve the speed of the method because MsFEM is well suited for parallel computation [48]. For the problems where $a_\epsilon(x, \eta, \xi) = a_\epsilon(x)b(\eta)\xi$ (see Sect. 4.2 and the next section for applications) our multiscale computations are very fast because the basis functions are built in the beginning of the computations. In this case, we have observed more than 95% CPU savings.

Applications of MsFEM to Richards' equation are presented in [33].

4.6 Generalizations of MsFEM and Some Remarks

Next, we present the framework of MsFEM for general equations. Consider

$$L_\epsilon u_\epsilon = f, \tag{35}$$

where ϵ is a small scale and $L_\epsilon : X \to Y$ is an operator. Moreover, we assume that L_ϵ G-converges to L^* (up to a sub-sequence), where u is a solution of

$$L^* u = f, \tag{36}$$

(we refer to [70], page 14 for the definition of G-convergence for operators). The objective of MsFEM is to approximate u in S^h. Denote S^h a family of finite dimensional space such that it possesses an approximation property (see [85], [71]) as before. Here h is a scale of computation and $h \gg \epsilon$. For (35) *multiscale mapping*, $E^{MsFEM} : S^h \to V_\epsilon^h$, will be defined as follows. For each element $v_h \in S^h$, $v_{\epsilon,h} = E^{MsFEM} v_h$ is defined as

$$L_\epsilon^{map} v_{\epsilon,h} = 0 \text{ in } K, \tag{37}$$

where L_ϵ^{map} can be, in general, different from L_ϵ and allows us to capture the effects of the small scales. Moreover, the domains different from the target coarse block K can be used in the computations of the local solutions. To solve (37) one needs to impose boundary and initial conditions. This issue needs to be resolved on a case by case basis, and the main idea is to interpolate v_h onto the underlying fine grid. Further, we seek a solution of (35) in V_ϵ^h as follows. Find $u_h \in S^h$ (consequently $u_{\epsilon,h} \in V_\epsilon^h$) such that

$$\langle L_\epsilon^{global} u_{\epsilon,h}, v_h \rangle = \langle f, v_h \rangle, \ \forall v_h \in S^h, \tag{38}$$

where $\langle u, v \rangle$ denotes the duality between X and Y, and L_ϵ^{global} can be, in general, different from L_ϵ. For example, for nonlinear elliptic equations we have $L_\epsilon u = -div a_\epsilon(x, u, \nabla u) + a_{0,\epsilon}(x, u, \nabla u)$, $L_\epsilon^{map} u = div a_\epsilon(x, \eta^u, \nabla u)$ in K, and $L_\epsilon^{global} = div a_\epsilon(x, \eta^u, \nabla u) + a_{0,\epsilon}(x, \eta^u, \nabla u)$ in K. The convergence of MsFEM is to show that $u_h \to u$ and $u_{\epsilon,h} \to u_\epsilon$, where $u_{\epsilon,h} = E^{MsFEM} u_h$ in appropriate space. The correct choices of L_ϵ^{map} and L_ϵ^{global} are the essential part of MsFEM and guarantees the convergence of the method.

In conclusion, we have presented a natural extension of MsFEM to nonlinear problems. This is accomplished by considering a multiscale map instead of the basis functions that are considered in linear MsFEM [48]. Our approaches share some common elements with recently introduced HMM [30], where macroscopic and microscopic solvers are also needed. In general, the finding of "correct" macroscopic and microscopic solvers is the main difficulty of the multiscale methods. Our approaches follow MsFEM and, consequently, finite element methods constitute its main ingredient. The resonance errors, that arise in linear problems also arise in nonlinear problems. Note that the resonance errors are the common feature of multiscale methods unless periodic problems are considered and the solutions of the local problems in an exact period are used. To reduce the resonance errors we use oversampling technique and show that the error can be greatly reduced by sampling from the larger domains. The multiscale map for MsFEM uses the solutions of the local problems in the target coarse block. This way one can sample the heterogeneities of the coarse block. If there is a scale separation and, in addition, some kind of periodicity, one can use the solutions of the smaller size problems to approximate the multiscale map. Note that a potential disadvantage of periodicity assumption is that the periodicity can act to disrupt large-scale connectivity features of the flow. For the examples similar to the non-periodic ones considered in this paper, with the use of the smaller size problems for approximating the solutions of the local problems, we have found very large errors (of order 50%).

5 Multiscale Simulations of Two-Phase Immiscible Flow in Adaptive Coordinate System

Previously, we discussed some applications of MsFEM to two-phase flows. In this section, we briefly discuss the use of adaptive coordinate system in multiscale simulations of two-phase porous media flows. In particular, we would like to present upscaling of transport equations and its coupling to MsFEM.

As we discussed earlier, the use of global information can improve the multiscale finite element method. In particular, the solution of the pressure equation at initial time is used to construct the boundary conditions for the basis functions. It is interesting to note that the multiscale finite element methods that employ a limited global information reduces to standard multiscale finite element method in flow-based coordinate system. This can be verified directly and the reason behind it is that we have already employed a limited global information in flow-based coordinate system. To achieve high degree of speed-up in two-phase flow computations, we also consider the upscaling of transport equation and its coupling to pressure equation.

We would like to derive an upscaled model for the transport equation. We will assume the velocity is independent of time, $\lambda(S) = 1$, and restrict ourselves to the two-dimensional case. Then using the pressure-streamline framework, one obtains

$$S_t^\epsilon + v_0^\epsilon f(S^\epsilon)_p = 0 \qquad (1)$$
$$S(p, \psi, t = 0) = S_0,$$

where ϵ denotes the small scale, v_0^ϵ denotes the Jacobian of the transformation and is positive, and p denotes the initial pressure. For simplicity, we assume $k(x)$ is diagonal and isotropic, and we have $\nabla \psi \cdot \nabla p = 0$, where ψ is the streamline function. For deriving upscaled equations, we will first homogenize (1) along the streamlines, and then to homogenize across the streamlines. The homogenization along the streamlines can be done following Bourgeat and Mikelic [15] or following Hou and Xin [54] and E [29]. The latter uses two-scale convergence theory and we refer to [76] for the results on homogenization of (1) using two-scale convergence theory. We note that the homogenization results of Bourgeat and Mikelic is for general heterogeneities without an assumption on periodicity, and thus, is more appropriate for problems considered in the paper. Following [15], the homogenization of (1) can be easily derived (see Proposition 3.4 in [15]). Here, we briefly sketch the proof.

For ease of notations, we ignore the ψ dependence of v_0^ϵ and S^ϵ, and treat ψ as a parameter. We consider

$$v_0^\epsilon(p) = v_0(p, \frac{p}{\epsilon}).$$

Moreover, we assume that the domain is a unit interval. Then, for each ψ, it can be shown that $S^\epsilon(p, \psi, t) \to \tilde{S}(p, \psi, t)$ in $L^1((0,1) \times (0,T))$, where $\tilde{S}$ satisfies

$$\tilde{S}_t + \tilde{v}_0 f(\tilde{S})_p = 0, \tag{2}$$

and where $\tilde{v}_0$ is harmonic average of v_0^ϵ, i.e.,

$$\frac{1}{v_0^\epsilon} \to \frac{1}{\tilde{v}_0} \quad \text{weak} * \text{ in } L^\infty(0,1).$$

The proof of this fact follows from Proposition 3.4. of [15] (see also [34]). In [34, 76], we prove a convergence rate of the fine saturation S^ϵ to the homogenized limit $\tilde{S}$ as $\epsilon \to 0$.

Theorem 5.1 *Assume that $v_0^\epsilon(p)$ is bounded uniformly*

$$C^{-1} \le v_0^\epsilon(p, \frac{p}{\epsilon}) \le D.$$

Denote by $F(t, T)$ the solution to $S_t + f(S)_T = 0$. The solution S^ϵ of (2) converges to $\tilde{S}$ (assuming initial conditions that don't depend on the fast scale) at a rate given by

$$\|S^\epsilon - \tilde{S}\|_\infty \le G\epsilon,$$

when F remains Lipschitz for all time, and

$$\|S^\epsilon - \tilde{S}\|_n \le G\epsilon^{1/n},$$

when F develops at most a finite number of discontinuities.

The homogenized operator given by (2) still contains variation of order ϵ through the fast variable $\frac{\psi}{\epsilon}$, however it does not contain any derivatives in that variable. Its dependence on $\frac{\psi}{\epsilon}$ is only parametric. We can homogenize the dependence of the partially homogenized operator on $\frac{\psi}{\epsilon}$ and arrive at a homogenized operator that is independent of the small scale. In the latter case, we will only obtain weak convergence of the partially homogenized solution. When we homogenized along the streamlines, the resulting equation was of hyperbolic type like the original equation. In a seminal and celebrated paper, Tartar [78] showed that homogenization across streamlines leads to transport with the average velocity plus a time-dependent diffusion term, referred to as macrodispersion, a physical phenomenon that was not present in the original fine equation. In particular, if the velocity field does not depend on p inside the cells, that is, $\tilde{v}(\psi, \frac{\psi}{\epsilon})$, then the homogenized solution, $\overline{\tilde{S}}$, (weak* limit of $\tilde{S}$, which will be denoted by $\overline{S}$), satisfies

$$\overline{S}_t + \overline{\tilde{v}}_0 \overline{S}_p = \int_0^t \int \overline{S}_{pp}(p - \lambda(t - \tau), \psi, \tau) d\mu_{\frac{\psi}{\epsilon}}(\lambda) d\tau. \tag{3}$$

Here, $d\nu_{\frac{\psi}{\epsilon}}$ the Young measure associated with the sequence $\tilde{v}_0(\psi, \cdot)$ and $d\mu_{\frac{\psi}{\epsilon}}$ is a Young measure that satisfies

$$\left(\int \frac{d\nu_{\frac{\psi}{\epsilon}}(\lambda)}{\frac{s}{2\pi i q} + \lambda} \right)^{-1} = \frac{s}{2\pi i q} + \overline{\tilde{v}}_0 - \int \frac{d\mu_{\frac{\psi}{\epsilon}}(\lambda)}{\frac{s}{2\pi i q} + \lambda}.$$

We have denoted by $\overline{\tilde{v}}_0$ the weak limit of the velocity. This equation has no dependence on the small scale and we consider it to be the full homogenization of the fine saturation equation. Efendiev and Popov [40] have extended this method for the Riemann problem in the case of nonlinear flux. Note that the homogenization across streamlines provides a weak limit of partially homogenized solution. Because the original solution S^ϵ strongly converges to partially homogenized solution for each ψ, it can be easily shown that $S^\epsilon \to \overline{S}$ weakly. We omit this proof here.

In numerical simulations, it is difficult to use (3) as a homogenized operator, and often a second order approximation of this equation is used. These approximate equations can be also derived using perturbation analysis. In particular, using the higher moments of the saturation and the velocity, one can model the macrodispersion. In the context of two-phase flow this idea was introduced by Efendiev et al. [39], Efendiev and Durlofsky [38], Chen and Hou [19] and Hou et al. [47]. In our case, the computation of the macrodispersion is much simpler because the transport equations have been already averaged along the streamlines, and thus we will be applying perturbation technique to one-dimensional problem.

We expand $\tilde{S}$, $\tilde{v}_0$ (following [39]) as an average over the cells in the pressure-streamline frame and the corresponding fluctuations

$$\begin{aligned} \tilde{S} &= \overline{S}(p, \psi, t) + S'(p, \psi, \zeta, t) \\ \tilde{v}_0 &= \overline{\tilde{v}}_0(p, \psi, t) + \tilde{v}_0'(p, \psi, \zeta, t) \end{aligned} \tag{4}$$

We will derive the homogenized equation for $f(S) = S$. Averaging (2) with respect to ψ we find an equation for the mean of the saturation

$$\overline{S}_t + \overline{\tilde{v}}_0 \overline{S}_p + \overline{\tilde{v}_0' S_p'} = 0.$$

An equation for the fluctuations is obtained by subtracting the above equation from (2)

$$S'_t + (\tilde{v}_0 - \overline{\tilde{v}_0})\overline{S}_p + \tilde{v}_0 S'_p - \overline{\tilde{v}'_0 S'_p} = 0.$$

Together, the equations for the saturation are

$$\overline{S}_t + \overline{\tilde{v}_0}\overline{S}_p + \overline{\tilde{v}'_0 S'_p} = 0 \tag{5}$$

$$S'_t + \tilde{v}'_0 \overline{S}_p + \tilde{v}_0 S'_p - \overline{\tilde{v}'_0 S'_p} = 0.$$

We can consider the second equation to be the auxiliary (cell) problem and the first equation to be the upscaled equation. We note that the cell problem for a hyperbolic equation is $O(1)$ whereas for an elliptic it is $O(\epsilon)$. We can obtain an approximate numerical method by solving the cell problem only near the shock region in space time, where the macrodispersion term is largest. In that case it is best to diagonalize these equations by adding the first to the second one

$$\overline{S}_t + \overline{\tilde{v}_0}\overline{S}_p = -\overline{\tilde{v}'_0}(\tilde{S}_p - \overline{S}_p)$$

$$\tilde{S}_t + \tilde{v}_0 \tilde{S}_p = \quad 0.$$

Compared to (5), it has fewer forcing terms and no cross fluxes, which leads to a numerical method with less numerical diffusion that is easier to implement.

The above derivation contains no approximation. Next, we follow the same idea as in the derivation to solve the equation for the fluctuations along the characteristics, but with the purpose of deriving an equation on the coarse grid. To achieve this, we will not perform analytical upscaling in the sense of deriving a continuous upscaled equation as in the previous section. We first discretize the equation with a finite volume method in space and then upscaled the resulting equation. Our upscaled equation will therefore be dependent on the numerical scheme.

We use the same definition for the average saturation and the fluctuations as in (4) and follow the same steps until (5). We discretize the macrodispersion term in the equation for the average saturation

$$\overline{\tilde{v}'_0 S'_p} = \frac{\overline{\tilde{v}'_0 S'}^{i+1} - \overline{\tilde{v}'_0 S'}^{i}}{\Delta p} + O(\Delta p).$$

A superscript $\cdot^i$ refers to a discrete quantity defined at the center of the conservation cell. Instead of solving the equation for the fluctuations on the fine characteristics as before, which would lead to a fine grid algorithm, we solve it on the coarse characteristics defined by

$$\frac{dP}{dt} = \overline{\tilde{v}_0}, \text{ with } P(p, 0) = p.$$

Compared to the equation that we obtained in the previous section for S', this equation for S' has an extra term, which appears second

$$S' =$$

$$-\int_0^t \left(\tilde{v}_0'(P(p,\tau),\psi)\overline{S}_p(P(p,\tau),\psi,\tau)+\tilde{v}_0'(P(p,\tau),\psi)S_p'(P(p,\tau),\psi,\tau)+\overline{\tilde{v}_0' S_p'}\right)d\tau$$

The second term is second-order in fluctuating quantities, and we expect it to be smaller than the first term so we neglect it. As before, we multiply by $\tilde{v}_0'$ and average over ψ to find

$$\overline{\tilde{v}_0' S'} = -\int_0^t \overline{\tilde{v}_0'\tilde{v}_0(P(p,\tau),\psi)\overline{S}_p(P(p,\tau),\psi,\tau)}d\tau.$$

In this form at time t it is necessary to know information about the past saturation in $(0,t)$ to compute the future saturation. Following [39], it can be easily shown that $\overline{S}_p(P(p,\tau)$ depends weakly on time, in the sense that the difference between $\overline{S}_p(P(p,\tau)$ and $\overline{S}_p(P(p,t)$ is of third-order in fluctuating quantities. Therefore we can take $\overline{S}_p(P(p,\tau)$ out of the time integral to find

$$\overline{\tilde{v}_0' S'} = -\int_0^t \overline{\tilde{v}_0'\tilde{v}_0'(P(p,\tau),\psi)}d\tau\,\overline{S}_p.$$

The term inside the time integral is the covariance of the velocity field along each streamline. The macrodispersion in this form can be computed independent of the past saturation.

The nonlinearity of the flux function introduces an extra source of error in the approximation. One can derive the upscaled equation by approximating nonlinear flux functions and this is presented in [34].

Extensive numerical studies are presented in [34, 76]. These numerical tests use multiscale finite volume element methods for two-phase flow. Note that global information is already incorporated into the multiscale basis functions and the standard MsFVEM is equivalent to MsFVEM using limited global information introduced earlier. In simulations, a moving mesh is used to concentrate the points of computation near the sharp front. Since the saturation equation is one-dimensional in the pressure-streamline coordinates, the implementation of the moving mesh is straightforward and efficient. We have presented the numerical results for different types of heterogeneities. All numerical results show that one can achieve very accurate results with low computational cost.

We present one set of numerical results for two-phase flow. The application of the proposed method to two-phase immiscible flow can be performed using the implicit pressure and explicit saturation (IMPES) framework. This procedure consists of computing the velocity and then using the velocity field in updating the saturation field. When updating the saturation field,

Fig. 26 *Left*: pressure and streamline function at time $t = 0.4$ in Cartesian frame. *Right*: pressure and streamline function at time $t = 0.4$ in initial pressure-streamline frame

we consider the velocity field to be time independent and we can use our upscaling procedure at each IMPES time step. First, we note that in the proposed method, the mapping is done between the current pressure-streamline and initial pressure-streamline. This mapping is nearly the identity for the cases when $\mu_o > \mu_w$. In Fig. 26, we plot the level sets of the pressure and streamfunction at time $t = 0.4$ in a Cartesian coordinate system (left plot) and in the coordinate system of the initial pressure and streamline (right plot). Clearly, the level sets are much smoother in initial pressure-streamline frame compared to Cartesian frame. In Fig. 27, we plot the saturation snapshots right before the breakthrough. In Fig. 28, the fractional flow is plotted. Again, the moving mesh algorithm is used to track the front separately. The convergence table is presented in Table 10. We see from this table that the errors decrease as first order which indicates that the pressure and saturation is smooth functions of initial pressure and streamline.

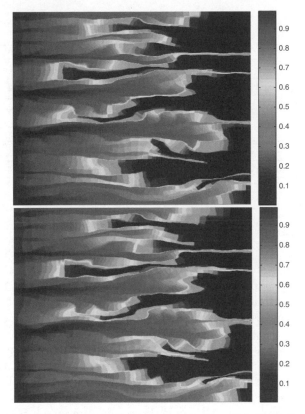

Fig. 27 *Left*: saturation plot obtained using coarse-scale model. *Right*: the fine-scale saturation plot. Both plots are on coarse grid. Variogram based permeability field is used. $\mu_o/\mu_w = 5$

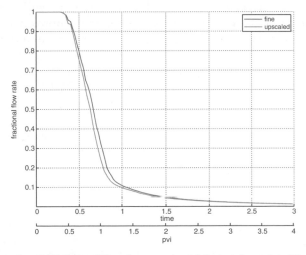

Fig. 28 Comparison of fractional flow for coarse- and fine-scale models. Variogram based permeability field is used. $\mu_o/\mu_w = 5$

Table 10 Convergence of the upscaling method for two-phase flow for variogram based permeability field

With $\tilde{S}$	50×50	100×100	200×200
L_2 pressure error at $t = \frac{3T_{final}}{4}$	0.0014	0.007	0.004
L_2 velocity error at $t = \frac{3T_{final}}{4}$	0.0235	0.0137	0.0072
L_1 saturation error $t = T_{final}$	0.0105	0.0052	0.0027

With $\bar{S}$	50×50	100×100	200×200
L_2 pressure error at $t = \frac{3T_{final}}{4}$	0.0046	0.0021	0.0008
L_2 velocity error at $t = \frac{3T_{final}}{4}$	0.0530	0.0335	0.0246
L_1 saturation error $t = T_{final}$	0.0546	0.0294	0.0134

6 Conclusions

In these lecture notes, we reviewed some of the recent advances in developing systematic multiscale methods with particular emphasis on multiscale finite element methods and their applications to fluid flows in heterogeneous porous media. In particular, the local approaches and their convergence properties for various flow problems are discussed. Moreover, improved subgrid capturing techniques through a judicious choice of local boundary conditions or through the use of limited global information are reviewed. Other topics, such as homogenization, the sampling techniques in numerical homogenization, and multiscale simulations of two-phase flows in heterogeneous porous media are also presented. Although the results presented in this paper are encouraging, there is scope for further exploration. These include the development and mathematical analysis of efficient numerical homogenization techniques for nonlinear convection–diffusion equations with various Peclet numbers (e.g., convection dominated), inexpensive approximations of multiscale basis functions, further exploration of accurate boundary conditions based on local multiscale solutions, the use of limited global information for nonlinear problems, development of adaptive criteria for multiscale basis functions (selection of coarse grid), applications of MsFEM to more multiphase/multi-component porous media flows, and etc.

Acknowledgements The authors gratefully acknowledge financial support from US DOE under grant DE-FG02-06ER25727. T. Hou would like also to acknowledge a partial support from NSF grant ITR Grant ACI-0204932.

References

1. J. Aarnes, *On the use of a mixed multiscale finite element method for greater flexibility and increased speed or improved accuracy in reservoir simulation*, SIAM MMS, 2 (2004), pp. 421–439.

2. J. Aarnes, Y. R. Efendiev, and L. Jiang, *Analysis of multiscale finite element methods using global information for two-phase flow simulations.* submitted.

3. J. Aarnes and T. Y. Hou *An Efficient Domain Decomposition Preconditioner for Multiscale Elliptic Problems with High Aspect Ratios,* Acta Mathematicae Applicatae Sinica, **18** (2002), 63-76.

4. R. A. ADAMS, *Sobolev spaces,* Academic Press [A subsidiary of Harcourt Brace Jovanovich, Publishers], New York-London, 1975. Pure and Applied Mathematics, Vol. 65.

5. T. Arbogast, *Implementation of a locally conservative numerical subgrid upscaling scheme for two-phase Darcy flow,* Comput. Geosci., 6 (2002), pp. 453–481. Locally conservative numerical methods for flow in porous media.

6. T. Arbogast and K. Boyd, *Subgrid upscaling and mixed multiscale finite elements.* to appear in SIAM Num. Anal.

7. M. AVELLANEDA AND F.-H. LIN, *Compactness methods in the theory of homogenization,* Comm. Pure Appl. Math., 40 (1987), pp. 803–847.

8. I. Babuska, U. Banerjee, and J. E. Osborn, *Survey of meshless and generalized finite element methods: A unified approach,* Acta Numerica, 2003, pp. 1-125.

9. I. Babuska, G. Caloz, and E. Osborn, *Special Finite Element Methods for a Class of Second Order Elliptic Problems with Rough Coefficients,* SIAM J. Numer. Anal., **31** (1994), 945-981.

10. I. Babuška and J. M. Melenk, *The partition of unity method,* Internat. J. Numer. Methods Engrg., 40 (1997), pp. 727–758.

11. I. Babuska and E. Osborn, *Generalized Finite Element Methods: Their Performance and Their Relation to Mixed Methods,* SIAM J. Numer. Anal., **20** (1983), 510-536.

12. I. Babuska and W. G. Szymczak, *An Error Analysis for the Finite Element Method Applied to Convection-Diffusion Problems,* Comput. Methods Appl. Math. Engrg, **31** (1982), 19-42.

13. A. Bensoussan, J. L. Lions, and G. Papanicolaou, *Asymptotic Analysis for Periodic Structures,* Volume 5 of Studies in Mathematics and Its Applications, North-Holland Publ., 1978.

14. A. Bourgeat, *Homogenized Behavior of Two-Phase Flows in Naturally Fractured Reservoirs with Uniform Fractures Distribution,* Comp. Meth. Appl. Mech. Engrg, **47** (1984), 205-216.

15. A. Bourgeat and A. Mikelić, *Homogenization of two-phase immiscible flows in a one-dimensional porous medium,* Asymptotic Anal., 9 (1994), pp. 359–380.

16. M. Brewster and G. Beylkin, *A Multiresolution Strategy for Numerical Homogenization,* ACHA, **2**(1995), 327-349.

17. F. Brezzi and M. Fortin, *Mixed and hybrid finite element methods,* Springer–Verlag, Berlin – Heidelberg – New-York, 1991.

18. F. Brezzi, L. P. Franca, T. J. R. Hughes and A. Russo, $b = \int g$, Comput. Methods in Appl. Mech. and Engrg., **145** (1997), 329-339.

19. Z. Chen and T. Y. Hou, *A mixed multiscale finite element method for elliptic problems with oscillating coefficients,* Math. Comp., 72 (2002), pp. 541–576 (electronic).

20. M. Christie and M. Blunt, *Tenth SPE comparative solution project: A comparison of upscaling techniques,* SPE Reser. Eval. Eng., 4 (2001), pp. 308–317.

21. M. E. Cruz and A. Petera, *A Parallel Monte-Carlo Finite Element Procedure for the Analysis of Multicomponent Random Media,* Int. J. Numer. Methods Engrg, **38** (1995), 1087-1121.

22. J. E. Dendy, J. M. Hyman, and J. D. Moulton, *The Black Box Multigrid Numerical Homogenization Algorithm,* J. Comput. Phys., **142** (1998), 80-108.

23. C. V. DEUTSCH AND A. G. JOURNEL, *GSLIB: Geostatistical software library and user's guide, 2nd edition,* Oxford University Press, New York, 1998.

24. M. Dorobantu and B. Engquist, *Wavelet-based Numerical Homogenization,* SIAM J. Numer. Anal., **35** (1998), 540-559.

25. J. Douglas, Jr. and T.F. Russell, *Numerical Methods for Convection-dominated Diffu-sion Problem Based on Combining the Method of Characteristics with Finite Element or Finite Difference Procedures,* SIAM J. Numer. Anal. **19** (1982), 871–885.

26. L. J. Durlofsky, *Numerical Calculation of Equivalent Grid Block Permeability Tensors for Heterogeneous Porous Media,* Water Resour. Res., **27** (1991), 699-708.

27. L.J. Durlofsky, R.C. Jones, and W.J. Milliken, *A Nonuniform Coarsening Approach for the Scale-up of Displacement Processes in Heterogeneous Porous Media,* Adv. Water Resources, **20** (1997), 335–347.

28. B. B. Dykaar and P. K. Kitanidis, *Determination of the Effective Hydraulic Con-ductivity for Heterogeneous Porous Media Using a Numerical Spectral Approach: 1. Method,* Water Resour. Res., **28** (1992), 1155-1166.

29. W. E, *Homogenization of linear and nonlinear transport equations,* Comm. Pure Appl. Math., XLV (1992), pp. 301–326.

30. W. E and B. Engquist, *The heterogeneous multi-scale methods,* Comm. Math. Sci., 1(1) (2003), pp. 87–133.

31. Y. R. Efendiev, *Multiscale Finite Element Method (MsFEM) and its Applications,* Ph. D. Thesis, Applied Mathematics, Caltech, 1999.

32. Y. Efendiev, V. Ginting, T. Y. Hou, and R. Ewing, *Accurate multiscale finite element methods of two-phase flow simulations.* J. Comput. Phys., **220** (2006), 155-174.

33. Y. Efendiev, T. Hou, and V. Ginting, *Multiscale finite element methods for nonlinear problems and their applications,* Comm. Math. Sci., 2 (2004), pp. 553–589.

34. Y. Efendiev, T. Hou, and T. Strinopoulos, *Multiscale simulations of porous media flows in flow-based coordinate system,* to appear in Comp. Geosciences.

35. Y. R. Efendiev, T. Y. Hou, and X. H. Wu, *Convergence of A Nonconforming Multiscale Finite Element Method,* SIAM J. Numer. Anal., **37** (2000), 888-910.

36. Y. Efendiev and A. Pankov, *Homogenization of nonlinear random parabolic operators,* Advances in Differential Equations, vol. 10,Number 11, 2005, pp., 1235-1260

37. Y. Efendiev and A. Pankov, *Numerical homogenization of nonlinear random parabolic operators,* SIAM Multiscale Modeling and Simulation, 2(2) (2004), pp. 237–268.

38. Y. R. Efendiev and L. J. Durlofsky, *Numerical modeling of subgrid heterogeneity in two phase flow simulations,* Water Resour. Res., 38(8) (2002), p. 1128.

39. Y. R. Efendiev, L. J. Durlofsky, S. H. Lee, *Modeling of Subgrid Effects in Coarse-scale Simulations of Transport in Heterogeneous Porous Media,* WATER RESOUR RES, **36** (2000), 2031-2041.

40. Y. R. Efendiev and B. Popov, *On homogenization of nonlinear hyperbolic equations,* Communications on Pure and Applied Analysis, 4(2) (2005), pp. 295–309.

41. R. Eymard, T. Gallouët, and R. Herbin, *Finite volume methods,* in Handbook of numerical analysis, Vol. VII, Handb. Numer. Anal., VII, North-Holland, Amsterdam, 2000, 713–1020.

42. J. Fish and K.L. Shek, *Multiscale Analysis for Composite Materials and Structures,* Composites Science and Technology: An International Journal, **60** (2000), 2547-2556.

43. J. Fish and Z. Yuan, *Multiscale enrichment based on the partition of unity,* Interna-tional Journal for Numerical Methods in Engineering, **62**, (2005), 1341–1359.

44. D. Gilbarg and N. S. Trudinger, Elliptic Partial Differential Equations of Second Order. Springer, Berlin, New York, 2001.

45. J. Glimm, H. Kim, D. Sharp, and T. Wallstrom *A Stochastic Analysis of the Scale Up Problem for Flow in Porous Media,* Comput. Appl. Math., **17** (1998), 67-79.

46. T. Hou, X. Wu, and Y. Zhang, *Removing the cell resonance error in the multiscale finite element method via a petrov-galerkin formulation,* Communications in Mathe-matical Sciences, **2**(2) (2004), 185–205.

47. T. Y. Hou, A. Westhead, and D. P. Yang, *A framework for modeling subgrid effects for two-phase flows in porous media.* to appear in SIAM Multiscale Modeling and Simulation.

48. T. Y. Hou and X. H. Wu, *A Multiscale Finite Element Method for Elliptic Problems in Composite Materials and Porous Media,* J. Comput. Phys., **134** (1997), 169-189.

49. T. Y. Hou and X. H. Wu, *A Multiscale Finite Element Method for PDEs with Oscillatory Coefficients,* Proceedings of 13th GAMM-Seminar Kiel on Numerical Treatment of Multi-Scale Problems, Jan 24-26, 1997, Notes on Numerical Fluid Mechanics, Vol. 70, ed. by W. Hackbusch and G. Wittum, Vieweg-Verlag, 58-69, 1999.

50. T. Y. Hou, X. H. Wu, and Z. Cai, *Convergence of a Multiscale Finite Element Method for Elliptic Problems With Rapidly Oscillating Coefficients,* Math. Comput., **68** (1999), 913-943.

51. T. Y. Hou, D.-P. Yang, and K. Wang, *Homogenization of Incompressible Euler Equation.* J. Comput. Math., **22** (2004), 220-229.

52. T. Y. Hou, D. P. Yang, and H. Ran, *Multiscale Analysis in the Lagrangian Formulation for the 2-D Incompressible Euler Equation,* Discrete and Continuous Dynamical Systems, **13** (2005), 1153-1186.

53. T. Y. Hou, D.-P. Yang, and H. Ran, *Multiscale analysis and computation for the 3-D incompressible Navier-Stokes equations,* Multiscale Modeling and Simulation, **6(4)** (2008), 1317-1346.

54. T. Y. Hou and X. Xin, *Homogenization of linear transport equations with oscillatory vector fields,* SIAM J. Appl. Math., 52 (1992), pp. 34–45.

55. Y. Huang and J. Xu, *A partition-of-unity finite element method for elliptic problems with highly oscillating coefficients,* preprint.

56. T. J. R. Hughes, *Multiscale Phenomena: Green's Functions, the Dirichlet-to-Neumann Formulation, Subgrid Scale Models, Bubbles and the Origins of Stabilized Methods,* Comput. Methods Appl. Mech Engrg., **127** (1995), 387-401.

57. T. J. R. Hughes, G. R. Feijóo, L. Mazzei, J.-B. Quincy, *The Variational Multiscale Method – A Paradigm for Computational Mechanics,* Comput. Methods Appl. Mech Engrg., **166**(1998), 3-24.

58. M. GERRITSEN AND L. J. DURLOFSKY, *Modeling of fluid flow in oil reservoirs,* Annual Reviews in Fluid Mechanics, 37 (2005), pp. 211–238.

59. P. Jenny, S. H. Lee, and H. Tchelepi, *Adaptive multi-scale finite volume method for multi-phase flow and transport in porous media,* Multiscale Modeling and Simulation, 3 (2005), pp. 30–64.

60. V. V. Jikov, S. M. Kozlov, and O. A. Oleinik, *Homogenization of Differential Operators and Integral Functionals,* Springer-Verlag, 1994, Translated from Russian.

61. Ioannis G. Kevrekidis, C. William Gear, James M. Hyman, Panagiotis G. Kevrekidis, Olof Runborg, and Constantinos Theodoropoulos, *Equation-free, coarse-grained multiscale computation: enabling microscopic simulators to perform system-level analysis,* Commun. Math. Sci. **1** (2003), no. 4, 715–762.

62. S. Knapek, *Matrix-Dependent Multigrid-Homogenization for Diffusion Problems,* in the Proceedings of the Copper Mountain Conference on Iterative Methods, edited by T. Manteuffal and S. McCormick, volume I, SIAM Special Interest Group on Linear Algebra, Cray Research , 1996.

63. P. Langlo and M.S. Espedal, *Macrodispersion for Two-phase, Immiscible Flow in Porous Media,* Adv. Water Resources **17** (1994), 297–316.

64. A. M. Matache, I. Babuska, and C. Schwab, *Generalized p-FEM in Homogenization,* Numer. Math. **86** (2000), 319-375.

65. A. M. Matache and C. Schwab, *Homogenization via p-FEM for Problems with Microstructure,* Appl. Numer. Math. **33** (2000), 43-59.

66. J. F. McCarthy, *Comparison of Fast Algorithms for Estimating Large-Scale Permeabilities of Heterogeneous Media,* Transport in Porous Media, **19** (1995), 123-137.

67. D. W. McLaughlin, G. C. Papanicolaou, and O. Pironneau, *Convection of Microstructure and Related Problems,* SIAM J. Applied Math, **45** (1985), 780-797.

68. S. Moskow and M. Vogelius, *First Order Corrections to the Homogenized Eigenvalues of a Periodic Composite Medium: A Convergence Proof,* Proc. Roy. Soc. Edinburgh, A, **127** (1997), 1263-1299.

69. ———, *Metric based up-scaling,* Comm. Pure and Applied Math., **60** (2007), 675-723.

70. A. Pankov, *G-convergence and homogenization of nonlinear partial differential operators*, Kluwer Academic Publishers, Dordrecht, 1997.
71. W. V. Petryshyn, *On the approximation-solvability of equations involving A-proper and pseudo-A-proper mappings*, Bull. Amer. Math. Soc., 81 (1975), pp. 223–312.
72. O. Pironneau, *On the Transport-diffusion Algorithm and its Application to the Navier-Stokes Equations*, Numer. Math. **38** (1982), 309–332.
73. R.E. Rudd and J.Q. Broughton, *Coarse-grained molecular dynamics and the atomic limit of finite elements* , Phys. Rev. B 58, R5893 (1998).
74. G. Sangalli, *Capturing Small Scales in Elliptic Problems Using a Residual-Free Bubbles Finite Element Method*, Multiscale Modeling and Simulation, 1 (2003), no. 3, 485–503
75. I. V. Skrypnik, *Methods for analysis of nonlinear elliptic boundary value problems*, vol. 139 of Translations of Mathematical Monographs, American Mathematical Society, Providence, RI, 1994. Translated from the 1990 Russian original by Dan D. Pascali.
76. T. Strinopoulos, *Upscaling of immiscible two-phase flows in an adaptive frame*, PhD thesis, California Institute of Technology, Pasadena, 2005.
77. T. Strouboulis, I. Babuška, and K. Copps, *The design and analysis of the generalized finite element method*, Comput. Methods Appl. Mech. Engrg., 181 (2000), pp. 43–69.
78. L. Tartar, *Nonlocal Effects Induced by Homogenization*, in PDE and Calculus of Variations, ed by F. Culumbini, et al, Birkhäuser, Boston, 925-938, 1989.
79. X.H. Wu, Y. Efendiev, and T. Y. Hou, *Analysis of Upscaling Absolute Permeability*, Discrete and Continuous Dynamical Systems, Series B, **2** (2002), 185-204.
80. P. M. De Zeeuw, *Matrix-dependent Prolongation and Restrictions in a Blackbox Multigrid Solver*, J. Comput. Applied Math, **33**(1990), 1-27.
81. S. Verdiere and M.H. Vignal, *Numerical and Theoretical Study of a Dual Mesh Method Using Finite Volume Schemes for Two-phase Flow Problems in Porous Media*, Numer. Math. **80** (1998), 601–639.
82. T. C. Wallstrom, M. A. Christie, L. J. Durlofsky, and D. H. Sharp, *Effective Flux Boundary Conditions for Upscaling Porous Media Equations*, Transport in Porous Media, **46** (2002), 139-153.
83. T. C. Wallstrom, M. A. Christie, L. J. Durlofsky, and D. H. Sharp, *Application of Effective Flux Boundary Conditions to Two-phase Upscaling in Porous Media*, Transport in Porous Media, **46** (2002), 155-178.
84. T. C. Wallstrom, S. L. Hou, M. A. Christie, L. J. Durlofsky, and D. H. Sharp, *Accurate Scale Up of Two Phase Flow Using Renormalization and Nonuniform Coarsening*, Comput. Geosci, **3** (1999), 69-87.
85. E. Zeidler, *Nonlinear functional analysis and its applications. II/B*, Springer-Verlag, New York, 1990. Nonlinear monotone operators, Translated from the German by the author and Leo F. Boron.

List of Participants

1. Alessandro Teta
   Università degli Studi de
   L'Aquila, L'Aquila
   Italy
   teta@univaq.it

2. Ali Faraj
   Laboratoire Mathématiques pour
   l'Industrie et la Physique
   Université Paul Sabatier
   Toulouse 3, Toulouse
   France
   faraj@mip.ups-tlse.fr

3. Anton Arnold
   Institut für Analysis und
   Scientific Computing
   Technische Universität
   Vienna, Austria
   anton.arnold@tuwien.ac.at

4. Antonio Greco
   Università di Palermo
   Palermo, Italy
   greco@math.unipa.it

5. Chiara Manzini
   Dipartimento di Matematica
   Applicata
   Università di Firenze
   Florence, Italy
   chiara.manzini@unifi.it

6. Claudia Negulescu
   Laboratoire CMI/LATP
   Université de Provence
   Marseille Cedex 13, France
   *claudia.negulesc@cmi.
   univ-mrs.fr*

7. David Sanchez
   Département de Mathématiques
   INSA de Toulouse, Toulouse
   France
   david.sanchez@insa-toulouse.fr

8. Domenico Finco
   Institut fur Angewandte
   Mathematik, Bonn, Germany
   finco@wiener.iam.uni-bonn.de

9. Elise Fouassier
   Ecole Normale Superieure de
   Lyon, UMPA, Lyon, France
   elise.fouassier@umpa.ens-lyon.fr

10. Evangelia Kalligiannaki
    Department of Mathematics
    University of Crete, Heraklion
    Greece
    euagelia@math.uoc.gr

11. Fanny Fendt
    IRMAR, Rennes, France
    fanny_fendt@yahoo.fr

12. Federica Pezzotti
Dipartimento di Matematica
Pura ed Applicata
Università de L'Aquila, L'Aquila
Italy
federica.pezzotti@univaq.it

13. Florian Mehats
Université de Rennes 1, Rennes
France
florian.mehats@univ-rennes1.fr

14. Francesco Vecil
Universitat Autonoma de
Barcelona, Barcelona
Spain
fvecil@mat.uab.es

15. Gianluca Panati
University of Rome
"La Sapienza", Rome, Italy
panati@mat.uniroma1.it

16. Gilberto Mongatti
Università di Firenze, Florence
Italy
mongatti@math.unifi.it

17. Giovanni Borgioli
Università di Firenze, Florence
Italy
giovanni.borgioli@unifi.it

18. Giovanni Frosali (editor)
Dipartimento di Matematica
Applicata,
Università di Firenze
Florence, Italy
giovanni.frosali@unifi.it

19. Giovanni Mascali
Università della Calabria
Cosenza, Italy
g.mascali@unical.it

20. Giuseppe Alí
CNR, IAC Napoli, Naples, Italy
g.ali@iac.cnr.it

21. Grégoire Allaire (lecturer)
Ecole Polytechnique, Palaiseau
France
gregoire.allaire@polytechnique.fr

22. Luigi Barletti
Università di Firenze
Florence, Italy
barletti@math.unifi.it

23. Maike Schulte
Westfälische Wilhelms -
Universität Münster, Münster
Germany
maikemg@math.uni-muenster.de

24. Marco Sammartino
University of Palermo, Palermo
Italy
marco@math.unipa.it

25. Naoufel Ben Abdallah (editor)
Laboratoire Mathématiques pour
l'Industrie et la Physique
Université Paul Sabatier
Toulouse 3, Toulouse, France
naoufel@mip.ups-tlse.fr

26. Omar Maj
Università di Pavia, Pavia, Italy
maj@fisicavolta.unipv.it

27. Omar Morandi
Dipartimento di Scienze
Matematiche,
Università Politecnica delle
Marche, Ancona, Italy
morandi@dipmat.univpm.it

28. Paolo Antonelli
Università degli Studi de
L'Aquila, L'Aquila, Italy
paolo.antonelli@univaq.it

29. Pierre Degond (lecturer)
IRMAR, Rennes, France
degond@mip.ups-tlse.fr

30. Josipa Pina Milišić
    Johannes Gutenberg University
    Mainz, Germany
    milisic@mathematik.uni-mainz.de

31. Riccardo Adami
    Dipartimento di Matematica e
    Applicazioni
    Università di Milano Bicocca
    Milan, Italy
    riccardo.adami@unimib.it

32. Roberto Beneduci
    Università della Calabria
    Cosenza, Italy
    rbeneduci@unical.it

33. Romina Gobbi
    Università degli Studi Roma Tre
    Rome, Italy
    gobbi@mat.uniroma3.it

34. Samy Gallego
    Laboratoire Mathématiques pour
    l'Industrie et la Physique
    Université Paul Sabatier
    Toulouse 3, Toulouse
    France
    gallego@mip.ups-tlse.fr

35. Silvia Palpacelli
    Università degli Studi Roma Tre
    Rome, Italy
    palpacel@mat.uniroma3.it

36. Tatiana Ryabukha
    Institute of mathematics
    NAS of Ukraine, Kiev, Ukraine
    vyrtum@imath.kiev.ua

37. Thomas Y. Hou (lecturer)
    Caltech, Pasadena, CA, USA
    hou@aem.caltech.edu

38. Valeria Ricci
    University of Palermo, Palermo
    Italy
    vricci@mat.uniroma1.it

39. Viacheslav Shtyk
    Institute of Mathematics
    NAS of Ukraine, Kiev
    Ukraine
    shtyk@imath.kiev.ua

40. Virginie Bonnaillie-Noël
    ENS Cachan
    antenne de Bretagne
    IRMAR, Rennes, France
    *virginie.bonnaillie@bretagne.
    ens-cachan.fr*

41. Vittorio Romano
    Università di Catania, Catania
    Italy
    romano@dmi.unict.it

42. Zhanna Artemichenko
    Institute of Mathematics
    NAS of Ukraine, Kiev, Ukraine
    artemich@imath.kiev.ua

LIST OF C.I.M.E. SEMINARS

Published by C.I.M.E

Published by Ed. Cremonese, Firenze

1966   39. Calculus of variations
        40. Economia matematica
        41. Classi caratteristiche e questioni connesse
        42. Some aspects of diffusion theory

1967   43. Modern questions of celestial mechanics
        44. Numerical analysis of partial differential equations
        45. Geometry of homogeneous bounded domains

1968   46. Controllability and observability
        47. Pseudo-differential operators
        48. Aspects of mathematical logic

1969   49. Potential theory
        50. Non-linear continuum theories in mechanics and physics and their applications
        51. Questions of algebraic varieties

1970   52. Relativistic fluid dynamics
        53. Theory of group representations and Fourier analysis
        54. Functional equations and inequalities
        55. Problems in non-linear analysis

1971   56. Stereodynamics
        57. Constructive aspects of functional analysis (2 vol.)
        58. Categories and commutative algebra

1972   59. Non-linear mechanics
        60. Finite geometric structures and their applications
        61. Geometric measure theory and minimal surfaces

1973   62. Complex analysis
        63. New variational techniques in mathematical physics
        64. Spectral analysis

1974   65. Stability problems
        66. Singularities of analytic spaces
        67. Eigenvalues of non linear problems

1975   68. Theoretical computer sciences
        69. Model theory and applications
        70. Differential operators and manifolds

Published by Ed. Liguori, Napoli

1976   71. Statistical Mechanics
        72. Hyperbolicity
        73. Differential topology

1977   74. Materials with memory
        75. Pseudodifferential operators with applications
        76. Algebraic surfaces

Published by Ed. Liguori, Napoli & Birkhäuser

1978   77. Stochastic differential equations
        78. Dynamical systems

1979   79. Recursion theory and computational complexity
        80. Mathematics of biology

1980   81. Wave propagation
        82. Harmonic analysis and group representations
        83. Matroid theory and its applications

Published by Springer-Verlag

Lecture Notes in Mathematics

For information about earlier volumes
please contact your bookseller or Springer
LNM Online archive: springerlink.com

Vol. 1809: O. Steinbach, Stability Estimates for Hybrid Coupled Domain Decomposition Methods (2003)

Vol. 1810: J. Wengenroth, Derived Functors in Functional Analysis (2003)

Vol. 1811: J. Stevens, Deformations of Singularities (2003)

Vol. 1812: L. Ambrosio, K. Deckelnick, G. Dziuk, M. Mimura, V. A. Solonnikov, H. M. Soner, Mathematical Aspects of Evolving Interfaces. Madeira, Funchal, Portugal 2000. Editors: P. Colli, J. F. Rodrigues (2003)

Vol. 1813: L. Ambrosio, L. A. Caffarelli, Y. Brenier, G. Buttazzo, C. Villani, Optimal Transportation and its Applications. Martina Franca, Italy 2001. Editors: L. A. Caffarelli, S. Salsa (2003)

Vol. 1814: P. Bank, F. Baudoin, H. Föllmer, L.C.G. Rogers, M. Soner, N. Touzi, Paris-Princeton Lectures on Mathematical Finance 2002 (2003)

Vol. 1815: A. M. Vershik (Ed.), Asymptotic Combinatorics with Applications to Mathematical Physics. St. Petersburg, Russia 2001 (2003)

Vol. 1816: S. Albeverio, W. Schachermayer, M. Talagrand, Lectures on Probability Theory and Statistics. Ecole d'Eté de Probabilités de Saint-Flour XXX-2000. Editor: P. Bernard (2003)

Vol. 1817: E. Koelink, W. Van Assche (Eds.), Orthogonal Polynomials and Special Functions. Leuven 2002 (2003)

Vol. 1818: M. Bildhauer, Convex Variational Problems with Linear, nearly Linear and/or Anisotropic Growth Conditions (2003)

Vol. 1819: D. Masser, Yu. V. Nesterenko, H. P. Schlickewei, W. M. Schmidt, M. Waldschmidt, Diophantine Approximation. Cetraro, Italy 2000. Editors: F. Amoroso, U. Zannier (2003)

Vol. 1820: F. Hiai, H. Kosaki, Means of Hilbert Space Operators (2003)

Vol. 1821: S. Teufel, Adiabatic Perturbation Theory in Quantum Dynamics (2003)

Vol. 1822: S.-N. Chow, R. Conti, R. Johnson, J. Mallet-Paret, R. Nussbaum, Dynamical Systems. Cetraro, Italy 2000. Editors: J. W. Macki, P. Zecca (2003)

Vol. 1823: A. M. Anile, W. Allegretto, C. Ringhofer, Mathematical Problems in Semiconductor Physics. Cetraro, Italy 1998. Editor: A. M. Anile (2003)

Vol. 1824: J. A. Navarro González, J. B. Sancho de Salas, $\mathscr{C}^\infty$ – Differentiable Spaces (2003)

Vol. 1825: J. H. Bramble, A. Cohen, W. Dahmen, Multiscale Problems and Methods in Numerical Simulations, Martina Franca, Italy 2001. Editor: C. Canuto (2003)

Vol. 1826: K. Dohmen, Improved Bonferroni Inequalities via Abstract Tubes. Inequalities and Identities of Inclusion-Exclusion Type. VIII, 113 p, 2003.

Vol. 1827: K. M. Pilgrim, Combinations of Complex Dynamical Systems. IX, 118 p, 2003.

Vol. 1828: D. J. Green, Gröbner Bases and the Computation of Group Cohomology. XII, 138 p, 2003.

Vol. 1829: E. Altman, B. Gaujal, A. Hordijk, Discrete-Event Control of Stochastic Networks: Multimodularity and Regularity. XIV, 313 p, 2003.

Vol. 1830: M. I. Gil', Operator Functions and Localization of Spectra. XIV, 256 p, 2003.

Vol. 1831: A. Connes, J. Cuntz, E. Guentner, N. Higson, J. E. Kaminker, Noncommutative Geometry, Martina Franca, Italy 2002. Editors: S. Doplicher, L. Longo (2004)

Vol. 1832: J. Azéma, M. Émery, M. Ledoux, M. Yor (Eds.), Séminaire de Probabilités XXXVII (2003)

Vol. 1833: D.-Q. Jiang, M. Qian, M.-P. Qian, Mathematical Theory of Nonequilibrium Steady States. On the Frontier of Probability and Dynamical Systems. IX, 280 p, 2004.

Vol. 1834: Yo. Yomdin, G. Comte, Tame Geometry with Application in Smooth Analysis. VIII, 186 p, 2004.

Vol. 1835: O.T. Izhboldin, B. Kahn, N.A. Karpenko, A. Vishik, Geometric Methods in the Algebraic Theory of Quadratic Forms. Summer School, Lens, 2000. Editor: J.-P. Tignol (2004)

Vol. 1836: C. Năstăsescu, F. Van Oystaeyen, Methods of Graded Rings. XIII, 304 p, 2004.

Vol. 1837: S. Tavaré, O. Zeitouni, Lectures on Probability Theory and Statistics. Ecole d'Eté de Probabilités de Saint-Flour XXXI-2001. Editor: J. Picard (2004)

Vol. 1838: A.J. Ganesh, N.W. O'Connell, D.J. Wischik, Big Queues. XII, 254 p, 2004.

Vol. 1839: R. Gohm, Noncommutative Stationary Processes. VIII, 170 p, 2004.

Vol. 1840: B. Tsirelson, W. Werner, Lectures on Probability Theory and Statistics. Ecole d'Eté de Probabilités de Saint-Flour XXXII-2002. Editor: J. Picard (2004)

Vol. 1841: W. Reichel, Uniqueness Theorems for Variational Problems by the Method of Transformation Groups (2004)

Vol. 1842: T. Johnsen, A. L. Knutsen, K₃ Projective Models in Scrolls (2004)

Vol. 1843: B. Jefferies, Spectral Properties of Noncommuting Operators (2004)

Vol. 1844: K.F. Siburg, The Principle of Least Action in Geometry and Dynamics (2004)

Vol. 1845: Min Ho Lee, Mixed Automorphic Forms, Torus Bundles, and Jacobi Forms (2004)

Vol. 1846: H. Ammari, H. Kang, Reconstruction of Small Inhomogeneities from Boundary Measurements (2004)

Vol. 1847: T.R. Bielecki, T. Björk, M. Jeanblanc, M. Rutkowski, J.A. Scheinkman, W. Xiong, Paris-Princeton Lectures on Mathematical Finance 2003 (2004)

Vol. 1848: M. Abate, J. E. Fornaess, X. Huang, J. P. Rosay, A. Tumanov, Real Methods in Complex and CR Geometry, Martina Franca, Italy 2002. Editors: D. Zaitsev, G. Zampieri (2004)

Vol. 1849: Martin L. Brown, Heegner Modules and Elliptic Curves (2004)

Vol. 1850: V. D. Milman, G. Schechtman (Eds.), Geometric Aspects of Functional Analysis. Israel Seminar 2002-2003 (2004)

Vol. 1851: O. Catoni, Statistical Learning Theory and Stochastic Optimization (2004)

Vol. 1852: A.S. Kechris, B.D. Miller, Topics in Orbit Equivalence (2004)

Vol. 1853: Ch. Favre, M. Jonsson, The Valuative Tree (2004)

Vol. 1854: O. Saeki, Topology of Singular Fibers of Differential Maps (2004)

Vol. 1855: G. Da Prato, P.C. Kunstmann, I. Lasiecka, A. Lunardi, R. Schnaubelt, L. Weis, Functional Analytic Methods for Evolution Equations. Editors: M. Iannelli, R. Nagel, S. Piazzera (2004)

Vol. 1856: K. Back, T.R. Bielecki, C. Hipp, S. Peng, W. Schachermayer, Stochastic Methods in Finance, Bressanone/Brixen, Italy, 2003. Editors: M. Fritelli, W. Runggaldier (2004)

Vol. 1857: M. Émery, M. Ledoux, M. Yor (Eds.), Séminaire de Probabilités XXXVIII (2005)

Vol. 1858: A.S. Cherny, H.-J. Engelbert, Singular Stochastic Differential Equations (2005)

Vol. 1859: E. Letellier, Fourier Transforms of Invariant Functions on Finite Reductive Lie Algebras (2005)

Vol. 1860: A. Borisyuk, G.B. Ermentrout, A. Friedman, D. Terman, Tutorials in Mathematical Biosciences I. Mathematical Neurosciences (2005)

Vol. 1861: G. Benettin, J. Henrard, S. Kuksin, Hamiltonian Dynamics – Theory and Applications, Cetraro, Italy, 1999. Editor: A. Giorgilli (2005)

Vol. 1862: B. Helffer, F. Nier, Hypoelliptic Estimates and Spectral Theory for Fokker-Planck Operators and Witten Laplacians (2005)

Vol. 1863: H. Führ, Abstract Harmonic Analysis of Continuous Wavelet Transforms (2005)

Vol. 1864: K. Efstathiou, Metamorphoses of Hamiltonian Systems with Symmetries (2005)

Vol. 1865: D. Applebaum, B.V. R. Bhat, J. Kustermans, J. M. Lindsay, Quantum Independent Increment Processes I. From Classical Probability to Quantum Stochastic Calculus. Editors: M. Schürmann, U. Franz (2005)

Vol. 1866: O.E. Barndorff-Nielsen, U. Franz, R. Gohm, B. Kümmerer, S. Thorbjønsen, Quantum Independent Increment Processes II. Structure of Quantum Lévy Processes, Classical Probability, and Physics. Editors: M. Schürmann, U. Franz, (2005)

Vol. 1867: J. Sneyd (Ed.), Tutorials in Mathematical Biosciences II. Mathematical Modeling of Calcium Dynamics and Signal Transduction. (2005)

Vol. 1868: J. Jorgenson, S. Lang, $Pos_n(R)$ and Eisenstein Series. (2005)

Vol. 1869: A. Dembo, T. Funaki, Lectures on Probability Theory and Statistics. Ecole d'Eté de Probabilités de Saint-Flour XXXIII-2003. Editor: J. Picard (2005)

Vol. 1870: V.I. Gurariy, W. Lusky, Geometry of Müntz Spaces and Related Questions. (2005)

Vol. 1871: P. Constantin, G. Gallavotti, A.V. Kazhikhov, Y. Meyer, S. Ukai, Mathematical Foundation of Turbulent Viscous Flows, Martina Franca, Italy, 2003. Editors: M. Cannone, T. Miyakawa (2006)

Vol. 1872: A. Friedman (Ed.), Tutorials in Mathematical Biosciences III. Cell Cycle, Proliferation, and Cancer (2006)

Vol. 1873: R. Mansuy, M. Yor, Random Times and Enlargements of Filtrations in a Brownian Setting (2006)

Vol. 1874: M. Yor, M. Émery (Eds.), In Memoriam Paul-André Meyer - Séminaire de Probabilités XXXIX (2006)

Vol. 1875: J. Pitman, Combinatorial Stochastic Processes. Ecole d'Eté de Probabilités de Saint-Flour XXXII-2002. Editor: J. Picard (2006)

Vol. 1876: H. Herrlich, Axiom of Choice (2006)

Vol. 1877: J. Steuding, Value Distributions of L-Functions (2007)

Vol. 1878: R. Cerf, The Wulff Crystal in Ising and Percolation Models, Ecole d'Eté de Probabilités de Saint-Flour XXXIV-2004. Editor: Jean Picard (2006)

Vol. 1879: G. Slade, The Lace Expansion and its Applications, Ecole d'Eté de Probabilités de Saint-Flour XXXIV-2004. Editor: Jean Picard (2006)

Vol. 1880: S. Attal, A. Joye, C.-A. Pillet, Open Quantum Systems I, The Hamiltonian Approach (2006)

Vol. 1881: S. Attal, A. Joye, C.-A. Pillet, Open Quantum Systems II, The Markovian Approach (2006)

Vol. 1882: S. Attal, A. Joye, C.-A. Pillet, Open Quantum Systems III, Recent Developments (2006)

Vol. 1883: W. Van Assche, F. Marcellàn (Eds.), Orthogonal Polynomials and Special Functions, Computation and Application (2006)

Vol. 1884: N. Hayashi, E.I. Kaikina, P.I. Naumkin, I.A. Shishmarev, Asymptotics for Dissipative Nonlinear Equations (2006)

Vol. 1885: A. Telcs, The Art of Random Walks (2006)

Vol. 1886: S. Takamura, Splitting Deformations of Degenerations of Complex Curves (2006)

Vol. 1887: K. Habermann, L. Habermann, Introduction to Symplectic Dirac Operators (2006)

Vol. 1888: J. van der Hoeven, Transseries and Real Differential Algebra (2006)

Vol. 1889: G. Osipenko, Dynamical Systems, Graphs, and Algorithms (2006)

Vol. 1890: M. Bunge, J. Funk, Singular Coverings of Toposes (2006)

Vol. 1891: J.B. Friedlander, D.R. Heath-Brown, H. Iwaniec, J. Kaczorowski, Analytic Number Theory, Cetraro, Italy, 2002. Editors: A. Perelli, C. Viola (2006)

Vol. 1892: A. Baddeley, I. Bárány, R. Schneider, W. Weil, Stochastic Geometry, Martina Franca, Italy, 2004. Editor: W. Weil (2007)

Vol. 1893: H. Hanßmann, Local and Semi-Local Bifurcations in Hamiltonian Dynamical Systems, Results and Examples (2007)

Vol. 1894: C.W. Groetsch, Stable Approximate Evaluation of Unbounded Operators (2007)

Vol. 1895: L. Molnár, Selected Preserver Problems on Algebraic Structures of Linear Operators and on Function Spaces (2007)

Vol. 1896: P. Massart, Concentration Inequalities and Model Selection, Ecole d'Été de Probabilités de Saint-Flour XXXIII-2003. Editor: J. Picard (2007)

Vol. 1897: R. Doney, Fluctuation Theory for Lévy Processes, Ecole d'Été de Probabilités de Saint-Flour XXXV-2005. Editor: J. Picard (2007)

Vol. 1898: H.R. Beyer, Beyond Partial Differential Equations, On linear and Quasi-Linear Abstract Hyperbolic Evolution Equations (2007)

Vol. 1899: Séminaire de Probabilités XL. Editors: C. Donati-Martin, M. Émery, A. Rouault, C. Stricker (2007)

Vol. 1900: E. Bolthausen, A. Bovier (Eds.), Spin Glasses (2007)

Vol. 1901: O. Wittenberg, Intersections de deux quadriques et pinceaux de courbes de genre 1, Intersections of Two Quadrics and Pencils of Curves of Genus 1 (2007)

Vol. 1902: A. Isaev, Lectures on the Automorphism Groups of Kobayashi-Hyperbolic Manifolds (2007)

Vol. 1903: G. Kresin, V. Maz'ya, Sharp Real-Part Theorems (2007)

Vol. 1904: P. Giesl, Construction of Global Lyapunov Functions Using Radial Basis Functions (2007)

Vol. 1905: C. Prévôt, M. Röckner, A Concise Course on Stochastic Partial Differential Equations (2007)

Vol. 1906: T. Schuster, The Method of Approximate Inverse: Theory and Applications (2007)

Vol. 1907: M. Rasmussen, Attractivity and Bifurcation for Nonautonomous Dynamical Systems (2007)

Vol. 1908: T.J. Lyons, M. Caruana, T. Lévy, Differential Equations Driven by Rough Paths, Ecole d'Été de Probabilités de Saint-Flour XXXIV-2004 (2007)

Vol. 1909: H. Akiyoshi, M. Sakuma, M. Wada, Y. Yamashita, Punctured Torus Groups and 2-Bridge Knot Groups (I) (2007)

Vol. 1910: V.D. Milman, G. Schechtman (Eds.), Geometric Aspects of Functional Analysis. Israel Seminar 2004-2005 (2007)

Vol. 1911: A. Bressan, D. Serre, M. Williams, K. Zumbrun, Hyperbolic Systems of Balance Laws. Cetraro, Italy 2003. Editor: P. Marcati (2007)

Vol. 1912: V. Berinde, Iterative Approximation of Fixed Points (2007)

Vol. 1913: J.E. Marsden, G. Misiołek, J.-P. Ortega, M. Perlmutter, T.S. Ratiu, Hamiltonian Reduction by Stages (2007)

Vol. 1914: G. Kutyniok, Affine Density in Wavelet Analysis (2007)

Vol. 1915: T. Bıyıkoğlu, J. Leydold, P.F. Stadler, Laplacian Eigenvectors of Graphs. Perron-Frobenius and Faber-Krahn Type Theorems (2007)

Vol. 1916: C. Villani, F. Rezakhanlou, Entropy Methods for the Boltzmann Equation. Editors: F. Golse, S. Olla (2008)

Vol. 1917: I. Veselić, Existence and Regularity Properties of the Integrated Density of States of Random Schrödinger (2008)

Vol. 1918: B. Roberts, R. Schmidt, Local Newforms for GSp(4) (2007)

Vol. 1919: R.A. Carmona, I. Ekeland, A. Kohatsu-Higa, J.-M. Lasry, P.-L. Lions, H. Pham, E. Taflin, Paris-Princeton Lectures on Mathematical Finance 2004. Editors: R.A. Carmona, E. Çinlar, I. Ekeland, E. Jouini, J.A. Scheinkman, N. Touzi (2007)

Vol. 1920: S.N. Evans, Probability and Real Trees. Ecole d'Été de Probabilités de Saint-Flour XXXV-2005 (2008)

Vol. 1921: J.P. Tian, Evolution Algebras and their Applications (2008)

Vol. 1922: A. Friedman (Ed.), Tutorials in Mathematical BioSciences IV. Evolution and Ecology (2008)

Vol. 1923: J.P.N. Bishwal, Parameter Estimation in Stochastic Differential Equations (2008)

Vol. 1924: M. Wilson, Littlewood-Paley Theory and Exponential-Square Integrability (2008)

Vol. 1925: M. du Sautoy, L. Woodward, Zeta Functions of Groups and Rings (2008)

Vol. 1926: L. Barreira, V. Claudia, Stability of Nonautonomous Differential Equations (2008)

Vol. 1927: L. Ambrosio, L. Caffarelli, M.G. Crandall, L.C. Evans, N. Fusco, Calculus of Variations and Non-Linear Partial Differential Equations. Cetraro, Italy 2005. Editors: B. Dacorogna, P. Marcellini (2008)

Vol. 1928: J. Jonsson, Simplicial Complexes of Graphs (2008)

Vol. 1929: Y. Mishura, Stochastic Calculus for Fractional Brownian Motion and Related Processes (2008)

Vol. 1930: J.M. Urbano, The Method of Intrinsic Scaling. A Systematic Approach to Regularity for Degenerate and Singular PDEs (2008)

Vol. 1931: M. Cowling, E. Frenkel, M. Kashiwara, A. Valette, D.A. Vogan, Jr., N.R. Wallach, Representation Theory and Complex Analysis. Venice, Italy 2004. Editors: E.C. Tarabusi, A. D'Agnolo, M. Picardello (2008)

Vol. 1932: A.A. Agrachev, A.S. Morse, E.D. Sontag, H.J. Sussmann, V.I. Utkin, Nonlinear and Optimal Control Theory. Cetraro, Italy 2004. Editors: P. Nistri, G. Stefani (2008)

Vol. 1933: M. Petković, Point Estimation of Root Finding Methods (2008)

Vol. 1934: C. Donati-Martin, M. Émery, A. Rouault, C. Stricker (Eds.), Séminaire de Probabilités XLI (2008)

Vol. 1935: A. Unterberger, Alternative Pseudodifferential Analysis (2008)

Vol. 1936: P. Magal, S. Ruan (Eds.), Structured Population Models in Biology and Epidemiology (2008)

Vol. 1937: G. Capriz, P. Giovine, P.M. Mariano (Eds.), Mathematical Models of Granular Matter (2008)

Vol. 1938: D. Auroux, F. Catanese, M. Manetti, P. Seidel, B. Siebert, I. Smith, G. Tian, Symplectic 4-Manifolds and Algebraic Surfaces. Cetraro, Italy 2003. Editors: F. Catanese, G. Tian (2008)

Vol. 1939: D. Boffi, F. Brezzi, L. Demkowicz, R.G. Durán, R.S. Falk, M. Fortin, Mixed Finite Elements, Compatibility Conditions, and Applications. Cetraro, Italy 2006. Editors: D. Boffi, L. Gastaldi (2008)

Vol. 1940: J. Banasiak, V. Capasso, M.A.J. Chaplain, M. Lachowicz, J. Miękisz, Multiscale Problems in the Life Sciences. From Microscopic to Macroscopic. Będlewo, Poland 2006. Editors: V. Capasso, M. Lachowicz (2008)

Vol. 1941: S.M.J. Haran, Arithmetical Investigations. Representation Theory, Orthogonal Polynomials, and Quantum Interpolations (2008)

Vol. 1942: S. Albeverio, F. Flandoli, Y.G. Sinai, SPDE in Hydrodynamic. Recent Progress and Prospects. Cetraro, Italy 2005. Editors: G. Da Prato, M. Röckner (2008)

Vol. 1943: L.L. Bonilla (Ed.), Inverse Problems and Imaging. Martina Franca, Italy 2002 (2008)

Vol. 1944: A. Di Bartolo, G. Falcone, P. Plaumann, K. Strambach, Algebraic Groups and Lie Groups with Few Factors (2008)

Vol. 1945: F. Brauer, P. van den Driessche, J. Wu (Eds.), Mathematical Epidemiology (2008)

Vol. 1946: G. Allaire, A. Arnold, P. Degond, T.Y. Hou, Quantum Transport. Modelling, Analysis and Asymptotics. Cetraro, Italy 2006. Editors: N.B. Abdallah, G. Frosali (2008)

Vol. 1947: D. Abramovich, M. Mariño, M. Thaddeus, R. Vakil, Enumerative Invariants in Algebraic Geometry and String Theory. Cetraro, Italy 2005. Editors: K. Behrend, M. Manetti (2008)

Vol. 1948: F. Cao, J-L. Lisani, J-M. Morel, P. Musé, F. Sur, A Theory of Shape Identification (2008)

Vol. 1949: H.G. Feichtinger, B. Helffer, M.P. Lamoureux, N. Lerner, J. Toft, Pseudo-differential Operators. Cetraro, Italy 2006. Editors: L. Rodino, M.W. Wong (2008)

Vol. 1950: M. Bramson, Stability of Queueing Networks, Ecole d' Eté de Probabilités de Saint-Flour XXXVI-2006 (2008)

Recent Reprints and New Editions

Vol. 1702: J. Ma, J. Yong, Forward-Backward Stochastic Differential Equations and their Applications. 1999 – Corr. 3rd printing (2007)

Vol. 830: J.A. Green, Polynomial Representations of GL_n, with an Appendix on Schensted Correspondence and Littelmann Paths by K. Erdmann, J.A. Green and M. Schoker 1980 – 2nd corr. and augmented edition (2007)

Vol. 1693: S. Simons, From Hahn-Banach to Monotonicity (Minimax and Monotonicity 1998) – 2nd exp. edition (2008)

Vol. 470: R.E. Bowen, Equilibrium States and the Ergodic Theory of Anosov Diffeomorphisms. With a preface by D. Ruelle. Edited by J.-R. Chazottes. 1975 – 2nd rev. edition (2008)

Vol. 523: S.A. Albeverio, R.J. Høegh-Krohn, S. Mazzucchi, Mathematical Theory of Feynman Path Integral. 1976 – 2nd corr. and enlarged edition (2008)

Vol. 1764: A. Cannas da Silva, Lectures on Symplectic Geometry 2001 – Corr. 2nd printing (2008)